Safety and Security Engineering IX

WITPRESS
WIT Press publishes leading books in Science and Technology.
Visit our website for new and current list of titles.
www.witpress.com

WITeLibrary
Home of the Transactions of the Wessex Institute.
Papers published in this volume are archived in the WIT eLibrary in volume 206 of
WIT Transactions on The Built Environment (ISSN 1743-3509).
The WIT eLibrary provides the international scientific community with immediate and
permanent access to individual papers presented at WIT conferences.
Visit the WIT eLibrary at www.witpress.com/elibrary.

Ninth International Conference on Safety and Security Engineering

SAFE 2021

Honorary Chairman

Conference Chairmen

International Scientific Advisory Committee

Sponsored by

WIT Transactions on the Built Environment

Organised by

Wessex Institute, UK
University of Rome "La Sapienza", Italy

WIT Transactions

Wessex Institute
Ashurst Lodge, Ashurst
Southampton SO40 7AA, UK

Senior Editors

Editorial Board

Safety and Security Engineering IX

Editors

Giorgio Passerini
Marche Polytechnic University, Italy
Member of WIT Board of Directors

Fabio Garzia
University of Rome "La Sapienza", Italy

Mara Lombardi
University of Rome "La Sapienza", Italy

WIT PRESS Southampton, Boston

Editors:

Giorgio Passerini
Marche Polytechnic University, Italy
Member of WIT Board of Directors

Fabio Garzia
University of Rome "La Sapienza", Italy

Mara Lombardi
University of Rome "La Sapienza", Italy

Published by

WIT Press
Ashurst Lodge, Ashurst, Southampton, SO40 7AA, UK
Tel: 44 (0) 238 029 3223; Fax: 44 (0) 238 029 2853
E-Mail: witpress@witpress.com
http://www.witpress.com

For USA, Canada and Mexico

Computational Mechanics International Inc
25 Bridge Street, Billerica, MA 01821, USA
Tel: 978 667 5841; Fax: 978 667 7582
E-Mail: infousa@witpress.com
http://www.witpress.com

British Library Cataloguing-in-Publication Data
A Catalogue record for this book is available
from the British Library

ISBN: 978-1-78466-443-5
eISBN: 978-1-78466-444-2
ISSN: 1746-4498 (print)
ISSN: 1743-3509 (on-line)

The texts of the papers in this volume were set individually by the authors or under their supervision. Only minor corrections to the text may have been carried out by the publisher.

Preface

This Volume contains the revised versions of selected papers presented at the 9th International Conference on Safety and Security Engineering, held online in November 2021, organised by Wessex Institute in the UK, University of Rome "La Sapienza" and University Politecnica delle Marche, Italy. The Conference was sponsored by the WIT Transactions on the Built Environment.

This ninth meeting in the Series was reconvened online, due to the criticality of being able to travel in the period scheduled for the conference and follows the success of the first eight meetings held in Rome (2005), Malta (2007), Rome (2009), Antwerp (2011), Rome (2013), Opatija in Croatia (2015), Rome (2017) and Ancona (2019). The purpose of the Conference is to provide a forum for the presentation and discussion of the most recent research and industrial developments in the theoretical and practical aspects of safety and security engineering.

The papers address the most recent developments in the theoretical and practical aspects of these important fields, which due to their special nature, brings together in a systematic way, many disciplines of engineering, from the traditional to the most technologically advanced. The papers cover areas such as risk analysis and assessment, safety engineering, disaster and emergency management, critical infrastructure, human factors and risk perception, earthquake and natural hazards, modelling and experiments, risk reduction in transport system in emergency conditions, counter terrorism issues, cyber security / E security, and many others.

The Editors are grateful to the reviewers, as well as to all authors for their contributions.

The Editors
2021

Contents

Section 7: Risk analysis and assessment

Section 8: Safety engineering

Section 9: Risk reduction in transport system in emergency conditions (Special session organised by F. Russo)

AS LOW AS REASONABLY PRACTICABLE: HOW IT WORKS IN THE RAIL AND ROAD TUNNELS IN ITALIAN RULES "RISK ACCEPTABILITY/ TOLERABILITY CRITERIA" – THE GU@LARP MODEL

MASSIMO GUARASCIO
Sapienza University of Rome, Italy

ABSTRACT

As Low As Reasonably Practicable, in short ALARP, arises from a UK court judgment (Edwards v. National Coal Board, 1949) and from a legislation act (1974) which requires "provisions and maintenance of plant and systems of work" that are "So Far As Reasonably Practicable", safe and without risks to health. The above requires that risk must be reduced to a level that is "As Low As Reasonably Practicable". After three catastrophic fire road tunnel accidents occurred in three major tunnels in France, Italy, Austria and Switzerland, the European Commission emitted a directive to ensure a minimum level of safety for road users in tunnels in the Trans European Road Network. In Italy in 2006 a decree adopted the European Directive introducing risk acceptability/tolerability criteria based upon ALARP principle.

Keywords: ALARP, risk acceptability, risk tolerability, fire risk, Gu@larp, tunnel.

1 INTRODUCTION

1.1 The origin of ALARP

The underlying fact at the origin of the acronym ALARP was a rock boulder fall in a coal mine roadway where Mr Edwards, a National Coal Board worker, was killed. The Edwards versus National Coal Board was an important case in the English Law. The above 1949 case revolved around whether it was "reasonably practicable" to prevent even the smallest possibility of a rock fall in a coal mine or in all similar scenarios. The judge, Mr Asquith, stated in his written judgment, *"**Reasonably practicable** is a narrower term than 'physically possible' and implies that a computation must be made [...] in which the **quantum of risk** is placed in one scale and the sacrifice involved in the **measures necessary for averting the risk** (whether in time, trouble or money) is placed in the other and that, if it be shown that there is a **great disproportion** between them – the risk being insignificant in relation to the sacrifice – the person upon whom the obligation is imposed discharges the onus which is upon him."* The above has established the concept of "**reasonable practicability**".

The Court of Appeal decided that "**reasonably practicable**" allowed the creation of "equation that measured the risk" present in a given situation against the "reasonable practicability" of mitigated that risk. In other words the equation asked if averting the risk was worth the effort it took to negate that risk. In addition the Court in Edwards determined that the size and wealth of the company should have no bearing on such decisions.

Mr Edwards was killed when an unsupported section of a travelling road in a mine gave way. Only about half the whole length of the road was shored up. The company argued that the cost of shoring up all roads in every mine was prohibitive when compared to the risk.

The question at issue was not the cost of shoring up all roads in every mine operated by the company. The issue was the cost of making safe the section of road that fell. Some roads are secure and show no signs of failing. Others may already have fallen and have already been repaired. The section in question was already supported by timber along half its length. The cost of making it safe was not great compared to the risk of injury and loss of life.

The Act requires employers to ensure "**so far as is reasonably practicable**" employees, and non-employees, are not exposed to risks to their health and safety from the employer's undertaking (i.e. business). The test for what is **reasonably practicable** was set out in this case. The case established the risk must be balanced against the '**sacrifice**', whether in money, time or trouble, needed to avert or mitigate the risk. By carrying out this exercise the employer can determine what measures are reasonable to take. This is effectively an implied requirement for risk assessment.

2 THE ALARP CONCEPT OF RISK ACCEPTABILITY/TOLERABILITY IN ITALIAN ROAD TUNNELS

2.1 The European Directive (2004) and the Italian Decree (2006)

As mentioned above, three major fire accidents occurred in 1999 and 2000 in three major European tunnels (Monte Bianco, Tauern and Gotthard) and have been at the origin of the European Directive (2004) of which "subject matter and scope" was aiming to ensure a minimum level of safety for road users in tunnels in the TERN by the prevention of critical events that may endanger human life, the environment and tunnel installations, as well as by the precision of protection in case of accident.

Two years later the Italian Parliament approved the European Directive with some textual modification in the Decree (2006). Instead of the formula "*minimum level of safety*" we read "*un livello minimo sufficiente di sicurezza*", which means the level has to be sufficient, eventually according to a given criterion. And further we are reminded that prevention and protection measures should be designed according to the hazard and risk conditions.

2.2 Rail and road tunnels: Risk acceptability and tolerability criteria in Italian rules

Fig. 1 shows the risk acceptability/tolerability criteria of the decree 25th October 2005, introduced in Italy for rail tunnels, where exceedance probability are "kilometre × year" units. In bi-log coordinates paper, the straight lines corresponding to the risk acceptability and tolerability limits have a slope greater than "1" and have origin at the value "2" fatalities. The rail tunnel Santa Croce case study is shown in Fig. 1 where the risk reduction effect of an additional protection system is verified at design level.

The road tunnel risk acceptability/tolerability criteria according to the EU Directive (2004) introduced in Italy by the 264 Decree (2006) are shown in Fig. 2. The broken lines in the bi-log coordinate paper, correspond to the integer values function exceedance probability distribution according to the Gu@larp model introduced later in the present paper. In the case of road tunnels, the slope of the equivalent straight lines of the risk acceptability and tolerability limits in bi-log coordinates diagram is exactly "1" and the origin point of the threshold axis (number of fatalities) is exactly N = 1, apart from the point N = 0 obviously.

Fig. 2 shows the Scampitella tunnel (Italy) risk-based design where three alternative solutions of protection systems availability is presented. The broken lines are respectively the acceptability and tolerability limits and the three solutions compared for optimization and compliance purposes.

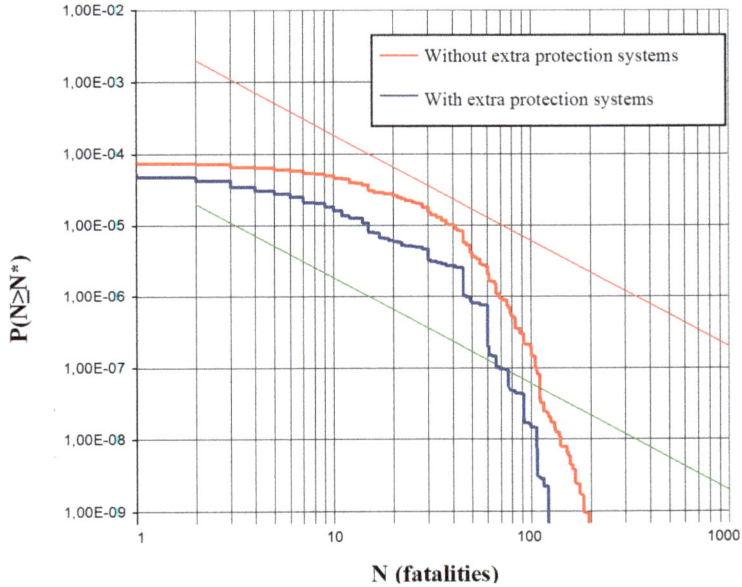

Figure 1: Risk acceptability/tolerability criteria of the decree 25th October 2005, introduced in Italy for rail tunnels.

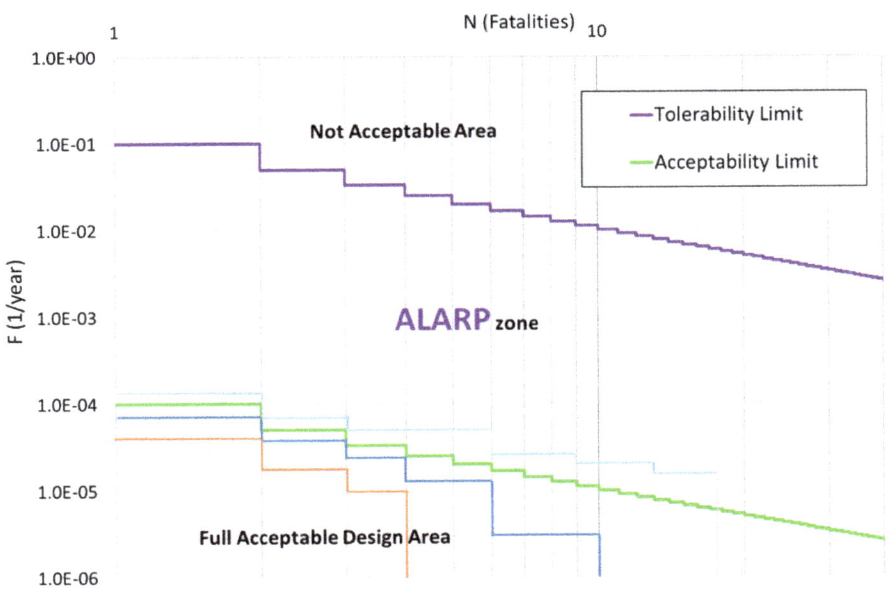

Figure 2: Risk acceptability/tolerability criteria according 264 Decree (2006) for road tunnel (broken lines for Scampitella tunnel).

3 THE RISK QUANTITATIVE CRITERIA

3.1 A comparative exam of the rail road tunnel risk type acceptability/tolerability criteria

Below a comparative analysis is discussed. Describing the major differences existing between the two criteria adopted for rail and road tunnels risk acceptability/tolerability according to the different parameters conditions of the respective characteristic equations. In the bi-log paper Fig. 3 presents different slope and origin in one case (red and yellow broken lines) corresponding to the rail tunnels and in the other (purple and green broken lines) to the road tunnels acceptability and tolerability criteria.

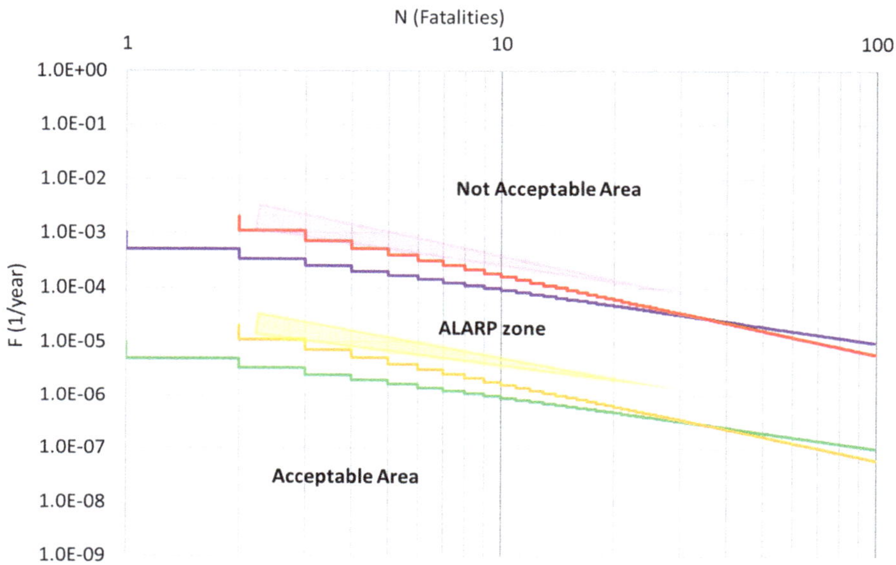

Figure 3: Comparison of the bi-log diagrams between road and rail tunnels.

Fig. 3 shows the highlighted areas where there are over-tolerated fatalities that correspond quantitatively to the measure of the shaded areas.

To better clarify the above concepts the same data are presented in Fig. 4, in arithmetical coordinates. In the case of the road tunnel model instead of a straight line type curve we get equilateral hyperbola equation, that shows the pure inverse proportionality link between exceedance probability, on the vertical axis, versus the number of fatalities threshold on the horizontal one. The above inverse proportionality property is effective from the social economical ethics point of view.

In order to highlight further the major differences between the rail and road tunnel risk criteria here is offered in Fig. 5 the behaviour of the contributions to the number of fatalities expected value "$E(N_i)$" corresponding to the all individual scenario each characterised by a different N_i (Fatalities estimated number).

The Ni value does depend upon the different hazard condition, protection system performances/availability and exposure map model that are assumed in risk analysis model. Model coupled with the used code for the fire and evacuation scenario simulation.

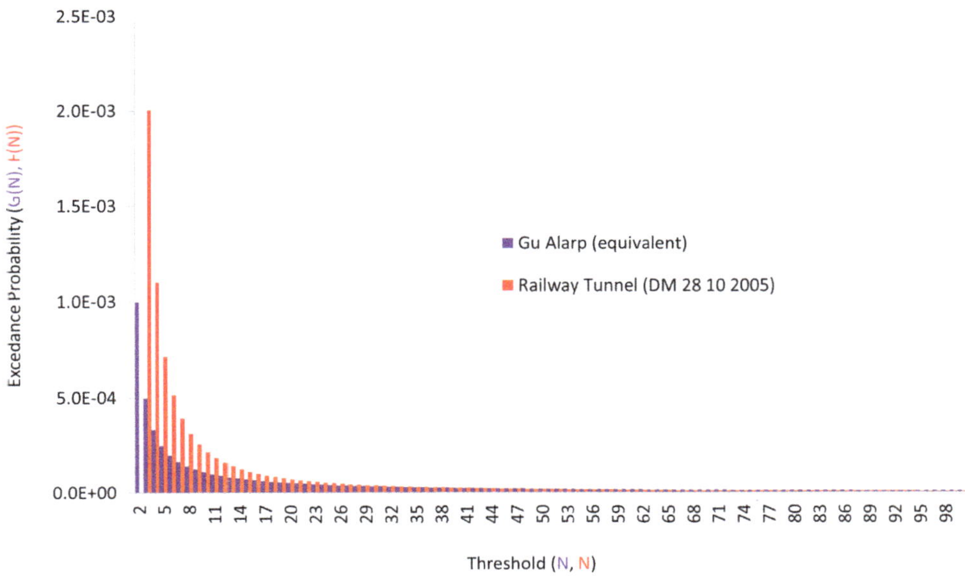

Figure 4: Exceedance probability ($G(N)$ e $F(N)$) as threshold level risk indicator.

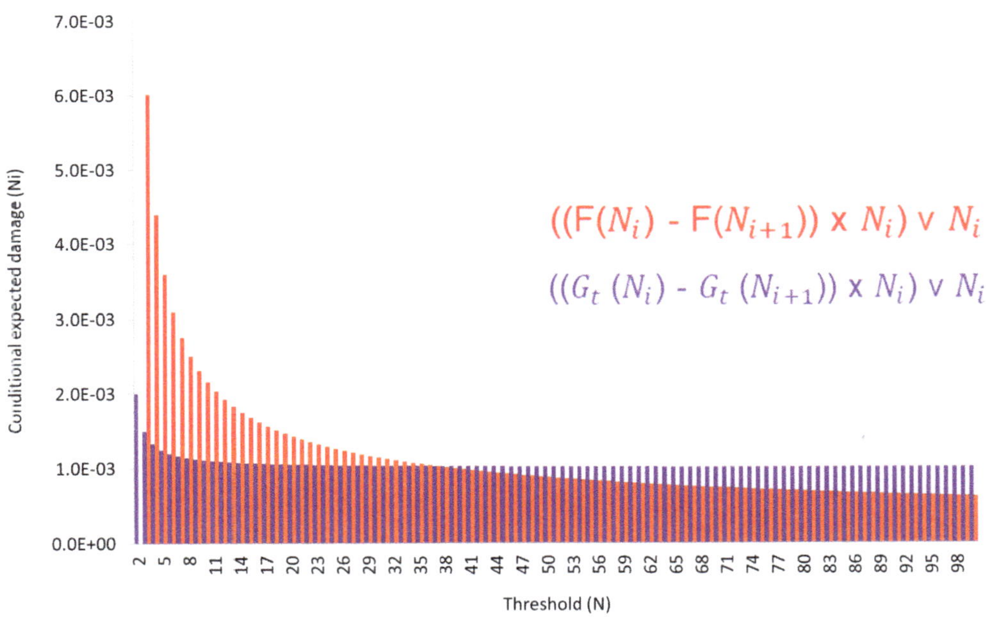

Figure 5: Single scenario expected damage (Ni fatalities) as scenario risk indicator.

Everybody can verify that, in the case of the road tunnel risk model, is lightly monotonic decreasing but substantially invariant with respect to the *Ni* values. The above property is exactly according the "As Low As Reasonably Practicable" spirit, but the mathematical properties of the *G(N)* integer values exceedance probability distribution model requires the assumption of a parent function defined in the continuous domain. The one suggested here has been inspired to the author by the Pareto type function where the $alpha$ shape parameter value is assumed equal to "1".

The Gu@larp density function is: $g[x\_, Gu\_] = (1 - Gu)\ DiracDelta[x] + If[x \geq 1, Gu/x^2, 0]$ and the corresponding exceedance probability $G(x) = Gu \cdot x^{-1}$.

4 HISTORICAL CONCLUSIVE REMARK

The journey of ALARP has been a long one. From the UK Court Occupational Safety Case in 1949 to the European Court Judgment (Curia, Luxemburg, 2007) where the philosophical and legal principle of "As Low As Reasonably Practicable" has been fully accepted.

How an individual fatality due to the rock fall in a coal mine in 1949 has been able to became a real "influencer" for many human activity sectors safety rules aimed to the risk reduction for the people? That's the question.

SECTION 1
CRITICAL INFRASTRUCTURE

APPLICATION OF SEISMIC GEOSTATISTICS TO IMPROVE THE TERRITORIAL INFRASTRUCTURE RESILIENCE AND SUSTAINABILITY

MASSIMO GUARASCIO, ANGELO LIBERTÀ, DAVIDE BERARDI,
ELEONORA DI BENEDETTO & MARA LOMBARDI
Sapienza University of Rome, Italy

ABSTRACT

The geostatistical methodology proposed in this article is able to provide quantitative data of the ground vibrational movements to the decision support systems to determine the priorities of the seismic interventions adaptation of existing transport infrastructures but also in the design phase of new infrastructures. This paper shows the use of geostatistics to estimate the arrival time and the value of the ground acceleration peak of the vertical and epicentral oscillation components of the Norcia earthquake of 2016. The validity of the non-stationary geostatistical methodology is confirmed by the calculated estimation errors in the 38 sites of the accelerometric stations and used to verify the representativeness of the estimated data. The histogram of the 38 normalized estimation errors is close to a Gaussian distribution with the frequency of the $\pm \sigma_{ck}$ class of 73.7% instead of 68.3%. This result was obtained for both seismic variables treated, arrival time of the maximum acceleration and peak ground acceleration value (PGA).

Keywords: geostatistical, seismic, resilience, sustainability, earthquake, seismicity, seismic hazard, historical earthquakes, seismic risk.

1 INTRODUCTION

1.1 Regionalized geostatistical variables

The knowledge with a low level of uncertainty of the directional seismic profiles of the ground oscillations following an earthquake is a relevant issue in active seismogenic areas. A uniform geographical coverage of quantitative seismic data is useful for decision support systems to determine the priorities of seismic adaptation interventions of existing transport infrastructures but also in the design phase of new infrastructures.

One response to engineering needs is the continuous monitoring of seismic quantities with an adequate geographical density of observation points to reduce as much as possible the level of uncertainty of the estimate in unobserved sites. At the same level of attention is the choice of efficient numerical methodologies for estimating the dynamic parameters of the vibrational movements of the ground, such as the Cartesian components of the amplitude, speed and acceleration vectors of the ground oscillations. This methodology should take into account the regionalized aspect of the seismic variables, strictly related to the geographic coordinates and the elevation of the site and not reducible to the epicentral distance alone.

The geostatistical methodology proposed in this article is able to provide the best local estimate of the seismic variables by making the most of existing data, but also to design seismic monitoring networks appropriate to the accepted degree of uncertainty. This paper shows the use of geostatistics to estimate at the nodes of a georeferenced grid the arrival time and the value of the ground acceleration peak of the vertical and epicentral oscillation components of the Norcia earthquake of 2016.

WIT Transactions on The Built Environment, Vol 206, © 2021 WIT Press
www.witpress.com, ISSN 1743-3509 (on-line)
doi:10.2495/SAFE210011

1.2 Seismic phenomenology and geostatistical model

In the epicentral area of an earthquake, the vibrations of the ground layers of engineering interest are originated mainly from the seismic energy released in the hypocentral area and transported to the surface by longitudinal (primary waves) and transverse (secondary waves) volume waves. In the shallow earthquakes, with a hypocentral area between 7 and 10 km deep, the extension of the ground vibrations in the second following the instant in which the first volume waves reach the earth's surface is of the order of 10 km. This is because in compact rocks the propagation velocity of volume seismic waves is between 3 and 6 km/s and therefore the amplitude of seismic cone after about one second is about 30°.

The vibrational movement of the ground in the epicentral area generate new classes of seismic waves that propagate radially for hundreds of kilometers in the most superficial soil layer of geostatigraphic system. The amplitude of the vibrations of the ground outside the epicentral area is the resulting vector of the vibrational movements carried by the surface waves and the vibrational movements carried by the volume waves rising from the deeper layers of the geostratigraphic system. Moving away from the hypocentral area: (i) the amplitude of the surface waves of Rayleigh and Love become predominant with respect to the amplitude of the longitudinal (P) and transverse (S) volume waves; (ii) the dynamic parameters of the vibrational movements of the ground tend to weaken. Otherwise, in the innermost area of the epicentral area, the amplitudes of the vibrational movements carried by the volume waves, rising almost vertically from the hypocentral area, are stationary and do not have a precise and detectable zoning. In other words, the different damages suffered by the infrastructures of the road networks and by the houses of the urban centers located in this seismic area are attributable more to the different construction techniques than to a different seismic stress. A first prediction of the geographical pattern of the vibrational parameters of the ground such as amplitude, velocity and acceleration is provided by the physical theory of elastic waves applied to the waves emitted radially by a point source in a homogeneous and isotropic material medium. The theoretical result highlights a damping law of the dynamic parameters of the oscillations inversely proportional to the distance from the source for the volume waves ($f(r) = a\, r^{-1}$) and a law inversely proportional to the square root of the distance from the source for the waves superficial ($f(r) = a\, r^{-0.5}$) (geometrical attenuation).

In geostratigraphic systems, the regionalization of the vibrational parameters of the soil differs from the geographical pattern predicted by the theory due to the heterogeneity of the rock formations crossed by seismic waves and the complex geometry of the fault systems and the stratigraphic contacts between lithological units:

- the different propagation velocity of seismic waves in rock formations deform the geometry of the wave surfaces with consequent directional variations of the geometrical attenuation;
- moreover, irregular local variations of the seismic oscillations are a consequence of the different inelastic behaviour of the rock formations. A part of the energy carried by the seismic waves is continuously and irregularly dissipated by friction (anelastic attenuation).

Despite this, over distances of tens of kilometers, the data of the seismograms recorded by the accelerometric stations show good consistency with the geographic pattern described by the function $a\, r^{-0.5}\, e^{-br}$ (first-guess); where a is the adjustment factor to the existing seismic data; r is the average distance from the epicentral zone of the maximum vibrations [1] and b is the inelastic damping coefficient of the seismic oscillations.

Fig. 1 shows the peak data of the epicentral component of ground acceleration (PGA) of the Norcia earthquake of 2016 and the attenuation profile predicted by the theoretical model. The graph highlights the presence of a directional anisotropy of the geographic pattern; the PGA data recorded by the accelerometric stations aligned in the 35° East direction with respect to the epicentral area (direction orthogonal to the Apennine axis) are positioned in the lower band of the scatter-diagram. Directional anisotropy is also present in the vertical component of the PGA.

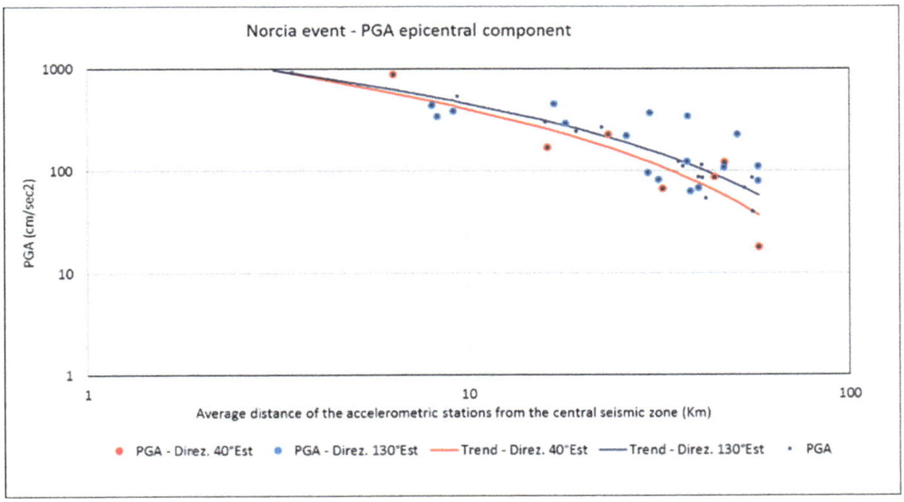

Figure 1: Theoretical model of geometric and inelastic damping of the acceleration peak vs PGA data recorded by the accelerometric stations during the Norcia earthquake (2016).

The goal of the work is to implement the best non-stationary linear estimator to estimate the arrival time and the local value of the acceleration peak at the nodes of a georeferenced grid with 500 m square cells. To this end it is:

- Assumed as a geographic pattern of the seismic variables: the function
  $a_p r^{-0.5} + e^{-b_p r}$ for the PGA
  $a_t + b_t\, r$ for the arrival time of the acceleration peak
- used the Universal CoKriging geostatistical estimator with External Drift to integrate the functions of the geographic pattern of the seismic variables and the seismic information recorded by the network of accelerometric stations.

The geostatistical estimator, proposed as an alternative to the statistical methods already used to reconstruct the geographical evolution of the dynamic parameters of the vibrational movements of the ground (ground motions), allowed to: (i) reconstruct the directional attenuation profiles of the seismic variables consistent with the data and directional variations modelled on the data itself; (ii) calculate the local estimation error for any point of the observed geographic domain conditioned by the position of the accelerometric stations around the estimated point and the directional correlation of the seismic variables. In this way the estimation error or the uncertainty in associating the estimated value to a geographical point will depend not only on the distance from the epicentral zone [1]–[3] but also from the local and directional variations of the seismic variables. The validation phase of the geostatistical estimator has been added to the procedure for estimating the seismic variables. The aim is to check with the existing data: (i) the correctness of the estimator *the average of the estimation errors is zero*; (ii) the representativeness of the estimate uncertainty indicator *the variance of the estimate calculated by the estimator is consistent with the variance of the estimation errors*.

2 GEOSTATISTICAL FRAMEWORK

In this section, the principles of the geostatistical Universal CoKriging estimator are described and the definition of regionalized variable is briefly introduced.

2.1 Natural phenomenon and regionalized variable

Initially geostatistics was developed to estimate mineral deposits [4] but the same methodologies were found to be valid and therefore currently applied in many fields of earth sciences (hydrogeology, meteorology, oceanography, geochemistry, environmental control, soil science, etc.). The basis of geostatistical methodologies is the observation that some natural phenomena have a geographical evolution and local states are closely related to their position. This typology of phenomena have been defined as "regionalized phenomena" and consequently the variables that describe them are defined as "regionalized variables" (V.R.) [5]. The set of states assumed by a V.R. at points x of a geographic domain they are represented by a function z(x). Locally, the z(x) function is highly irregular and the variations from one point to another are often unpredictable, while over greater distances the function is more continuous and shows the structural characteristics of the physical phenomenon that underlies the observed regionalized phenomenon. This feature suggested developing the function z(x) into a sum of two functions; the first, indicated by m(x), which represents the geographical trend of the regionalized phenomenon and the second, indicated by y(x), which reflects the irregular local variations of the observed phenomenon: z(x) = m(x) + y(x). As they are defined, the two components are statistically independent. This double aspect is found in numerous geophysical and geochemical variables such as: temperature of the layer of atmosphere in contact with the earth's surface (temperature decreases with latitude and topographical altitude), concentration of pollutants in the soil (concentration increases near an oil plant) or vibrations of the ground following a seismic event (vibrations of the ground decrease with the distance from the epicentral area of the earthquake).

2.2 Azimuth directional variography

The irregular geographical evolution of natural phenomena has suggested the conceptual choice of considering the V.R. z(x) of the realizations of Random Functions (F.A.) Z(x) defined on a precise geographical domain D of the Euclidean space $\mathbb{R}^n$. The presence of a geographical pattern in some regionalized phenomena leads to the following model:

$$Z(x) = m(x) + Y(x). \tag{1}$$

The function m(x) represents at each point x the expected value of the F.A. Z(x): E|Z(x)|=m(x). The residual Y(x) of the FA Z(x) is a locally stationary FA with E|Y(x)|=0 and covariance Cov|Y(x_1), Y(x_2)| = E|Y(x_1) Y(x_2)|<∞. This model is generalized by assuming that the geographical variations of the FA Y(x) are stationary and depend only on the distance and direction (intrinsic hypothesis):

$$\gamma(\vec{h}) = \frac{1}{2}\,\boldsymbol{E}\left|[(Y_i(x+\vec{h}) - Y_i(x)]^2\right|. \qquad (2)$$

The function $\gamma(\vec{h})$ is defined semivariogram of the F.A. stationary Y(x) and all the F.A. non-stationary defined as $Z_k(x) = m_k(x) + Y(x)$ [5]. In the study of the natural phenomena described by several V.R. related, information on the geographical structure of the F.A. $Z_i(x)$ e $Z_j(x)$ is completed by the cross-variogram of the F.A. of the residuals $Y_i(x)$ and $Y_j(x)$:

$$\gamma_{i,j}(\vec{h}) = \frac{1}{2}\,\boldsymbol{E}|[(Y_i(x+\vec{h}) - Y_i(x)][(Y_j(x+\vec{h}) - Y_j(x)]|. \qquad (3)$$

The semivariogram or variogram and the experimental cross-variogram are calculated using existing observations from which the theoretical models are subsequently adjusted. This operation is done taking into account both the existing data and the phenomenological characteristics of the regionalized variables studied [6].

2.3 Universal CoKriging with external drift

Universal CoKriging with external drift is a specialization of Universal CoKriging (UCoK) applied when the functions that describe the geographic pattern of the regionalized variables studied (external drift) are known. UCoK is widely used in earth sciences to map geophysical or geochemical variables such as the minimum and maximum air temperature or the concentration of metals or pollutants in the soil. The UCoK estimator is numerically equivalent to a multivariate geographic regression conditional on the directional variogram and cross-variogram functions of the target variable and the auxiliary variables. The estimate of the target variable Z_o at point x is obtained from the linear combination (weighted average) of the data of the target variable Z_o and the auxiliary variables Z_i which have a geographical correlation with the target variable:

$$Z_0^*(x) = \sum_{k=0}^{m} \sum_{i=1}^{n} \lambda_{k,i}\, Z_k(x_i), \qquad (4)$$

where $\lambda_{0,i}$ are the weight coefficients of the target variable and $\lambda_{k,i}$ are the weight coefficients of the auxiliary variables (index k> 0) and x_j are the coordinates of the n observation points (x_j represents the pair of geographic or metric coordinates necessary to locate the *i*th point in the geographic domain).

The universal CoKriging with external drift assumes that (i) the average of the F.A. objective and the F.A. auxiliaries are in every point of the geographic domain equal to less than one or more multiplying factors to the functions that describe the respective geographic patterns, i.e. $E|Z_i(x)| = \sum_{l=0}^{L} a_{i,l}\, f_i^l(x)$; (ii) for each pair of points in the geographic domain, the mean of the squared differences of the F.A. and the average of the product of the difference between two F.A. are determined by the variogram and cross-variogram model:

For each point to be estimated the estimation coefficients $\lambda_{k,i}$ are obtained by setting the following conditions:

(i) average estimation error of the zero target variable: $\boldsymbol{E}|[(Z_0^*(x) - Z_0(x)]| = 0$.
The condition of unbiased for the UCoK estimator becomes: $\boldsymbol{\lambda}_i^T\, \boldsymbol{F}_{i,l} = \delta_{i,l}\, \boldsymbol{f}_l^T$

(ii) variance estimator error of the minimum target variable:

$$\boldsymbol{Var}|[(Z_0^*(x) - Z_0(x)]| = 2\,\sum_i \boldsymbol{\lambda}_i^T\, \boldsymbol{\Gamma}_{i,0} - \sum_i \sum_j \boldsymbol{\lambda}_i^T\, \boldsymbol{\Gamma}_{i,j}\, \boldsymbol{\lambda}_j = \min.$$

These two conditions are implemented in the following linear system [6]:

$$\begin{cases} \sum_{k=0}^{m} -\boldsymbol{\Gamma}_{i,j}\, \boldsymbol{\lambda}_j \; + \; \boldsymbol{F}_i\, \boldsymbol{\mu}_i = -\boldsymbol{\Gamma}_{i,0} & i = 0 \ldots m \\ \boldsymbol{F}_i^T\, \boldsymbol{\lambda}_i = \boldsymbol{f}_{l,0}\, \delta_{i,l} & i = 0 \ldots m \end{cases} \tag{5}$$

where $\boldsymbol{\Gamma}_{i,j}$ are the nxn matrices of variograms or cross-variograms of F.A. Z_i and Z_j between each observation points pair; $\boldsymbol{F}_{i,l}$ are the nxL matrix of the values of the first-guess functions of the V.R. ith at the observation points; $\boldsymbol{\Gamma}_{i,0}$ are the vectors of the variograms and cross-variograms between the point to be estimated (target-point) and the observation points of the F.A. to be estimated with the F.A. Z_i and $\boldsymbol{f}_{l,0}\, \delta_{i,l}$ are the vectors of the values of the first-guess function of the V.R. to be estimated and sold set to zero the conditions of unbiased of the V.R. auxiliaries.

The variance of the estimation error or cokriging variance is given by:

$$\sigma_{CK}^2 = \sum_{k=0}^{m} \boldsymbol{\lambda}_k^T\, \boldsymbol{\Gamma}_{k,0} \; + \; \boldsymbol{\mu}_l^T\, \boldsymbol{f}_{i,0}, \tag{6}$$

where $\boldsymbol{\lambda}_{i,0}$ are the vectors of the F.A. Z_i and $\boldsymbol{\mu}_{i,0}$ are the vectors of the Lagrange parameters associated with the F.A. Z_i. The variance of the estimate quantifies the uncertainty of the estimate and, as defined, depends: (i) on the number of observation points; (ii) on the geometry of the observation points; (iii) on the position of the observation points with respect to the point to be estimated and (iv) on the local variability of the V.R. treated.

3 NORCIA EARTHQUAKE OF 30 OCTOBER 2016

This section presents the results of the seismic data processing of the 2016 Norcia earthquake (seismogenic zone of the Central Italian Apennines). The UCoK with External Drift estimator was used to estimate the time of arrival seismic variables and the peak value of the vertical (PGA_V) and epicentral (PGA_E) component of ground acceleration around 60 km from the epicentral area. The cartesian components peak values of the ground vibrational movements acceleration vector are some of the indicators most used in the design and seismic evaluation of road infrastructures and urban buildings [2], [7]. For this reason it was decided to test the UCoK with external drift geostatistical estimator on the two aforementioned seismic variables of the Norcia earthquake. To estimate the two seismic variables, the following were used: (i) the seismograms recorded by the accelerometric stations; (ii) the geographic coordinates and the quote of the accelerometric stations; (iii) the geo-referenced altimetric model with 500 m cells (DTM). The phases of the local estimation procedure of the seismic variables at the DTM nodes are:

A. localize the epicentral seismic area and the first-guess functions coefficients of the seismic variables;
B. determine the directional variogram of the seismic variables residuals not explained by the first-guess functions: difference between the real seismic data and the forecast provided by the theoretical model;
C. validate the geostatistical estimator using: the existing seismic data, the first-guess functions (point A), the directional variogram (point B) and the digital terrain model (DTM);
D. use the validated geostatistical estimator (point C) to estimate the seismic variables at the nodes of the geo-referenced grid of the DTM.

3.1 Ground motion data

The two seismic variables data were determined from the seismograms of the vertical oscillation component and of the N–S and E–W oscillation components. To this end, the seismograms recorded by the accelerometric stations around 60 km of the geographical area with the largest number of buildings collapsed in the Norcia earthquake were examined.

In each accelerometer station the PGA_V data are defined as follows:

$$z_v(x) = \max\left[z'_v(x,t);\ t_0 \leq t \leq t_f\right], \tag{7}$$

where $z'_K(x,t)$ are the acceleration data recorded in the vertical seismogram in the interval time between t_0 and t_f referable to the seismic event. The arrival time of the vertical oscillations peak coincides with the instant t of recording of the maximum data. The arrival time and the epicentral oscillations peak value on the epicentral seismogram produced by combining the data of the N–S and E–W seismograms were similarly determined.

Table 1: Quartile and max value of PGA_V and PGA_E (Norcia earthquake)

		PGA vertical component			PGA epicentral component		
Seismic event	Number of data	I Quartile (cm/s^2)	II Quartile (cm/sec^2)	Max value (cm/s^2)	I Quartile (cm/s^2)	II Quartile (cm/s^2)	Max value (cm/s^2)
Norcia (2016)	40	44.19	249.12	893.50	85.04	298.91	919.54

3.2 First-guess of seismic variables

The first-guess fields of the two seismic variables are described by continuous mathematical functions defined at all points of the geographical survey domain:

Arrival time of the acceleration peak:

$$m_t(x) = a_t(x_0) - b_t(x_0)\,\bar{r}_0(x), \tag{8}$$

where x_0 identifies the coordinates pair of the center of gravity of the seismic event epicentral area; $a_t(x_0)$ and $b_t(x_0)$ are the coefficients of the regression line; $\bar{r}_0(x)$ is the average distance of the x coordinate point from the epicentral area (average of the distances between the point and the points of the circle with a radius of 5 Km centred at the point x_0).

Peak ground acceleration (PGA):

$$m_p(x) = \frac{a_p(x_0)}{\sqrt{\bar{r}_0(x)}}\, e^{-b_p(x_0)\,\bar{r}_0(x)}, \tag{9}$$

where x_0 identifies the pair of coordinates of the center of gravity of the epicentral area of the seismic event; $a_p(x_0)$ and $b_p(x_0)$ are the coefficients of the theoretical pattern of the geographical evolution of the PGA; $\bar{r}_0(x)$ is, as for the seismic variable time of arrival, the average distance of the point of coordinates x from the epicentral area.

In the two first-guess functions the distance between any pair of points x' and x" is defined as: $d(x',x'') = \sqrt[2]{\boldsymbol{X}^T\boldsymbol{H}\,\boldsymbol{X}}$ where $\boldsymbol{X} = (x''_1 - x'_1, x''_2 - x'_2)^T$ is the difference vector of the metric coordinates of the two points, and $\boldsymbol{H} = \boldsymbol{\psi}\,\boldsymbol{M}\,\boldsymbol{\psi}^T$ is the product of the rotation matrix of the cartesian reference system and of the metric tensor $\boldsymbol{M}$ used to introduce in the first-guess functions the directional anisotropy of the seismic variables:

$$\boldsymbol{M} = \begin{vmatrix} (1+\alpha)^2 & 0 \\ 0 & 1 \end{vmatrix}. \tag{10}$$

The coefficient α is the expansion factor of the metric coordinates of the Cartesian axis orthogonal to the Apennine alignment (oriented at 35° East).

The position of the epicentral area must be compatible with the geographic pattern described by the $m_p(x)$ function and with the PGA data at the points of the accelerometric stations. To this end, the least squares estimator was first applied to estimate the coefficients of the first-guess function at the grid nodes of the DTM and then the grid node with the minimum root mean square error (RMSE) was selected as the centre of gravity of the epicentral area. In each node the error was calculated as the difference between the observed PGA data $z_p(x_i)$ and the value of the first-guess function at the point: $e_p(x_i) = z_p(x_i) - m_p(x_i)$.

The estimate of the coefficients of the first-guess function of the PGA_V and the PGA_E was obtained considering the quote *q(x)* of the accelerometric stations and the grid nodes and assuming that the average of the F.A. $Z_p(x)$ is expressed as:

$$m_p(x) = c_0 + c_q q(x) + \sum_{l=1}^{3} \frac{c_l\,(-1)^{l-1}}{\sqrt{\bar{r}_0(x)}} \frac{[\bar{r}_0(x)]^{l-1}}{(l-1)!}, \tag{11}$$

and solving three linear systems to calculate the estimation coefficients of $c_l^*(x) = \sum_{i=1}^{n} \lambda_{l,i}\, z(x_i)$:

$$\begin{cases} \boldsymbol{I}\,\boldsymbol{\lambda}_l + \boldsymbol{F}\,\boldsymbol{\mu} & = 0 \\ \boldsymbol{F}^T\,\boldsymbol{\lambda}_l & = \delta_{k,l} \end{cases}, \tag{12}$$

where $\boldsymbol{I}$ is the identity matrices *nxn*; $\boldsymbol{F}$ is the *n*x5 matrix of the values of the development terms of $m_p(x)$ in the observation points; $\boldsymbol{\mu}$ is the vector of the five Lagrange multipliers and $\delta_{k,l}$ is the operator to set equal to 1 the *k*th term of the coefficient to be estimated. The coefficients of the first-guess function are calculated starting from the three coefficients $c_l^*(x)$. The coefficients of the first-guess time of arrival function are estimated by minimizing the error $e_t(x_i) = z_t(x_i) - a_t(x_0) - b_t(x_0)\,\bar{r}_0(x)$ on the n accelerometric stations.

3.3 Seismic variables variography

The variographic analysis of the two seismic variables of the vertical and epicentral oscillations of the Norcia earthquake was performed on the residual data calculated as the difference between the observed data and the value of the first-guess function at the point. For the seismic variable PGA it was observed that the variance of local dispersion of the residuals decreases with the distance from the epicentral area (Fig. 3), for this reason the experimental variograms and cross-variograms were calculated using the residuals of the time of arrival of the acceleration peak $y_t(x_i) = z_t(x_i) - m_t(x_i)$ and the normalized residuals of the PGA:

$$y_P'(x_i) = \frac{z_P(x_i) - m_P(x_i)}{\sigma_P(x_i)}, \tag{13}$$

where $\sigma_P(x_i)$ is the standard deviation of the residuals of PGA around the point x_i.

The structure analysis was performed by processing the pairs of accelerometric stations in the four directions: 35° East (direction orthogonal to the Apennine alignment), 80° East, 125° East (direction of the Apennine alignment) and 170° East. of accelerometric stations have been grouped in the four directions and for multiple lengths of 10 km, applying respectively a directional tolerance of ± 20° and a distance tolerance of ± 5 km. For each

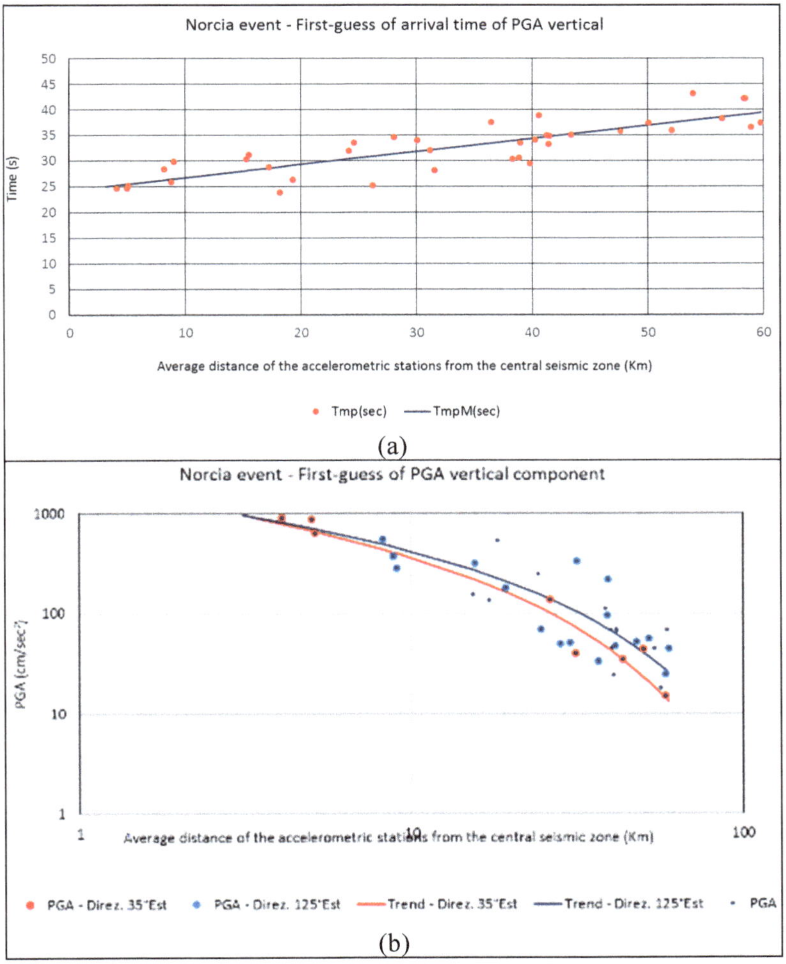

Figure 2: (a) First guess function of the arrival time vs (b) Vertical PGA during the Norcia earthquake (2016).

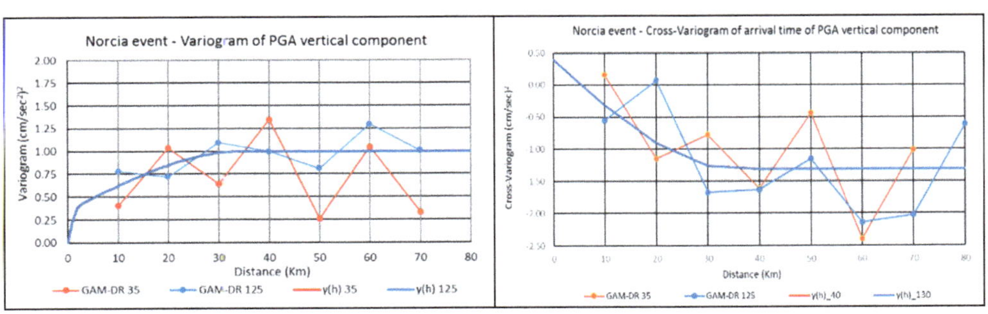

Figure 3: Variogram and cross-variogram of the peak value of the vertical acceleration with the arrival time of the Norcia event.

direction–distance the variogram and the cross-variogram of the two seismic variables; for example, the PGA variogram of a direction–distance is calculated as:

$$\gamma_p(h_\alpha) = \frac{1}{2\,N(h_\alpha)}\sum\left[y'_p(x+h_\alpha) - y'_p(x)\right]^2, \tag{14}$$

where $N(h_\alpha)$ is the number of pair of accelerometric stations of the class classe h_α.

The spherical scheme [6] defined by the following parameters was used to model the experimental variograms and cross-variograms: maximum correlation distance (range) and local dispersion variance. The variograms and the cross-variogram of the two seismic variables of the vertical oscillation components of the Norcia earthquake are shown in Fig. 3.

The variograms and the cross-variogram confirm the hypothesis of local stationarity of the two seismic variables of both the vertical and epicentral oscillation component. The variograms of the seismic variables of the vertical oscillation component detect the presence of a geographical correlation over distances of less than 30 km. In particular, the geographical variations of the PGA are isotropic while for the arrival time of the acceleration peak there is a greater correlation in the direction 125° East (Fig. 3).

For the epicentral oscillation component, the geographical correlation is detected only on the arrival time of the acceleration peak while the geographical variations of the PGA do not detect correlations over distances greater than 10 km.

3.4 Cross-validation

Once the coefficients of the first-guess functions have been calculated and the variogram and cross-variogram models of the variables to be estimated have been chosen, the application of the estimation process is a sequence of numerical calculations well tested in practice.

Verifying the representativeness of these choices is therefore the phase to confirm the applicability of the estimator implemented to the project. For this purpose, the arrival time of the acceleration peak and the PGA were estimated at the sites of the accelerometric stations by applying the UCoK with external drift estimator and imposing on the local average of the two F.A. the following form:

- Arrival time of PGA: $m_t(x) = c_{t,0} + c_{t,1}\,\overline{r}_0(x) + c_{t,q}q(x)$
- PGA: $m_p(x) = c_{p,0} + c_{p,1}\,\frac{a_p^*}{\sqrt{\overline{r}_0(x)}}\,e^{-ba_p^*\,\overline{r}_0(x)} + c_{p,q}q(x)$

The estimates were made using the data of both seismic variables, but taking one as an auxiliary variable from time to time. The estimation process was applied to all the accelerometric stations, not considering each time the data of the station to be estimated. In this way, the estimation error of each was calculated as the difference between the real data and the estimated data and $(x_i) = z^*(x_i) - z(x_i)$.

The results reveal an estimation error very close to zero and a variance of the experimental error close to the estimation variance calculated by the applied estimator (Table 2). In the scatter-diagram of Fig. 4, the estimated data are compared to the detected data of the two seismic variables.

4 ESTIMATION BY UNIVERSAL COKRIGING WITH EXTERNAL DRIFT

The validated UCoK with external drift estimator was used to estimate the arrival time and the peak value of the vertical and epicentral components of the ground acceleration of the

Table 2: Validation of the seismic variable estimation model (Norcia event).

		PGA vertical component			PGA epicentral component		
Seismic event	Number of data	Average errors	Variance errors	Variance estimate	Average errors	Variance errors	Variance estimate
PGA	38	–0.52	7111.45	7026.15	–0.60	10520.75	10835.20
Arrival time	38	0.08	4.51	4.44	0.105	7.98	8.08

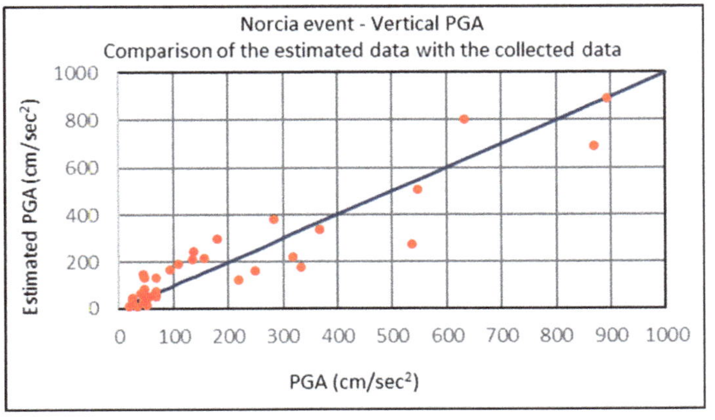

Figure 4: Cross-validation of the geostatistical model for estimating seismic variables.

Norcia earthquake (Fig. 5). The estimation was performed using data and first-guess functions of both seismic variables, variogram and cross-variogram models and digital terrain model (DTM). The value and variance of the two seismic variables were estimated in each node of the DTM grid.

5 CONCLUSIONS

The study highlights how the mapping of the estimated data represents the seismic phenomenology over the entire geographic domain. The local absence of observed data generated by a non-uniform distribution of the accelerometric stations is compensated by the physics law of damping of elastic oscillations in a homogeneous and isotropic material medium.

The network of observed data was used to adjust the physical law to the geostratigraphic system affected by the earthquake. This first result is representative of the geographical evolution of the two seismic variables over distances of tens of kilometers.

The non-stationary geostatistical estimator was applied to the data observed around each grid node of the DTM for a more precise evaluation of the local seismic state. The validity of the non-stationary geostatistical methodology is confirmed by the estimation errors calculated in the 38 sites of the accelerometric stations, which were used to verify the representativeness of the estimated data. The histogram of the 38 normalized estimation errors is close to a Gaussian with the frequency of the $\pm \sigma_{ck}$ class of 73.7% instead of 68.3%.

This result was obtained for both seismic variables treated: arrival time of the maximum acceleration and peak ground acceleration value (PGA).

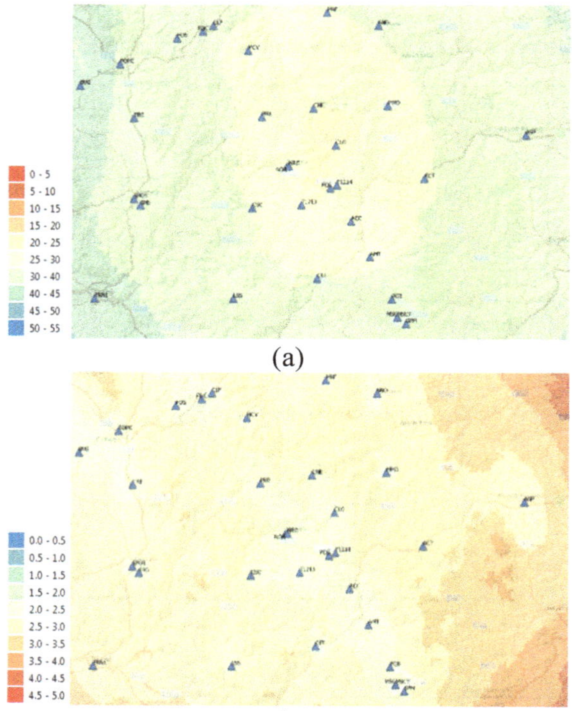

(a)

(b)

Figure 5: (a) Arrival time; and (b) Variance of estimation (Norcia earthquake).

REFERENCES

[1] Campbell, K.W., Strong motion attenuation relations: A ten-year perspective. *Earthquake Spectra*, **1**(4), pp. 759–804, 1985. DOI: 10.1193/1.1585292.

[2] Sabetta, F. & Pugliese, A., Estimation of response spectra and simulation of nonstationary earthquake ground motions. *Bulletin of the Seismological Society of America*, **86**(2), pp. 337–352, 1996. DOI: 10.1785/BSSA0860020337.

[3] Hernandez, B., Cocco, M., Cotton, F., Stramondo, S., Scotti, O., Courboulex, F. & Campillo, M., Rupture history of the 1997 Umbria-Marche (Central Italy) main shocks from the inversion of GPS, DInSAR and near field strong motion data. *Ann. Geophys.*, **47**(4), 2004. DOI: 10.4401/ag-3349.

[4] Guarascio, M., Huybrechts, C.J. & David, M., Advanced geostatistics in the mining industry. *Proceedings of the NATO Advanced Study Institute* , Istituto di Geologia Applicata of the University of Rome, Italy, 13–25 October, 1975.

[5] Matheron, G., Random sets theory and its applications to stereology. *Journal of Microscopy*, **95**, pp. 15–23, 1972. DOI: 10.1111/j.1365-2818.1972.tb03708.x.

[6] Chilés, J. & Delfiner, P., *Geostatistics: Modeling Spatial Uncertainty*, Wiley Inter-Science Publication, New York, 2012.

[7] Guarascio, M., Lombardi, M., Rossi, G. & Sciarra, G., Geostatistics/reliability based risk analysis of the Vajont landslide, pp. 607–615, 2009. DOI: 10.2495/SAFE090561.

STRUCTURAL CONDITION ASSESSMENT OF SEWERS AND ITS APPLICATION ON A CASE STUDY IN TOULOUSE, FRANCE

GEORGIA KANELLAKI[1], MARTA DEL PALACIO[1] & IDRISS BENSLIMANE[2]
[1]Infraneo, Madrid, Spain
[2]Infraneo, Paris, France

ABSTRACT

Urban densification and transformation of cities have induced an interconnection and superposition between new and old structures. Thus, it is important to verify the structural compatibility and to evaluate the risk induced on the underground structures. The goal is to avoid generating an irreversible dysfunction with disproportionate costs on urban functioning vital elements, such as wastewater and rainwater collectors. These structures are frequently impacted by current construction projects, such as tunnels, underground parking and buildings. For this reason, their resistance should be assessed. Meanwhile, most of the sewage systems of metropolitan cities have reached their design working life or have been severely damaged. Structural health monitoring of collectors is therefore an ongoing field and a useful solution to extend service life of the structures. This paper aims to provide some insights toward structural condition assessment methods of sewage systems, such as visual inspection, radar auscultation, internal hydraulic jacking tests and modelling of the structural stability due to external parameters. Furthermore, a case study of a 150/100 cm ovoid-shaped combined sewer in Toulouse, France, will be analysed applying the aforementioned methods. It is a non-reinforced concrete structure which collects wastewater and rainwater from three main conduits of the city upstream of the purification station. In regard to the tunnelling works for the metro passage underneath the collector, it was necessary to assess the impact of this operation on its integrity. 3D soil-structure interaction model has been performed using the finite element method. Structure parameters and hardening soil ones have been given according to the diagnostic. The results of numerical simulation are given in this paper and they were used as constraints for the tunnelling operation.
Keywords: structural condition assessment, structural auscultation, underground structures, sewage systems, tunnel construction impact, internal hydraulic jacking tests.

1 INTRODUCTION

Modern urban planning imposes the superposition between old and new infrastructures. Thus, it is indispensable to evaluate the health condition of the existing structures and to verify their stability due to external parameters. Sewage structures are key-elements in the functioning of daily life in cities. They are underground structures the majority of which dates back from the beginning of the 20th century in European metropolitan cities. Most of the times, they have reached their service life or have been seriously damaged. Meanwhile, they are often impacted by the construction of new infrastructures like tunnels, underground parking, buildings, road backfills, etc. Any sever dysfunction of the structure could have harmful consequences, like flooding or pollution of the groundwater. Structural condition assessment of collectors is, therefore, a field that water and sewage stakeholders have developed the last decades in order to control the health state of their structures and to privilege trenchless maintenance works. Sewage stakeholders have classified the collectors as visitable, semi-visitable and non-visitable and they have adapted the types of auscultation depending on the dimension of the pipeline.

This paper aims to offer some insights of structural condition assessment of visitable collectors. Typical auscultation methods like visual inspection and ground penetration radar

WIT Transactions on The Built Environment, Vol 206, © 2021 WIT Press
www.witpress.com, ISSN 1743-3509 (on-line)
doi:10.2495/SAFE210021

(GPR), but also innovative solutions, explicitly developed in the context of sewage auscultation, as internal hydraulic jacking tests will be presented. The goal is to estimate qualitatively and quantitatively the current state of the structure and the ground casing and bedding the collector. These two entities often work as a system and influence the stability of the sewer. Auscultations results are used to carry out a diagnostic study of the structure and according to the conclusions propose recommendations of maintenance works. They can also be the input data to assess the stability of the structure due to construction of new projects in the proximity.

In this context, a case study of a 150/100 cm ovoid-shaped combined sewer collector in Toulouse, France, will be analysed applying the aforementioned methods. It is a representative inspection, as the collector has been subjected to multiple types of emergency diagnostics due to collapsing of a part of the duct next to the national highway. In addition, metro construction underneath the collector is expected several meters upstream the collapsing zone, so the stability of the structure should be evaluated. This example will allow understanding how conclusions are extracted from each auscultation and are combined to define a degradation state of the structure. Subsequently, these conclusions for the structure and the soil will be interpreted into material properties to produce a finite element model and to define the thresholds of the settlements curve imposed by the tunnel on the structure.

2 STRUCTURAL CONDITION ASSESSMENT IN SEWAGE SYSTEMS

Diagnosis studies use different methods, such as visual inspection, non-destructive auscultations, core sampling and geotechnical essays, to divide the surveyed length of the collector into homogeneous sections of similar characteristics. Subsequently, the sewer and its surrounding ground are classified in an acceptable, medium and poor state according to diagnosis results. In this study, only non-destructive methods will be assessed.

2.1 Visual inspection

Visual inspection is carried out within the context of technical standards and recommendations, based on experience feedback [1]. This experience has grouped the degradations of sewers into five distinct families, ranked in descending order of potential structural risk, and therefore in relation to restructuring needs: breaks, deformations, leaks, punctual anomalies, degradation of coatings. According to the frequency and the significance of anomalies, the condition of the structure can be classified good, medium or poor. The visual inspection frequently allows to do pre-diagnostics studies in order to determine the cause of the damage and the adequate type of auscultation for the system soil/structure. Meanwhile this method is not sufficient enough to do a complete diagnosis.

2.2 Ground penetration radar auscultation

Radar auscultations allow detecting the existence, the position, the depth or the nature of an object through the study of the reflection of electromagnetic waves. GPR is a non-destructive technique (NDT) which detects interfaces between materials (structure, geological levels) or heterogeneity presence (cavities, flow paths, etc.). The wave frequencies used for this kind of auscultations are between 10 MHz to 5 GHz [2]. Radar auscultation is feasible by surface or from inside buried structures. Underground infrastructures are studied in a frequency range of 100 MHz to 1 GHz which is used to investigate on the meter scale, but specific information acquisition as concrete reinforcement detection (investigation depth less than 0.5 m) uses frequencies until 2 GHz. Sewage systems condition assessment uses widely GPR auscultations to determine information about:

- **Structure:** Nature of structure (concrete reinforced or non-reinforced, masonry), thickness, quality and reinforcing bars location.
- **Interface between structure and soil:** Presence of timbering and metal frames from construction shielding and quality assessment (voids presence, decompressions, detachments, etc.).
- **Ground (cover/casing/bedding soil):** Ground nature, geological levels and quality of the ground around the sewer, determination of nearby infrastructures presence (urban networks, underground structures, foundations, etc.).

2.3 Internal hydraulic jacking tests

Internal hydraulic jacking tests aim to calculate the soil micro-deformation modulus and determine a quality index for the underground structure by measuring a global stiffness of the system soil-structure. The jacking device contains a hydraulic jack and three rods placed between the walls (with a longitudinal spacing of approximately an equivalent diameter of the section) and just below the vault. The principles of the test are presented in Fig. 1(a). This auscultation method has been developed in France. Water and wastewater stakeholders have widely acknowledged this method during the last two decades.

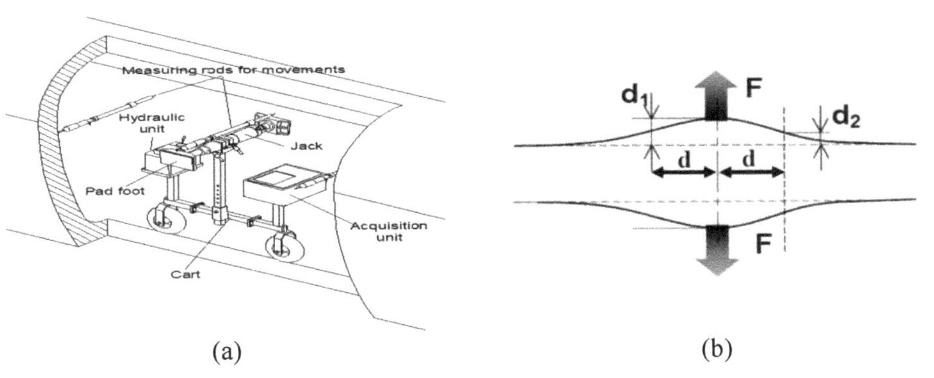

Figure 1: (a) Jacking device presentation; and (b) internal hydraulic jacking tests principles.

The test consists in applying an ovalization loading on the conduit by means of a horizontal jack. Three displacements are extracted with help of the measuring rods, one on level of the jack and two others at a certain distance from this last one (Fig. 1(b)). The measures remain always in the elasticity domain (micro-deformations). Each test has two loading cycles and each parameter is measured and calculated twice. The values obtained are generally averaged to constitute an aggregated result.

The calculation is inspired by a simple continuous beam on elastic supports (spring stiffness k), which represents the system soil-structure [3]. The stiffness of this system in longitudinal direction (K_g) is a combination of the bending rigidity of the structure (EI) and the modulus of reaction of the ground (k). Increasing relative stiffness of the soil results in a concentration of the deformation around the loading application point. Thus, the test consists in assessing simultaneously the total stiffness of the system (eqn (1)) and the localisation of the response (eqn (2)) to determine the stiffness of the beam and its supports distinctly, by using the displacements measurements (d_1, d_2) and the applied load (*F*).

Global stiffness:

$$K_g = \frac{F}{d_1/2}, \tag{1}$$

Deformation transmission ratio:

$$\Omega = \frac{d_2}{d_1}. \tag{2}$$

The goal of this procedure is to achieve decoupling the system, i.e. to represent the factors *EI* and *k* with functions related directly to the measured values of K_g and Ω. Decoupling of the system is done with the help of a finite element (FE) model and iterative calculations in order to obtain a polynomial expression for each variable. The model is based on the assumptions that the two parameters are distinguished and their mechanical behaviour is elastic, linear, homogeneous and isotropic. Each material is characterised by a longitudinal deformation modulus (E_i) and a Poisson coefficient (ν). The interface between the two materials does not consider slippage and detachment and the only charge considered is the internal jack load. The iterative procedure calculates the theoretical stiffness and transmission rate for eleven representative cases, from the unsupported structure (K_0, without soil contribution) until a system with a very compact soil contribution. The resolution is based on a dimensionless problem that defines the curves of the soil and structure modulus in relation to total stiffness in function of the transmission ratio ($E_c/K_g - \Omega$ and $E_s/K_g - \Omega$). Afterwards, the cubic degree polynomial adjustment of these curves allows obtaining their analytical resolution (eqns (3) and (4)).

$$\frac{E_s}{K_g} = f_1\left(\frac{\Omega}{\Omega_0}\right) \cong \sum_{n_1}^{k=0} \alpha_k \left(\frac{\Omega}{\Omega_0}\right)^k \tag{3}$$

$$\frac{K_c}{K_g} = f_2\left(\frac{\Omega}{\Omega_0}\right) \cong \sum_{n_2}^{k=0} \overline{\alpha}_k \left(\frac{\Omega}{\Omega_0}\right)^k \tag{4}$$

where E_s and K_c are the soil and structure stiffness respectively, k is the polynomial expression degree (limited at 3) and $\alpha_0, \alpha_1, \alpha_2, \alpha_3, \overline{\alpha}_0, \overline{\alpha}_1, \overline{\alpha}_2, \overline{\alpha}_3$ are the polynomial constants. Ω_0 is the transmission ratio for the case of the unsupported structure. The quality index of the structure for each measure is obtained as $IQ = K_c/K_0$ where K_0 is the unsupported structure stiffness.

The transmission ratio takes its maximal value for the case of the unsupported structure (Ω_0), beyond this value, decoupling is not possible. This value Ω_{max} depends exclusively on structure's thickness and geometry. The relation E_c/K_g is an increasing function, while E_s/K_g is a decreasing function which depends quasi-linearly on Ω in a wide range of variation, except for small values of the three-dimensional effect (Fig. 2). This case reflects rather the compression (crushing) of the collector in the thickness direction than an overall response of the structure in bending. It should therefore also be possible to deduce a minimum value Ω_{min} below which the test no longer falls within the proposed interpretation scheme.

The curves tendency allow a qualitative interpretation of the tests results for a measurements section. When K_g and Ω vary in the same direction, sewer's stiffness follows total stiffness evolution. For the opposite case, it is observed the same behaviour for soil stiffness. Thus, it is possible to distinguish zones that qualitatively have different geometrical or mechanical characteristics.

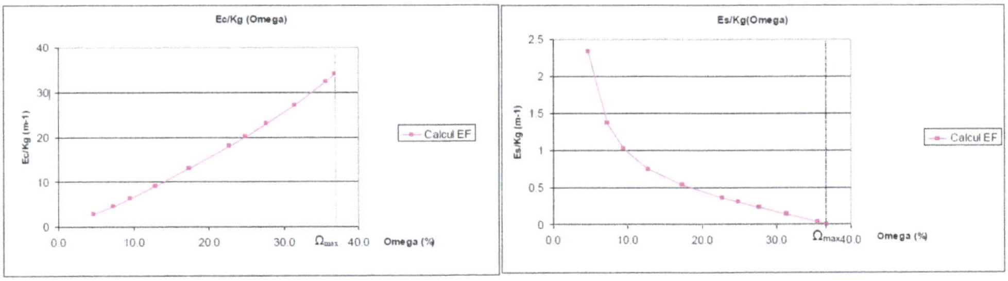

Figure 2: Evolution of E_c/K_g and E_s/K_g in relation to Ω.

The calculated soil modulus (E_s) corresponds to the microdeformation domain of deformations (less than 10^{-5}). The relationship between the soil modulus and the deformation domain is defined as logarithmic (Fig. 3). Using the results of the internal hydraulic jacking tests, the modulus (E_s) is translated into the deformation domain of underground structures (tunnels) as its value divided by 2 or 3.

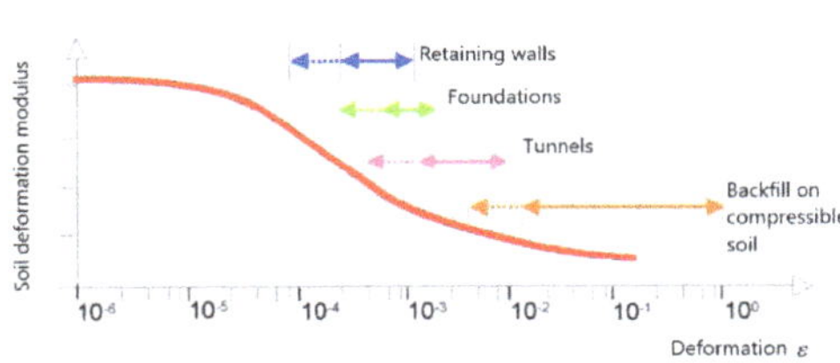

Figure 3: Evolution of the soil modulus in relation to the referenced deformation domain [4].

3 ABC COLLECTOR IN TOULOUSE

ABC collector in Toulouse, France is a representative study case of sewage systems condition assessment. It is a 150/100 cm ovoid-shaped combined sewer built with non-reinforced concrete. It collects wastewater and rainwater from three main conduits of the city upstream the purification station, it is thus under important hydraulic charge and should be rarely non-operational. The collector has been subjected to multiple types of emergency diagnostics on different parts along its length, due to collapsing of a part of the conduit. In addition, the passage of the upcoming metro line 3 and the fragility of the structure pointed out the necessity of a tunnelling impact study to determine the admissible settlement curve induced by the metro construction underneath the sewer (Fig. 4).

3.1 Collapsing of the collector, structural condition assessment and rehabilitation

In February 2021, the sewer subjected to an emergency diagnostics on a 450 lm part of the structure linear (RV-1 to RV3 and RV3.4 to RV4 + 40 m) due to collapsing of a part of the conduit (RV3.1 to RV3.4), situated next to the national highway. The structural condition assessment included visual inspection, internal hydraulic jacking tests, radar auscultation, core sampling and resistance tests of the structure and the soil behind. Visual inspection revealed poor degradation of the structure. Thus, radar auscultation and internal hydraulic jacking tests allowed investigating further the cause of the collapsing.

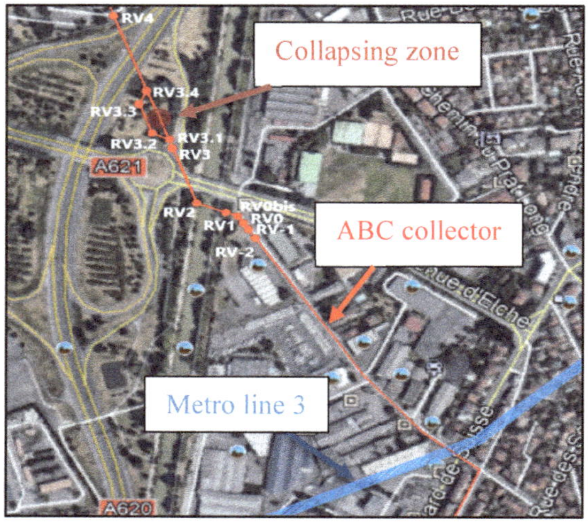

Figure 4: Disposition of the ABC collector in the studied zone.

Radar devices took measures from the intrados of the sewer on the vault and the base of the sidewalls (Fig. 5(a)). Radar auscultation determined the quality of the soil, the structure and the interface between them. Radar survey indicated heterogeneities at the interface and in the ground on both sides and the vault of the collector for the zone next to the collapsing (Fig. 5(b)). These results can be correlated with the presence of water leakage due to structure degradation. These water exfiltration are the cause of soil decompression by entrainment of fines. In addition, the survey estimated the medium thickness of the structure to 25 cm.

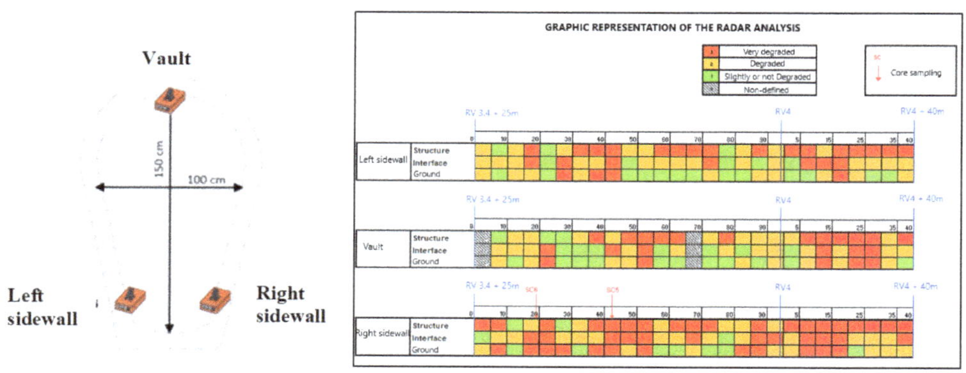

(a) (b)

Figure 5: (a) Radar position; and (b) Results of the radar survey for the part between RV3.4 + 25 m and RV4 + 40 m of ABC collector.

Internal hydraulic jacking tests obtained initially the total stiffness and the transmission ratio of the system (Fig. 6). A model of the ovoid structure was produced according to its

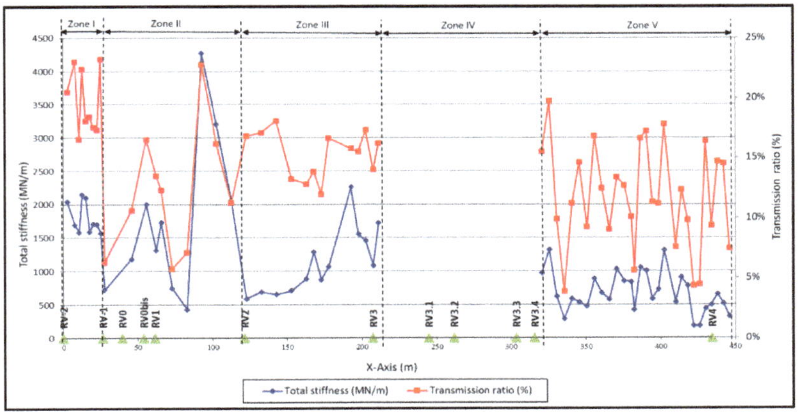

Figure 6: Internal hydraulic jacking tests measures.

Table 1: Zone definition for internal hydraulic jacking tests measurements and mean results.

Zone	Total stiffness (K_g)	Transmission ratio (Ω)	Covering high	Soil stiffness (E_s)	Quality index (IQ)
I	155–200 MN/m	18–24%	3.5 m	210 MN/m	60%
II	500–4200 MN/m	5–24%	3–9 m	670 MN/m	33%
III	600–2300 MN/m	12–18%	9 m	340 MN/m	33%
IV	Not auscultated	Not auscultated	–	–	
V	100–1200 MN/m	3–21%	3.6 m	318 MN/m	16%

geometrical characteristics and a theoretical modulus of 35 GPa. The parameters of the unsupported structure were determined to $\Omega_0 = 24\%$ and $K_0 = 2194$ MN/m. The analysis pointed out five zones where the characteristics could be grouped qualitatively (Table 1).

The decoupling of the system complemented radar results by defining a poor to medium state of the structure in the zone around the collapsing (Zone III and V, Fig. 7). In the zone I, more than 100 m upstream the collapse, the structure and ground charts present a more homogenous behaviour. The structure is in a medium to acceptable state, while the soil is loose to compact (Fig. 8). The difference of the covering high of the structure due to the highway backfills disposition (escalation from 3 m to 9 m at RV1 + 40 m) produce heterogeneous results concerning the compactness of the soil in the zone II.

3.2 Metro construction impact on the structure: Tunnelling impact study

As the structure is sensible, it was indispensable to assess the impact induced by the tunnel construction and to define thresholds for its integrity. The metro line 3 will be constructed underneath the collector. The sewer will be impacted by the settlements of the bedding ground due to its deconfinement during the passage of the tunnel boring machines (TBM). That procedure induces tensile stresses at the invert. The goal was to determine the acceptable thresholds for the sewer in terms of maximum settlement / maximum slope, induced by the TBM underneath the structure. These thresholds are set so as not to exceed the serviceability limit of the material constituting the collector structure. In the case of exceeding these

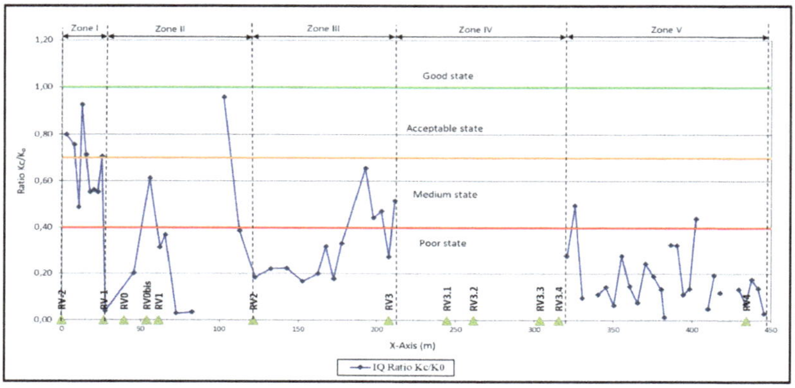

Figure 7: Quality index resulted from internal hydraulic jacking tests results.

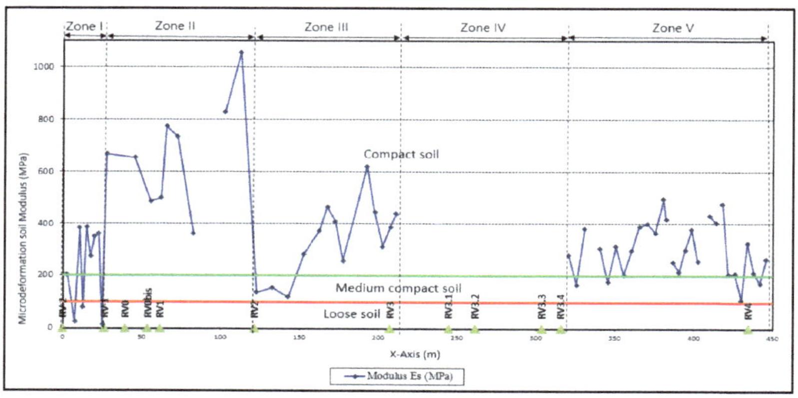

Figure 8: Soil modulus resulted from internal hydraulic jacking tests results.

admissible settlements, a reinforcement solution should be applied on the structure. A 3D soil-structure interaction model was performed using CESAR 3D LCPC software [5] in order to produce the tunnelling impact study. The structural condition assessment on the collector a few linear meters upstream the tunnelling impact zone was an important tool for the definition of the material properties of the structure and the ground.

3.2.1 Settlement curve underneath the collector

The settlements induced underneath the structure were defined based on recent bibliography. Most of the current research has focused on measuring and determining the settlements generated on the surface [6], which present the shape of a Gaussian curve. It has been observed that the settlements at a given depth follow this curve. The goal of an impact study is to evaluate the stresses generated at the underground structure from the existing empirical formulas for the determination of settlements and from numerical modelling methods. Peck's formula [7] is one of the most acknowledged to calculate le settlement curve of a tunnel (Fig. 9). According to this formula the displacement on each point of the transversal axis of the

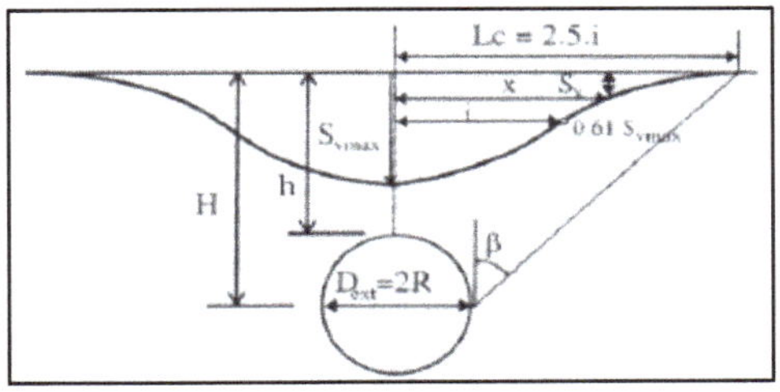

Figure 9: Settlement curve due to tunnel passage according to Peck's formula [7].

passage is a Gaussian curve (eqn (5)). It depends on the maximum settlement (S_{vmax}, eqn (6)), the volume loss due to the excavation (V_L), the tunnel diameter (D_e) and depth (H). The distance of the inflexion point from the centre of the hollow is determined in this case by the O'Reilly and New approach [8] and depends on a parameter (K) of the soil nature on each ground layer (eqn (7)).

$$S(x) = S_{vmax}\, e^{\left(-\frac{x^2}{2\,i^2}\right)}, \tag{5}$$

$$S_{vmax} = 0{,}313\,\frac{V_L D_e^2}{(0{,}5H)}, \tag{6}$$

$$i = K_i H_i, \tag{7}$$

where K_i and H_i the K parameter and the depth for each ground layer.

The effect of the impact depends on the relative position between the tunnel and the sewer. A transverse passage will produce displacements along the sewer and thus longitudinal stresses. A longitudinal passage will produce stresses in the cross section.

3.2.2 Finite element model

TBM will cross the sewer in an oblique angle (Fig. 4). The model considered that it is perpendicular to the collector in order to assess the case corresponding to the steepest Peck's curve slope. Several theoretical settlement curves of different amplitudes applied to the model, according to different excavation depths with maximum settlements of 5 mm, 10 mm and 15 mm (Fig. 10). TBM characteristics and geotechnical parameters were based on experience feedback report (REX) for the construction of the metro line B in Toulouse [9]. The excavation parameters D_e = 7.8 m and H = 7.54 to 22.3 m and the settlements measured in the REX (S_{vmax} ~ –3 mm) for a tunnel excavated with slurry pressure balanced (SPB) tunnel boring machines conditioned the variables K = 1.12 and the loss of volume V_L = 0.20%.

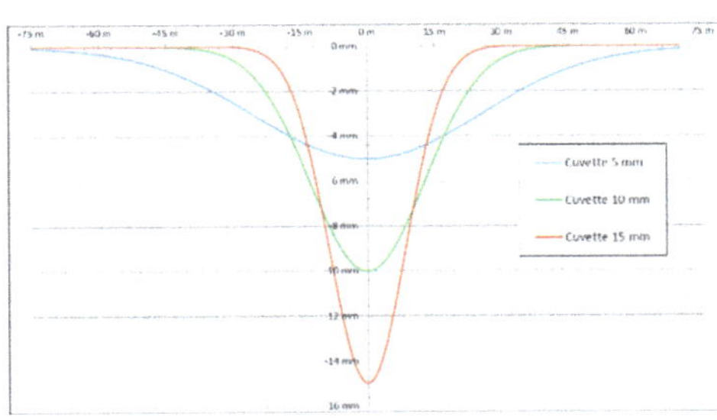

Figure 10: Figure settlement curves for maximum settlement of 5, 10, and 15 mm.

$$S_{vmax} = 5\ mm, H = 22.3\ m, i = 0.12‰\ ,$$

$$S_{vmax} = 10\ mm, H = 11.3\ m, i = 0.47‰\ ,$$

$$S_{vmax} = 15\ mm, H = 7.54\ m, i = 1.07‰.$$

The finite element model used the static soil-structure interaction theory. The calculations carried out without amplification coefficients of the loads so that the soil elasto-plastic behaviour is unaffected. The ultimate limit state (ULS) of the structure is obtained by applying the magnification factors to the results. 3D model represented a section of the soil around the sewer with a width of five equivalent diameters of the pipe (to neglect the edge effects) and a depth of 1 m underneath the invert base. It assessed only one quarter of the problem due to the symmetry of the settlement curve as well as the axial symmetry of the structure. The model only covered the longitudinal half of the settlement basin and the transverse half of the collector (Fig. 11). The length of the numerical models was set to 70 m (larger than half the width of the longest settlement basin) in order to neglect as much as possible the edge effects on the boundary conditions of the model. The boundary conditions defined as blocked displacements assigned to the sides of the soil volume.

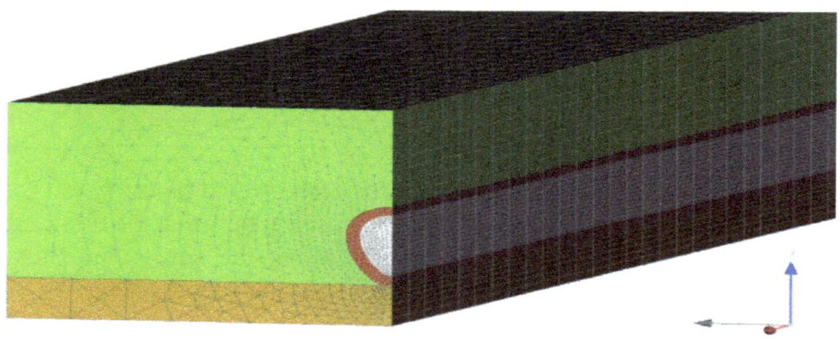

Figure 11: 3D mesh model.

3D model considered the structure as geometrically continuous. No defect, discontinuity or damage was taken into account. Thus, the mechanical properties of the materials describe a global behaviour and they were defined by the auscultations. The behavioural models considered for the materials are isotropic linear elasticity for the sewer and hardening soil model (HSM) [10] in elasto-plasticity for the soil (casing, cover, bedding and interface). Soil characteristics estimated from typical geotechnical parameters for each layer and the radar observations that characterised a terrain overly medium quality (Table 2). Soil loading modulus for the casing terrain determined by the internal hydraulic jacking tests results. It considered that the model was into the tunnelling domain of deformations. Compression essays produced on the core sampling of the structure and measured fluctuating resistances. Concrete thresholds were fixed by considering a conservative modulus of elasticity at 25 GPa and calculating the associated compression and tensile resistances for SLS according to the EC2. The deformation threshold was determined on the basis of the AFTES recommendations [11], which indicate a limit of 0.3‰ at SLS for underground concrete networks. Taking into account the sensitive nature of sewerage networks (exfiltration), this value was limited to 0.15‰.

Table 2: Geotechnical parameters of the model.

Terrain	Soil parameters					
	Nature	Density (γ)	Poisson coefficient (ν)	Cohesion (c')	Friction angle (φ')	Loading Modulus ($E_{50\,ref}$)
Casing	Backfills	20 kN/m^3	0.33	0 kPa	25°	50 MPa
Bedding	Alluvium	20 kN/m^3	0.33	10 kPa	30°	100 MPa

The loads were applied in distinct phases according to the construction mode of the structure. The first phase consisted on the initialisation of the soil stresses by applying its self-weight. Afterwards, the phase of the collector construction followed, with the application of its own weight as well as the soil deconfinement forces around the structure (λ = 1, collector built in trench). Finally, the excavation of the TBM under the collector was modelled. In this phase, the settlement (Gaussian curve) was imposed as a displacement on the z-axis on the ground one meter below the invert of the collector. This modelling convention took into account the possible influence of the presence of the structure for the reduction of the settlements generated by the excavation. In this phase, a hydrostatic pressure was also considered inside the collector to take into account the weight of the water in case the pipe is full at the time of the passage of the TBM (Fig. 12).

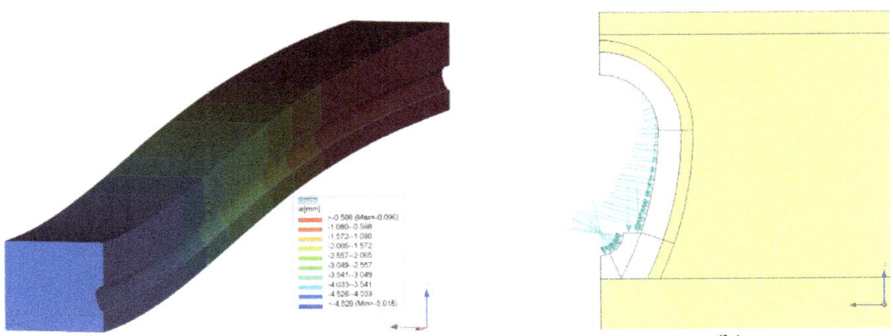

(a) (b)

Figure 12: (a) Settlement curve imposed on the model; and (b) hydraulic load into the sewer.

Several models were carried out to apply the different theoretical settlements proposed. The maximum normal stresses generated in the collector in the longitudinal direction occur at the point of maximum settlement. The tensile stress is the dimensioning parameter as its threshold is exceeded when the maximum settlement of 10 mm is applied (Table 3). The

Table 3: Modelling results.

Curve	S_{vmax} = 5 mm	S_{vmax} = 10 mm	Thresholds
Compression max σ_{yy}	–0.283 MPa	–0.665 MPa	$\sigma_{c\,ELS}$ = 0.6 f_{ck} = –4.39 MPa
Tensile max σ_{yy}	0.067 MPa	0.805 MPa	$\sigma_{t\,ELS}$ = 0.6 f_{ctm} = 0.68 MPa
Deformation max ε_{yy}	0.01‰	0.07‰	0.15‰

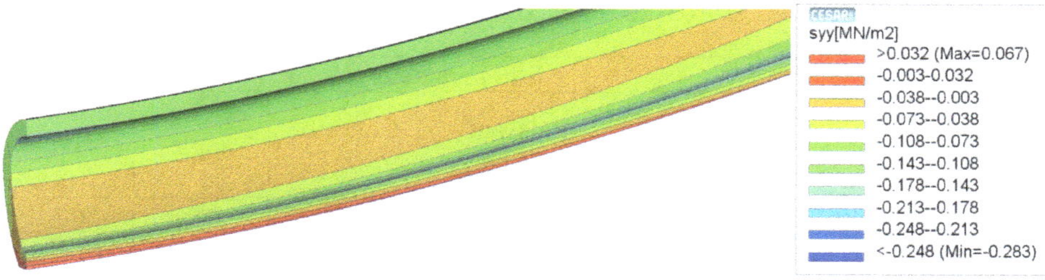

Figure 13: Normal stresses in the longitudinal direction (MPa) for the case of a 5 mm settlement curve.

threshold curve applied under the sewer is a maximum settlement of 5 mm. For this case, the tensile stresses are located in the invert and compressive stresses in the vault. They extend over a length of about 15 ml on either side of the tunnel axis (Fig. 13).

4 CONCLUSION

The paper provides a case study for the structural condition assessment of visitable sewers. It is based on existing methods like visual inspection and non-destructive auscultations, such as radar and internal hydraulic jacking tests, explicitly developed for this type of underground infrastructures. Through these methods a tunnelling impact study has been undertaken and the risk induced on the underground structures due to the construction of new metro lines has been determined. The admissible settlement curve imposed on the collector calculated by a soil-structure interaction finite element model. Calculated admissible settlements will serve as constraints for the tunnelling operation. Carrying out internal hydraulic jacking tests and visual inspection after the tunnel construction could allow to detect the evolution of the structure cracking and of the soil compactness. If necessary, a reinforcement solution should be applied.

REFERENCES

[1] Projet national rehabilitation des réseaux d'assainissement urbains (RERAU), Restructuration des collecteurs visitables, 2004.

[2] Quinta-Ferreira, M., Ground penetration radar in geotechnics: Advantages and limitations. *World Multidisciplinary Earth Sciences Symposium (WMESS)*, 2018.

[3] Thepot, O., Prise en compte des caractéristiques en petites deformations des sols dans l'étude du comportement des collecteurs enterrés. *Ecole Nationale des Ponts et Chaussées*, 2004.

[4] Khiatine, M., Reiffsteck, P. & Ramdane, B., Evaluation of dynamic soil properties for alluvial plain of Bejaia using field data and laboratory tests. *Geotechnical and Geological Engineering*, **37**, pp. 4707–4730, 2019.

[5] CESAR 3D – LCPC, Computer software, *itech*, 2014.

[6] Viggiani, G. & Soccodato, F.M., Predicting tunnelling-induced displacements and associated damage to structures. *Rivista italiana di geotecnica*, **4**, 2004.

[7] Peck, R.B., Deep excavations and tunnelling in soft ground. *7th International Conference of Soil Mechanics and Foundations*, pp. 226–290, 1969.

[8] O'Reilly, M.P. & New, B.M., Settlements above tunnels in the United Kingdom: Their magnitude and prediction. *Proc. Tunnelling*, **82**, pp. 387–400, 1982.

[9] Vanoudheusden, E., Impact de la construction de tunnels urbains sur les mouvements de sol et bâti existant: Indice du mode de pressurisation du front. L'Institut National des Sciences Appliquées de Lyon, 2006.
[10] Schaz, T., Vermeer, P.A. & Bonnier, P.G., The hardening soil model: Formulation and verification. *Beyond 2000 in Computational Geotechnics – 10 Years of Plaxis*, 1999.
[11] Recommendations de l'Association Française des Tunnels et de l'Espace Souterrain (AFTES), Tassements liés au creusement des ouvrages en souterrain, p. 13, 2018.

SECTION 2
CYBER SECURITY/
E SECURITY

ASSET MANAGEMENT TOOLS IN THE DIGITAL ENVIRONMENT

LAZAR D. GITELMAN, TATYANA B. GAVRILOVA & MIKHAIL V. KOZHEVNIKOV
Academic Department of Energy and Industrial Enterprises Management Systems,
Ural Federal University, Russia

ABSTRACT
The paper summarizes the changing approaches to asset management of manufacturing companies amid the unprecedented uncertainty in markets, progressing digital transformation and production facilities being enriched with intelligent information and communication systems. A conceptual model has been designed that reflects the new principles and the key conditions of asset management systems' operation in the digital industry. Having examined the cases of the energy sector and high-tech industries, the authors identify the main trends in the innovative development of production asset management: cutting-edge reliability-focused repair and maintenance methods, the digital twin technology, and platform solutions for business customization. The results of the study could be used for asset management strategy development in manufacturing and for improvement of the relevant legal and regulatory frameworks, including the international standards PAS 55 and ISO 55000.
Keywords: production and intellectual assets, unstable environment, digital transformation, repair and maintenance, digital twin, platform solutions, customization.

1 INTRODUCTION

ISO 55000:2014 defines asset management as "the coordinated activity of an organization to realise value from assets" [1]. This kind of activity implies reaching the balance of asset-related performance, cost and risk and should accompany the whole lifecycle of an asset because with time, different components of the asset may change due to diverse factors. Competent asset management is crucial for meeting stakeholders' expectations within the existing budget constraints and/or ensuring compliance with the absolute (e.g. legal) restrictions.

Traditionally, asset management was focused on production assets, in particular fixed capital – long-term physical assets such as facilities, buildings, machinery and equipment, vehicles, and so on. In recent years, however, more attention has been given to intangibles – assets that are not physical in nature – such as information and knowledge, results of R&D, and intellectual property [2]. To perform well in a fast-changing and unpredictable environment, a company needs to adjust its strategy regularly, which, in its turn, requires a synchronous reaction on the part of the asset management system [3]. Digital transformation has made data, systems of data processing and analysis crucial for companies wishing to keep their competitive edge in the long run, thus increasing the significance of intangibles in business [4]. These systems support the exchange of data, communications and transactions for better engagement between different stakeholders. Moreover, although the significance of industrial assets in infrastructure sectors still remains very high, it is now not the fact of asset ownership as such that matters the most but the maximization of value for the consumer that could be achieved with the help of these assets.

Digital transformation and modern IT-systems have become pivotal to the success of businesses seeking to create value from assets in a turbulent world.

This study aims at systematizing the most significant changes in the approaches to asset management of industrial companies amid the unprecedented uncertainty of the market

www.witpress.com, ISSN 1743-3509 (on-line)
doi:10.2495/SAFE210031

environment, digital transformation and enrichment of industrial production with intelligent data and communications systems.

2 NEW PRINCIPLES OF ASSET MANAGEMENT

The general requirements to the asset management system are described in the well-known international standard PAS 55 [5], [6]. PAS 55 outlines the types of activities that are necessary for efficient asset management and for evaluation of the assets' current condition. The Standard is of non-regulatory nature: it provides guidelines on how to improve the asset management policy and eventually optimize all management processes in the organization as a whole.

PAS 55 is based on the use of the Plan-Do-Check-Act (PDCA) cycle (see Fig. 1): the development of an asset management system should start with formulating the principles of the technical policy, devising a long-term strategy and setting goals regarding the results, risks and costs, and drawing long-term plans, which should correspond to the expectations of asset owners and other stakeholders.

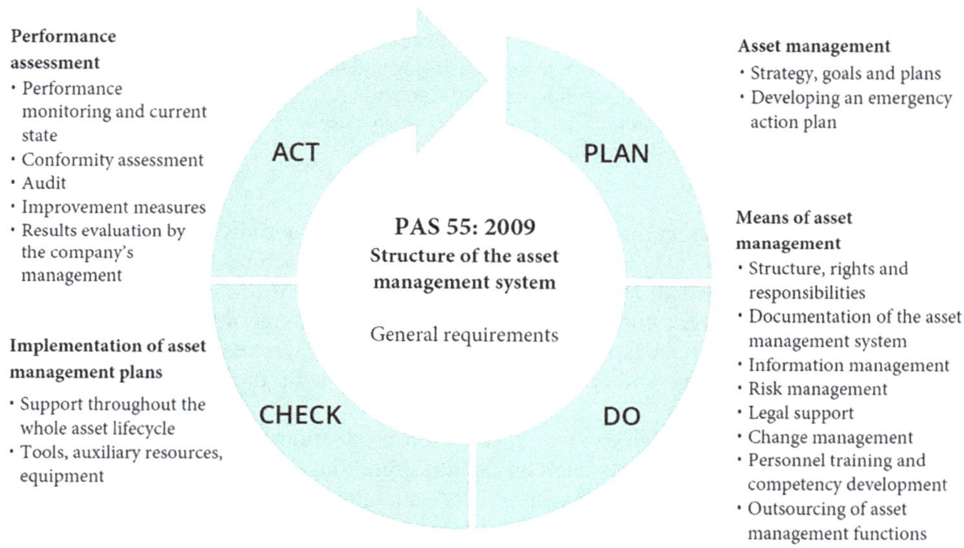

Figure 1: PDCA cycle according to PAS 55.

In our view, as digital transformation is becoming faster and more large-scale, the range of requirements to asset management is becoming broader and new principles and tools are being introduced to asset management activities.

1. Within an organization, the asset management system should be working in sync with other management systems (see Fig. 2). If an asset management system comprises assets of all types (that is, apart from physical assets, it includes intangible, financial, information and human resources), it is necessary to ensure effective coordination of all these subsystems and communication between them. The latter is based on an information infrastructure that needs to be constantly improved. Such infrastructure should include the knowledge management function, that is, the knowledge of the company's assets and the related processes, operations and regulations.

2. To select the main assets for production and improvement of the key products and services, the platform and modular principles are applied. Assets whose impact on the technological architecture of products is minimal should be replaceable easily and cheaply.
3. The asset management system maintains real-time access to the data on the retrospective and current state of the company's assets. Predictive analytics tools, big data technologies and robotic tools can be used for prediction and prevention of machine failure.
4. For optimization of the overall performance – the value gained from assets, it is necessary to identify the trade-offs between different factors (results, costs and risks) considered throughout the whole asset lifecycle. The potential growth in the complexity of assets throughout their life, the increasing uncertainty of the external environment make expert communications even more important. Expert communications include specialized research, continuous personnel training, and organization of interdisciplinary teams.

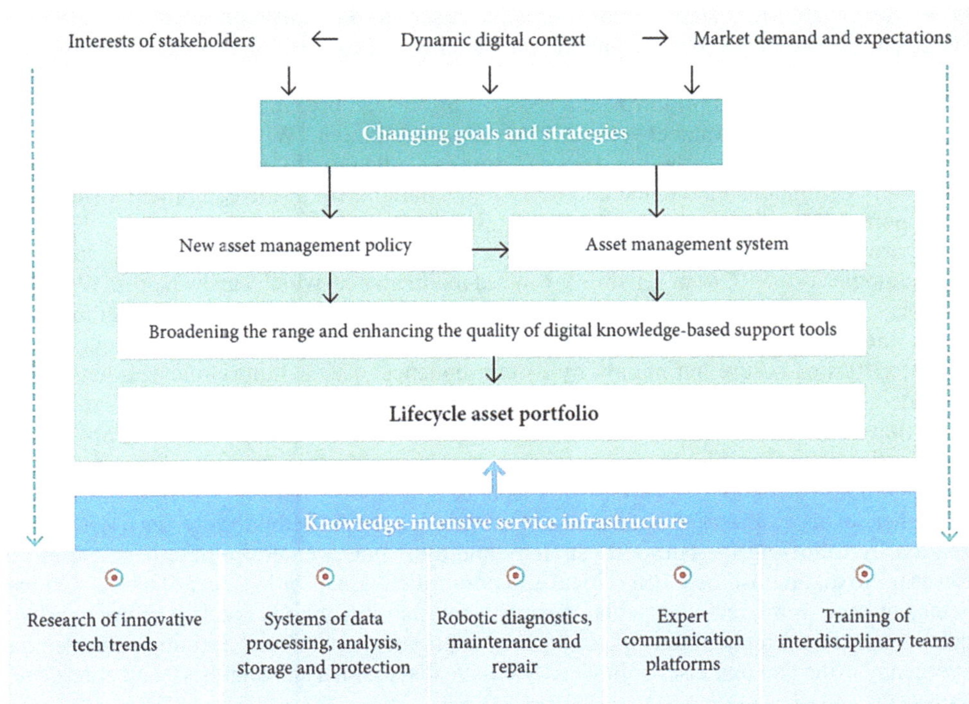

Figure 2: Connections between the key elements of the asset management system.

The main conditions that determine the successful operation of an asset management system are as follows:

- Effective coordination of the asset management policy and ensuring its compliance with the company's general strategy and digital transformation goals.

- Flexible organization culture; the company's ability to ensure efficient communication with consultants, stakeholders and professional communities.
- Coordination of plans for development and digitization of asset subsystems of various types.
- Good internal communications between specialists of different subsystems; cross-functional coordination, including platform-based coordination.
- High levels of competency and sufficient experience of the staff working in the system of asset management; organization of staff training for non-standard problem-solving.
- High quality of information support and its compliance with the decision justification requirements, including big data analytics and availability of information about the state of the assets, their characteristics, risks and costs.
- Complete data accessibility; the speed of data transfer and keeping the knowledge management system updated from different sources.

3 EMERGING OPPORTUNITIES IN ASSET MANAGEMENT TECHNIQUES

At the lower level (management of individual assets throughout their lifecycle), the most advanced approach is to switch from the scheduled preventive maintenance (SPM) system to condition-based maintenance (CBM). The CBM approach results in reduced maintenance and repair costs, it helps minimize the risks of unforeseen breakdowns between overhauls, and improve the performance of machines and equipment. With regard to the above-mentioned considerations, the key factor of success is the quality of the diagnostics system (preferably continuous automated control over the parameters of the equipment's state) and IT-support.

Regarding the development of strategies of asset use, one of the most advanced methodologies is the RCM or reliability-centred maintenance, which can be helpful to choose the most suitable failure management policy. The primary idea behind the RCM methodology is that the rules of equipment maintenance are determined not only by the nature and characteristics of failure but mainly by its consequences [7]. A failure management policy may include predictive or preventive maintenance, changes in the operational rules and regulations, changes in the design or configuration of the system, and other measures aimed at mitigating the effect of a technical failure [8]. These measures may also include changes in the methods of conducting certain maintenance and repair works and personnel training.

Higher levels of the asset management system require a methodology for asset system creation by following the corporate strategy and asset management policy. It is, therefore, necessary to distinguish between critical and non-critical assets or systems of assets. Critical assets, in their turn, require special means for sustaining their smooth performance and include ensuring preparedness of emergency equipment, personnel training (e.g. regular emergency drills and exercises followed by on-site discussions and analysis), and emergency plans.

The asset management system implies design, implementation and support of the information infrastructure, whose purpose is to provide all the key stakeholders with access to all the necessary data. These data may include registers of assets, drawings, agreements, licenses, and documents specifying regulatory requirements, policies, standards, directives, data on the state of assets and their productivity, and employees' undocumented knowledge. Information about this or that asset is usually communicated from the lowest level of asset operations.

There is intensive information exchange both inside the organization and between the organization and external stakeholders (consultants, specialist communities, and partners), based on a wide range of tools and methods.

Companies can benefit from the technologies for real-time monitoring of assets, which allow to gain situational awareness much faster than before. Through analysis of retrospective and current data and the use of special prognostic tools, it is now possible to predict the failure of machines or their parts and thus optimize the costs of maintenance and repair in the future. As a result, risk- and efficiency-based asset management strategies are becoming more widely spread [9], [10]. Such strategies are particularly relevant in relation to critical infrastructure, where emergency situations, such as unexpected asset failure may lead to huge financial and reputation losses and in some cases have even more disastrous consequences of the technogenic, social and economic nature.

To implement a *risk-based asset management strategy*, modern risk management analysis tools are necessary, such as robotic devices, gaining popularity in power engineering for intellectual monitoring of electrical power and heat networks [11]. For example, drones equipped with sensors, cameras and intelligent software can be used for inspection of high voltage electrical lines. They are able to detect damages and send this information to the dispatcher. This way the difficult and time-consuming work of powerline inspection is done remotely with drones being able to access the most hard-to-rich places, capturing a wealth of data, finding and assessing damage. The use of drones has increased the accuracy of powerline inspections since many damaged spots are usually hidden from human vision.

Together with the growing digitization of the asset management sphere, there is a growing need for *digital management asset (DAM) systems*, combining data processing, storage and distribution, management of knowledge and digital content [12]. DAM systems encompass not only equipment and fixed assets but can also apply to asset management in the broadest sense – to organizations in different sectors (e.g. industry, public services, finance and investment), for control and regulation on the level of individual companies or on the national and even global scale [13].

To create and develop a DAM system, substantial IT-support is required, including software and hardware and qualified personnel. The majority of organizations go for outsourcing as the most evident solution, that is, they buy the services of SaaS data centres (SaaS – Software as a Service). Large organizations which make extensive use of information technologies and offer IT-services to their customers create data centres of their own. In order to provide an adequate level of data management, companies purchasing cloud services employ qualified IT-specialists to maintain efficient communication with service providers, ensure cyber security, and develop IT-strategies.

A real breakthrough in digital management was the development of the *digital twin*. The digital twin is a virtual interactive copy of a real physical object, product, technological process or production in general [14]. For example, a digital twin of a manufacturing facility can model the location of the machines, staff flow, operational processes and emergencies. The functions of a digital twin are not limited only to the collection of data acquired during the development and manufacturing of a system (product). In fact, the digital twin continues gathering and analysing information throughout the whole lifecycle of the real object, for example, with the help of IoT technologies.

This tool is quite expensive to implement, although for large corporations engaged in long and capital-intensive investment projects digital twins pay off multiple times. For example, thanks to the interactive analysis of the data from the digital twin by Schneider Electric, one of the European oil refineries had received a warning of potential compressor failure 25 days before it actually happened. The company thus managed to save several millions of dollars through the digital twin concept married with technologies of big data analysis, AI, machine learning, and industrial IoT [15]. Another example is General Electric's Digital Twin – a corporate system for control over a power generation company's performance. The principle behind this system is the development of a behavioural digital twin of an asset (usually equipment): a mathematical model is created for analysis of the asset's behaviour [16].

There are different types of digital twins developed for specific processes. For instance, ABB's system of data acquisition and processing Transformer Electronic Control is applied in electric power engineering to monitor temperature and currents and create thermal models for transformers and voltage regulation. These models calculate the most heated points in accordance with the standards of the International Electrotechnical Commission and Institute of Electrical and Electronics Engineers. They are also used for modelling of entire thermal behaviour, which allows to compare the measured and expected values.

A novel perspective on asset management is offered by the developing *platform approach*. Platform is understood as a set of various assets (equipment, processes, knowledge, people and relationships) used to create a portfolio of products and services able to change their value, consumer properties and functions as new product components are promptly added or old and irrelevant are removed [17].

To construct the necessary set of assets, the modular approach is applied, which distinguishes between two types of assets – basic and variable. Basic assets are important for the creation of a unique technological architecture of products – the key competitive advantage of the company. Variable assets are pivotal to the production of additional (complementary) services and also maintain the overall stability and efficiency of the asset system.

Pioneers of the platform model are the companies which in fact possess no physical assets such as Uber, Airbnb and Facebook. Online platforms are now making deep inroads into such industries as heavy engineering, aerospace engineering, mining and electronics, and power engineering [18]. The main reason behind this growing interest in platforms in the industrial sector is the increasing demand for customization. Enterprises that used to be orientated towards serial, standardized production are now seeking to manufacture unique products, in other words, to become original equipment manufacturers.

For companies transferring to platform solutions, however, there is a risk of making the wrong decisions regarding the choice of assets that would enable them in the long-term to manufacture sufficient quantities of high-tech, adaptive products without having to make large investment or attract considerable external funding [19]. The choice of the assets that would enable the company to adopt a platform model is a crucial moment which happens at the very beginning of the platform's lifecycle. It is important that such a model should align together the strategy of asset management, corporate and product strategies.

The process of improvement of an asset management system with regard to the innovations described above is illustrated in Fig. 3. This process comprises two parts. First, the asset management system is formed. It goes hand in hand with the continuous strategic process to ensure active knowledge exchange and communications between various stakeholders as well as non-stop monitoring of the external environment. This stage results in the use of proactive training solutions to encourage employees to participate in innovation activities. Second, it is necessary to ensure expanded and more intensified use of digital instruments in asset management, including platform technologies and interdisciplinary team interactions. In other words, the focus is made on the stimulation of active interactions between manufacturers and consumers within value networks. The organization thus becomes more adaptable and responsive to the changes in the turbulent external environment, bringing about a whole host of competitive advantages.

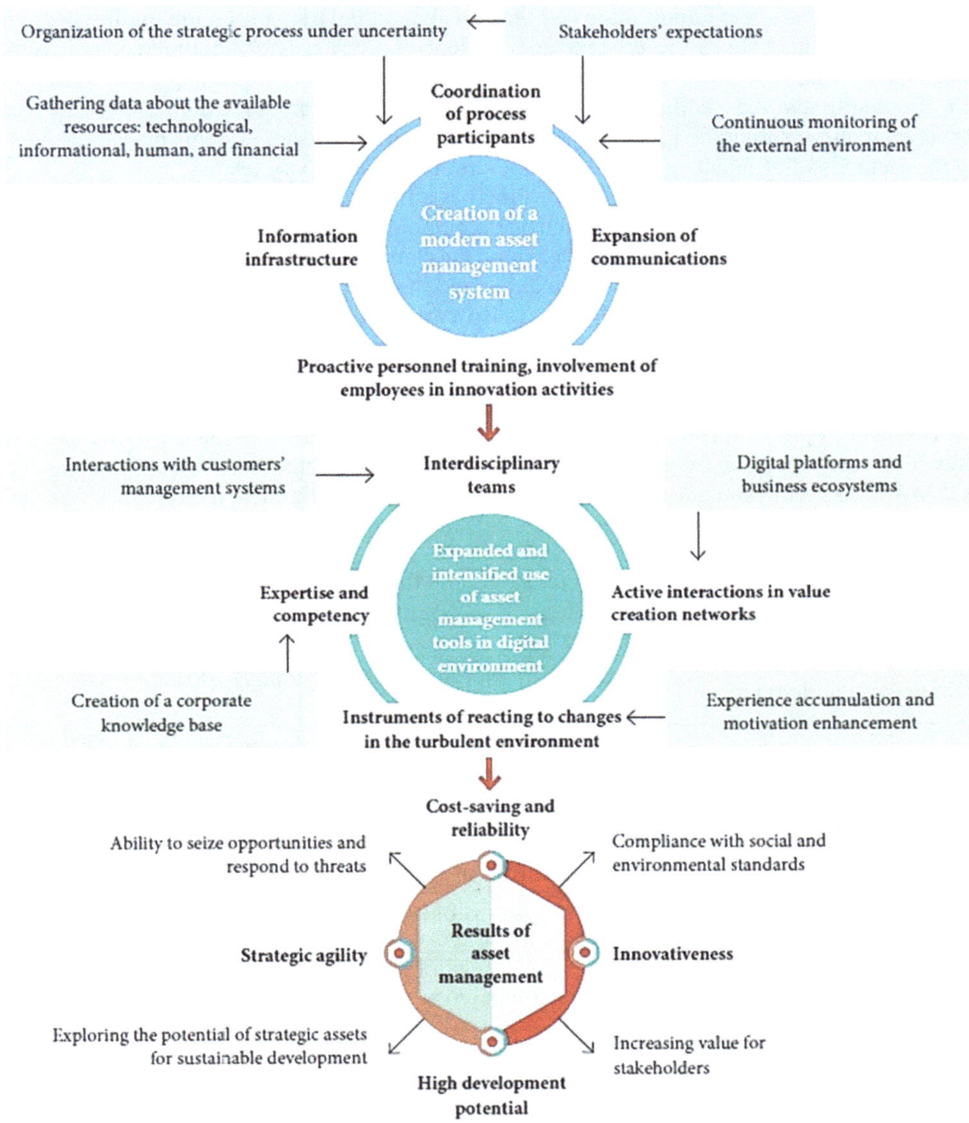

Figure 3: Improvement of the asset management system in a changing environment.

The creation and development of an advanced asset management system puts heavy demands on the personnel's skills and knowledge. Organization of asset management education for the staff, intensive internal communications, formation of interdisciplinary teams, continuous training, analysis of the lessons learnt from previous experience, learning from external sources, creation and development of the knowledge management system constitute the core of asset management practices.

It is best that training in asset management should follow the logic described below.

At the first stage, it is important to demonstrate that digital transformation implies keeping constantly updated about the context and content of the necessary competencies, which means mastering new knowledge, practices and methods of work and using them to replace the obsolete ones. All of the above is a key to successful work in a digital environment. Special attention should be given to the skills of systems thinking, systems engineering and to the methods of efficient communication within interdisciplinary teams.

The asset management system cannot function efficiently without reliable and extensive information support and intensive knowledge accumulation and exchange. The second stage, therefore, includes the disciplines that deal with the characteristics of information, knowledge and experience necessary for decision-making and the modern information support systems for asset management processes. Training also covers modes of communication, platform tools, tech eco-systems, interactions with IT-providers and matters of cyber security.

The third stage focuses on the methods of forecasting and planning under uncertainty, risk assessment and means of enhancing asset reliability and adaptability.

The list of options recommended for personnel training depending on the particular jobs they do was proposed by the Global Forum on Maintenance and Asset Management (GFMAM). The list contains 6 groups of subjects and 39 disciplines that together form the structure of asset management competencies (Table 1).

Table 1: Examples of disciplines included into the field of asset management [20].

Groups of subjects	Disciplines
Strategy and planning	Asset management policy/Demand analysis/Principles of strategic and operational planning
Decision-making	Lifecycle value realization/Resourcing strategy, Capital investment/Shutdown and outage strategy
Lifecycle delivery	Legislation and technical regulations/Systems engineering/ Configuration management/Operations, maintenance and repair/Modernization and decommissioning of assets
Information support	Information systems of asset management/Asset information standards
Organization structure and personnel	Leadership/Competency and behaviour management
Risks and analysis	Fault and incident response/Management of change/ Sustainable development/Stakeholder engagement

4 CONCLUSIONS

Companies embracing digital transformation can maximize the efficiency of their value chains based on asset lifecycle management: from planning and design through development to processes of operation and maintenance. Digital solutions enable companies to ensure asset observability and relatively predictable operational failure-free performance. Cloud-based computing, industrial Internet of Things, and AI are transforming traditional industrial operations and offer companies unprecedented opportunities of business growth. This, of course, also brings about new risks and uncertainty to companies' activities on the market, such as cyber security risks and breaches of confidentiality. Enterprises will face the challenging task of balancing their production processing, from operations planning to real-time equipment and facility management, to meet the existing operational performance constraints in order to achieve the best possible investment payoff.

Our experience with large energy companies has led us to the conclusion that modernization of asset management systems should include the following key components.

1. Assessment of the company management's readiness to digitize their business, which includes estimation of the capital investments required to implement smart production technologies.
2. Asset inventory, assessment of the assets' actual state (including physical wear and tear and obsolescence) and of their compatibility with the modern digital solutions.
3. Analysis of different scenarios of asset digitization and estimation of the impact that digitization will have on various business processes; estimation of the company's overall performance, including cost-saving, operating profit, emergency repair costs.
4. Proactive personnel training, including training aimed at developing engineering management and engineering economic competencies.

ACKNOWLEDGEMENT

The work was supported by Act 211 Government of the Russian Federation, contract No. 02.A03.21.0006.

REFERENCES

[1] ISO 55000:2014, Asset management – Overview, principles and terminology. https://www.iso.org/obp/ui/#iso:std:iso:55000:ed-1:v2:en. Accessed on: 20 May 2021.

[2] Gavrikova, E., Volkova, I. & Burda, Y., Strategic aspects of asset management: An overview of current research. *Sustainability*, **12**, p. 5955, 2020.

[3] Maletič, D., Maletič, M., Al-Najjar, B. & Gomišček, B., Development of a model linking physical asset management to sustainability performance: An empirical research. *Sustainability*, **10**, p. 4759, 2018.

[4] Asset Management 2020, A brave New World. https://www.pwc.com/gx/en/asset-management/publications/pdfs/pwc-asset-management-2020-a-brave-new-world-final.pdf. Accessed on: 20 May 2021.

[5] PAS 55-1:2008, Asset management. Part 1: Specification for the optimized management of physical assets, http://www.irantpm.ir/wp-content/uploads/2014/01/pass55-2008.pdf. Accessed on: 20 May 2021.

[6] PAS 55-2:2008, Asset management. Part 2: Guidelines for application of PAS 55-1. http://www.irantpm.ir/wp-content/uploads/2014/01/pass55guide.pdf. Accessed on: 20 May 2021.

[7] Moubray, J., *Reliability-Centered Maintenance.* 2nd ed., Industrial Press Inc.: New York, 1997.

[8] Gitelman, L.D., Kozhevnikov, M.V., Chebotareva, G.S. & Kaimanova, O.A., Asset management of energy company based on risk-oriented strategy. *WIT Transactions on Ecology and the Environment*, vol. 246, WIT Press: Southampton and Boston, pp. 125–135, 2020.

[9] Candón, E. et al., Implementing intelligent asset management systems (IAMS) within an Industry 4.0 manufacturing environment. *IFAC-PapersOnLine*, **52**(13), pp. 2488–2493, 2019.

[10] Aronson, K.E., Murmansky, B.E., Murmanskii, I.B. & Brodov, Y.M., An expert system for diagnostics and estimation of steam turbine components' condition. *International Journal of Energy Production and Management*, **5**(1), pp. 70–81, 2020.

[11] Gitelman, L.D., Kozhevnikov, M.V. & Kaplin, D.D., Asset management in grid companies using integrated diagnostic devices. *International Journal of Energy Production and Management*, **4**(3), pp. 230–243, 2019.

[12] Digital asset management – A closer look at the literature. http://citeseerx.ist.psu.edu/viewdoc/download?doi=10.1.1.369.7294&rep=rep1&type=pdf. Accessed on: 20 May 2021.

[13] Haberly, D., MacDonald-Korth, D., Urban, M. & Wójcik, D., Asset management as a digital platform industry: A global financial network perspective. *Geoforum*, **106**, pp. 167–181, 2019.

[14] Macchi, M., Roda, I., Negri, E. & Fumagalli, L., Exploring the role of digital twin for asset lifecycle management. *IFAC-PapersOnLine*, **51**(11), pp. 790–795, 2018.

[15] Digital twin: Bring your data to life for better performance. https://www.se.com/us/en/download/document/DT_eguide_09-10-19AR0_EN/. Accessed on: 20 May 2021.

[16] GE Digital Twin. Analytic engine for the digital power plant. https://www.ge.com/digital/sites/default/files/download_assets/Digital-Twin-for-the-digital-power-plant-.pdf. Accessed on: 20 May 2021.

[17] Robertson, D. & Ulrich, K., Planning for product platforms. *Sloan Management Review*, **39**(4), pp. 19–31, 1998.

[18] Gitelman, L.D., Kozhevnikov, M.V. & Sandler, D.G., Technology platforms as a tool for solving complex innovation problems. *International Journal of Design and Nature and Ecodynamics*, **11**(4), pp. 584–592, 2016.

[19] André, S. & Elgh, F., Modeling of transdisciplinary engineering assets using the design platform approach for improved customization ability. *Advanced Engineering Informatics*, **38**, pp. 277–290, 2018.

[20] *The Asset Management Landscape*, 2nd ed., 2014. https://www.gfmam.org/sites/default/files/2019-05/GFMAMLandscape_SecondEdition_English.pdf. Accessed on: 20 May 2021.

CYBERSECURITY: PANORAMA AND IMPLEMENTATION IN 2021

ADEL I. G. IBRAHIM
RMG, Kingdom of Saudi Arabia

ABSTRACT

Cyber threats have changed the universe of enterprise security. These threats are often difficult to determine and locate particularly in the digital and mobile era. Cybercriminals behind these threats possess knowledge, intelligence, creativity, flexibility, and resilience, which increase with time. Security measures taken to mitigate these threats require the cooperation of multiple security disciplines, because a single discipline cannot address the issues of growing threats. This paper first presents the panorama of cybersecurity and its components; subsequently, it reviews the cybersecurity landscape in terms of various frameworks, models, and recommendations issued from specialized organizations and institutions such as NIST, ISO, CSI, and ISACA. This paper attempts to guide enterprises to navigate this supercharged landscape as well as to implement a sound cybersecurity model that is suitable for a specified industry and situation. A methodology was developed in this study. The methodology recommends a framework that is based on a cross section of standard frameworks but adapted to the levels of decision making in the enterprise. The proposed methodology was applied to an institution in the public sector, whereby the cybersecurity panorama was explored, and the best practices suitable for the activity and the processes of the institution were implemented. Thus, a project with defined phases was executed. The methodology also suggests a sense of continuity, as cybersecurity is a never-ending endeavour.

Keywords: cybersecurity, cybercrime, malicious software, cybersecurity standards, cybersecurity frameworks, NIST 800, CIS controls, cybersecurity governance, cybersecurity strategy, cybersecurity implementation.

1 INTRODUCTION

In the 21st century, almost all companies use some type of digital systems (at least email, which implies internet access) and, therefore, are vulnerable to cyber-attacks because their data are not protected. Information breaks occur on a practically regular basis, uncovering email addresses, passwords, credit card details, and other exceptionally sensitive information.

Several users do not comprehend the severity of the issue until it "bites" them. Individuals and organizations alike are attacked fraudulently, usually by experts.

A credit scoring company, Experian, has published statistics for late 2020, showing that 31% of stolen personal information led to complete identity thefts.

Data theft in companies are more subtle, as the effect of cyber breach can manifest itself few months after the occurrence of the breach, rarely or never noticeable immediately to address the issues.

Published lists of confidential customers' data or R&D top-secret product development data on the dark web are disastrous for the companies' operations.

Living with the discomfort of these information leaks has apparently become an inescapable fact presently. One significant approach for handling these leaks is to prepare for cybersecurity.

The preparation for cybersecurity is a construct that includes policies, some hardware, software, and an awareness of personnel.

The next few sections describe some notable incidents that occurred in 2020/2021, followed by discussions of the most common cyber threat. Subsequently, widely used

WIT Transactions on The Built Environment, Vol 206, © 2021 WIT Press
www.witpress.com, ISSN 1743-3509 (on-line)
doi:10.2495/SAFE210041

standards are discussed, followed by a description of a methodology that is currently being applied.

2 CYBERSECURITY INCIDENTS

An internet search can reveal the information of some major cybersecurity breaches. Table 1 summarizes some of the notable cybersecurity breaches in 2020/2021.

Table 1: Notable cybersecurity breaches 2020/2021.

#	Company	Date discovered	Damage	Exposed
1	Channel Nine (Australian broadcaster)	28/3/2021	• broadcast • internet • publishing	Inability to broadcast news and shows, no internet access, publishing tools down
2	Harris Federation	3/2021	Ransomware attack	Voluntary shut down, disabling accesses to email and coursework for 37,000 students
3	CNA Financial (cyber insurance)	21/3/2021	Ransomware attack	Voluntary shut down for three days, disrupting operations for customers and employees
4	Florida water system	11/2/2021	Poisoning water supply	Increase in the amount of sodium hydroxide to a potentially dangerous level, from 100 ppm to 110,000 ppm
5	Bombardier (airplane manufacturer)	2/2021	Compromise of confidential data	Compromise of the confidential data of suppliers, customers. and around 130 employees located in Costa Rica
6	Facebook (social media)	3/4/2021	533 million users exposed	Phone numbers, DOB, locations, past locations, full name, and in some cases, email addresses
7	One Class (online learning)	29/6/2020	Over 1 million	Students' full names, email addresses, schools/universities, phone numbers, account details, and school enrolment details
8	BlueKai (web usage tracking data (for marketing))	19/6/2020	Over 2 billion entries	Names, residential addresses, email addresses, and other identifiable data including web browsing activity
9	Postbank (financial institution)	14/6/2020	At least 8 million	8–10 million beneficiaries who receive social grants
10	Keepnet (security company)	9/6/2020	5 billion	*(Second breach in a few months)*
11	CPA Canada (Professional Association)	4/6/2020	329,000	Members' data
12	Truecaller (caller id service in India)	27/5/2020	47.5 million	Phone numbers, service providers, names, genders, and more information
13	Pentagon hacking	3/6/2021	Undisclosed	Thousands of emails

Note: The list is not exhaustive. For further information, see [1]–[3].

3 CYBERSECURITY RISK TYPES AND ANSWERS

The risks on information and communications technology (ICT) operation stem from two main threats:

1. Sabotage: Sabotage refers to an internal human resource problem to be resolved by human resources practices. There exist few reports in the literature on this issue [4]. The best safeguards are in the Human Resource Policy proposed by ISO 27001 but still cannot fully guard from cyber threats, even if they are thoroughly implemented. companies must trust their employees up to a certain extent and ensure that their backup procedure is sound and diversified.
2. Internet: The internet is considered a window to the world, and it is the most dangerous window that is open to hackers and other cyber criminals.

3.1 Cybersecurity threats

Cybersecurity threats can be classified into several categories as summarized in Table 2.

Table 2: Main cybersecurity threats.

#	Type of cybersecurity threat	Description
1	Distributed denial of service (DDOS)	Denying access to bona fide users by bombarding the resource that they want to access, to the extent that the resource stops functioning (the hammer technique).
2	Man in the middle (MitM)	Attack targeting a user that remotely accesses a server, which is implemented by intercepting the communication and mimicking the bona fide user to steal their credentials and sensitive information. It can also be used to confuse the user by returning different responses.
3	Malware and spyware	A series of attacks that penetrate the system and install themselves to monitor user activity and steal confidential data and any other malicious intent.
4	Advanced persistent threat (APT)	This sophisticated attack stems from gaining unauthorized access to a system and remaining undetected for a period of time to exfiltrate sensitive data.
5	Password attacks	Gaining access to the password of a bona fide user either by knowledge, educated guesses, or brute force.
6	Social engineering	This threat attacks the weakest links in a network – human targets.

Social engineering attacks are targeted toward normal users, who are lured to perform a seemingly inoffensive action – business as usual – that triggers a chain reaction of offensive results planned by hackers. The social security attacks are so common that they deserve a further development, as summarized in Table 3, for which the data are taken from PA Knowledge [5] and Zielinski [6]. Software and appliances products that are mainly associated to firewalls can be used, but the safe solution to address most of these threats is to make users aware so that they practice caution.

Table 3: Social engineering attacks [5], [6].

#	Type of social engineering attack	Description
1	Phishing	Sending fraudulent emails from a seemingly official source, asking to click on a link that introduces the hacker's malicious program (usually providing the hacker a backdoor access to data and operations).
2	Homograph	A user accesses a fake website that seems to have a genuine URL and then submits sensitive information on this website.
3	Trojan virus	A user downloads a file either from a fake website or from an email attachment that will create a backdoor that can be used by the attackers.
4	Ransomware	When introduced, it prevents a user from accessing the user system and data unless a ransom is paid.
5	Malvertising or adware	Online advertising; when opened, it introduces the hacker's programs.
6	Wiper malware	When introduced in a target system, it wipes systems and data.
7	Drive by download	Hacking genuine websites, inserting malicious software that will be installed in the device of the website user.
8	Rogue security	A software that warns a user of the presence of a virus on their device. The user genuinely believes it and submits financial information to pay for the removal of viruses.

3.2 Usual software for handling the security risks

The cybersecurity industry is thriving, and the developers of many products in the market claim that the products address many of the cybersecurity risks. Some products are almost necessary, and some are purchased depending on the budget of companies and the sensitivity of their systems. Table 4 lists the features of some of these software.

3.3 Policies and procedures

Software deployment is insufficient to handle cybersecurity. A set of policies and procedures related to the discipline of the manpower should be implemented. Most frameworks (described in the next section) recommend a set of these policies.

4 CYBERSECURITY FRAMEWORKS (STANDARDS)

Various organizations – associations, standards, and institutes – have developed frameworks for cybersecurity. These frameworks are a set of documentation policies or procedures that state the best practices for handling cybersecurity in any entity. They need to be applied, and, if the standard (e.g. ISO) has a certificate, evidence must be provided for certification that these best practices are implemented in the entity.

These frameworks are usually based on access control, audit and accountability, awareness and training, cloud services safeguards, configuration management, continuous vulnerability management, email browser and web services protection, identification and

Table 4: List of software used to counter cyber threat [7].

#	Software	Function
1	Penetration testing	Known as ethical hacking, whereby a hacker tries to penetrate a system externally and prepares a report on its vulnerability from the outside.
2	Vulnerability assessment	This software allows to identify (from the inside) possible penetration points.
3	Assets management	An inventory of all information assets (usually classified as well); these assets are servers, network devices, databases, and people.
4	Patch management software (includes bug fixing and security patches)	Software vendors usually discover bugs and potential security vulnerabilities in their software and prepare patches to be distributed to their clients, in order to plug the security gap or bug fix. This software manages and deploys these patches.
5	Configuration hardening	Some associations propose the best configuration for a piece of software (operating system, database, etc.). Cybersecurity personnel implement these recommendations and harden the configurations.
6	Multifactor authentication	This software allows for a second personal authentication for the user in addition to a password; the most common is a one-time passcode (OTP) on an individual's cell phone.
7	Data leakage system	This software allows to encrypt any item of data to circulate in the network so that it can only be seen by a recipient who has the decryption; any other person cannot read it in clear.
8	Security information and event management (SIEM)	The organizations install a security operating centre (SOC) whose main objective is to identify any attack (event) and raise an alarm.
9	User and entity behaviour analytics (UEBA)	The software creates a baseline of user behaviour using avant-garde techniques such as machine learning. It alerts when there is a change in behaviour, which implies that a different user may have gained access.
10	Security orchestration automation and response (SOAR)	These series of software automate and save the workflow and human intervention for the first time when responding to a cybersecurity attack, so that it is triggered automatically in the case of a recurrence of the same attack without human intervention.
11	File integrity monitoring	This software checks that any software (apart from data) has been unchanged or officially changed.
12	Intrusion detection system/ intrusion prevention system (IDS/IPS)	Mainly a feature in the firewalls, allowing the detection of known signatures of a cybersecurity attack and preventing the packet to go through.

Table 4: Continued.

#	Software	Function
13	Network access control	This software allows for setting a policy (who can access, which device can access, etc.) for accessing the network. It repudiates any other attempt of access outside the set policy.
14	Network detection and response (NDR)	This software allows for monitoring the traffic for malicious actions and suspicious behaviour and react in response to network threats.
15	Applications whitelisting	An index of allowed applications and software scripts to be run in a system. It does not allow for running any other application, thus limiting access to potentially harmful applications.
16	Endpoint detection and response	This software allows for detecting and responding to malicious activity on endpoint devices as well as responding to some such threats. It shortens the response time of security teams.
17	Malware protection	Malware = malicious software; types are viruses, worms, Trojan horses, spyware, malvertising, and ransomware.
18	Data encryption – standing data	Encryption of records or specific fields in a database so that it is accessible only to persons having the decryption keys.
19	Data encryption – transmitted data	Encrypts the network traffic to guard from "sniffing" the traffic.

authentication, incident response management, malware protection, media protection, monitoring of cyber-attacks, patch management, personnel security, physical and environmental protection, penetration testing, system and communications protection, and system and information integrity.

Two types of standards (frameworks) are used for cybersecurity [8]:

- Specific industry/trade/sector cybersecurity standards/frameworks.
- Generic or universal cybersecurity standards that are applicable anywhere.

4.1 Specific industry/trade/sector cybersecurity standards

These standards are developed by specific industries and sectors; they include many features that cover the operations of the sector. They are built based on a set of compliance requirements or cybersecurity controls that must be adhered to in order to be compliant and/or accredited.

4.1.1 Health Insurance Portability and Accountability Act (HIPAA), 1996 [9]

As the name suggests, this act is applicable in the medical field, banning the disclosure of patients' medical data without the consent of the person. The standard developed contains 64 compliance requirements and is applicable to all health providers: medical institutions (hospitals, clinics, nursing homes, pharmacies, etc.), professionals (doctors, psychologists, dentists, chiropractors, etc.), and medical insurance plans (insurance companies or employers). These health providers all need to be HIPAA-compliant by applying strict rules pertaining to the confidentiality, integrity, and privacy of electronically protected health information (e-PHI) when it is created, stored, received, updated, or transmitted via electronic channels. It is mainly applicable in the USA but has a wider applicability in the Medical Profession.

4.1.2 Payment Card Industry – Data Security Standard (PCI-DSS) [10]

PCI-DSS is a security framework aimed at safeguarding cardholders' data when they use credit, debit, and cash cards. It is aimed at financial institutions, merchants, and individuals who create, use, transmit, accept, and honour payment with any such cards. It has 245 compliance requirements. It can lead to a PCI certification after a thorough audit. It is a worldwide standard championed by credit/debit cards companies such as Visa and Mastercard.

4.1.3 SOC 2 [11]

SOC 2 is a framework applicable to ICT service companies or software as a service (SaaS) companies. It safeguards customer data when they are uploaded into a cloud. It has 61 compliance requirements to ensure that organizations can control the privacy and security of customer and client data. It is an auditing procedure developed by the American Institute of CPAs (AICPA) and is based on a trust principle.

4.1.4 Others

- EU General Data Protection Regulation (GDPR) was effective from 25 May, 2018.
- It is an EU regulation with 98 compliance requirements to protect the personal data and privacy of EU citizens during transactions occurring within EU member states.
- NIST 800-53 is a framework developed by the National Institute of Standards in the USA (NIST), specific to the federal information systems, with 931 compliance requirements.
- NIST 800-171 lays out information security guidelines for the Department of Defense (DoD), USA. It is a subset of NIST 800-53.

Note: NIST has developed guidelines for specific purposes as above, but it has also developed a general guideline, NIST CSF (Cybersecurity Framework), which is described below.

4.2 General cybersecurity standards

These standards are to be used by any organization irrespective of its sector or trade from the government to SMEs or academic institution; they are applicable in any sector.

4.2.1 NIST CSF National Institute of Standards Cybersecurity Framework [12]

NIST CSF is one of the most popular frameworks in the world, and many national frameworks follow its recommendations. Fig. 1 shows its features.

NIST CSF has 108 subcategories of cybersecurity outcomes and controls, leading to 23 categories covered by the five functions shown in Fig. 1, which are applicable to all types of entities.

4.2.2 ISO/IEC 27001 [13]

ISO/IEC 27001 is an international standard to manage information security; it is published jointly by the International Organization for Standardization and the International

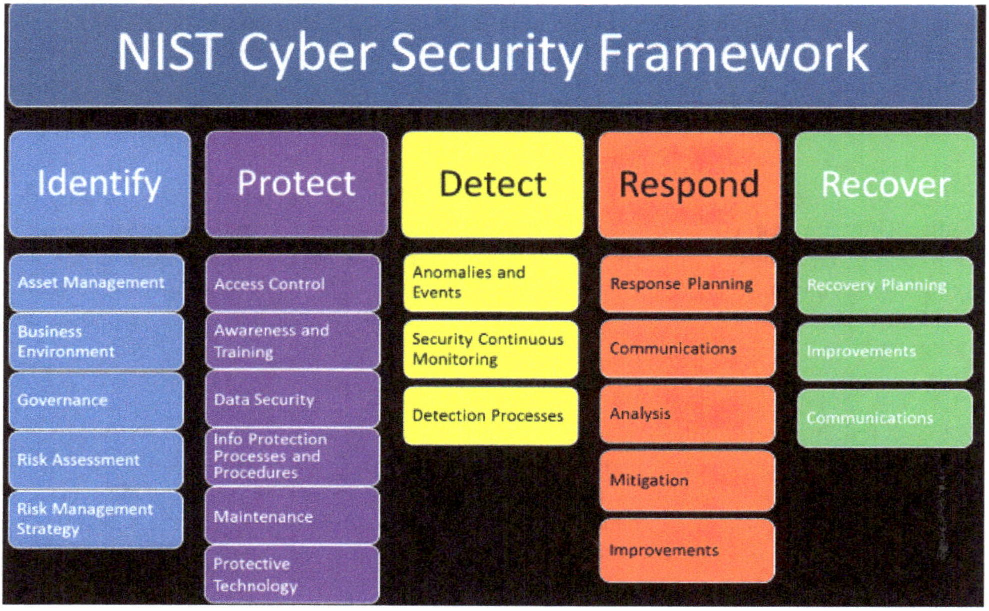

Figure 1: NIST Cybersecurity Framework. *(Source: Security Affairs.co.)*

Electrotechnical Commission. It has 114 compliance requirements for entities of any size in any sector. The standard is used for developing an information security management system and has an Annexure A for a series of policies and procedures. It has been widely applied across the world except in the USA.

4.2.3 CIS V8 Critical Security Controls (CSC) [14]

CIS V8 Critical Security Controls (CSC) refers to a framework developed by the Center of Internet Security. Fig. 2 shows the top 18 of these controls. The CSCs contain 153 safeguards. The CIS had divided the application of the standard according to the maturity level in implementation groups (IGs) orders from simple to the most complicated. These IGs are as follows:

- IG1 is called hygiene (to denote entry-level), having 56 cyber defence safeguards against the most common threats. It can be implemented by normal Information technology persons.
- IG2 refers to an intermediate level, having 74 additional safeguards, built upon the ones in IG1 but needs specialists (cybersecurity dedicated specialist) for implementation.
- IG3 refers to the expertise level and contains 23 additional safeguards to the above 45 and 74. This level requires experts for implementation. It caters to repel attacks by a sophisticated adversary.

5 WHAT TO DO: METHODOLOGY

The following points are extracted from all the frameworks mentioned above, and the methodology is proposed to adopt them in an organized manner by implementing a group of policies detailed in Fig. 3, whereby every block represents a policy containing the set of directives aimed to fulfil its purpose.

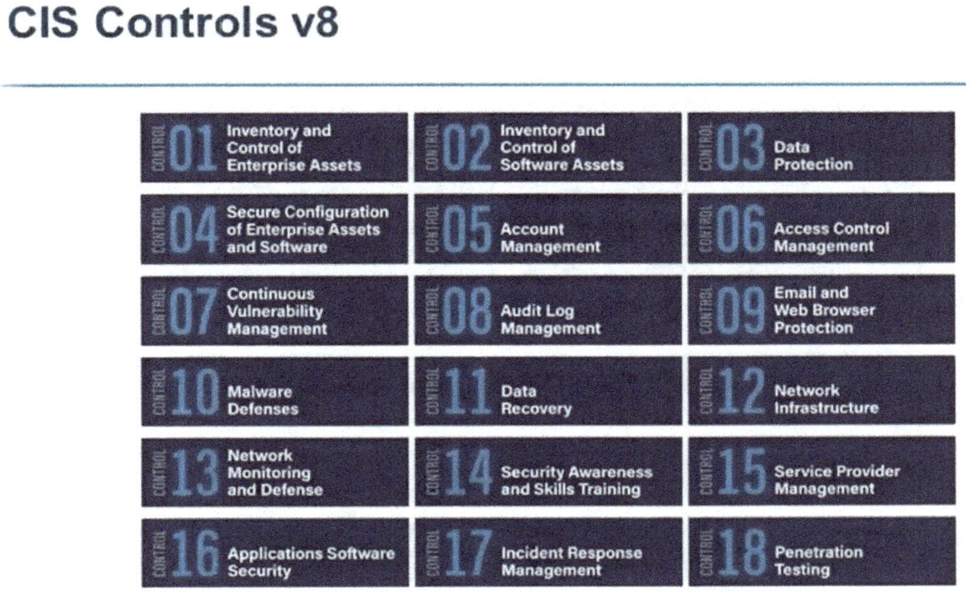

Figure 2: Top 18 security controls of CIS v8.

The methodology can be implemented through two groups within the organization:

- The leadership group, which is concerned with the strategy, governance, training, and definition of the responsibilities. These tasks are performed by a committee, which is composed of senior managers in the business units and the CTO, to oversee and monitor changes and advances over time. The committee can be called the ICT governance committee because it is responsible for the strategic direction of ICT services.
- The CIO group is concerned with operational aspects, policies and third-party relationships. This group is composed of the ICT hierarchy and relevant contracts management group (from purchasing). For protection, this group is responsible for identifying and acquiring the necessary software and tools performing the ICT protection. Budget is obtained and negotiated with the governance committee.

Fig. 3 illustrates the basic methodology as a framework.

5.1 Cybersecurity leadership

5.1.1 Cybersecurity governance

The organization must define a cybersecurity strategy, risk management framework, structured processes, roles and responsibilities, and top management oversight.

5.1.2 Cybersecurity strategy

The organization must define, document, approve, and implement a cybersecurity strategy, including the various initiatives, projects, programs, and references to internal and external relevant stakeholders. It should ensure reaching the desired and targeted cybersecurity maturity level with an alignment of objectives across all sectors.

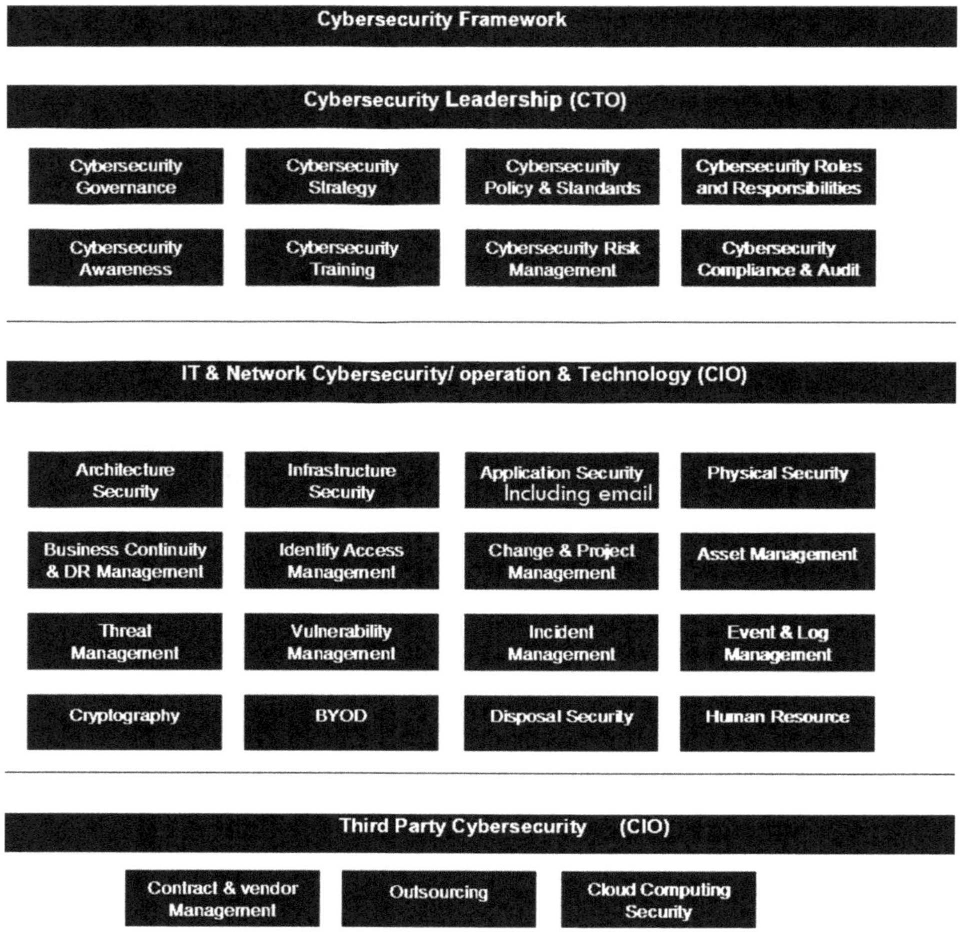

Figure 3: Methodology components.

5.1.3 Cybersecurity policies and standards
The organization must define, document, approve, implement, and communicate a set of cybersecurity policies and standards applicable to the organization setups and digital footprint.

5.1.4 Cybersecurity roles and responsibilities
The organization must define, document, approve, and implement roles and responsibilities within the context of the organization. These definitions shall be communicated to ensure that all relevant personnel and stakeholders understand their roles and responsibilities within the framework of cybersecurity.

5.1.5 Cybersecurity awareness
This topic concerns all organization's employees. It entails training (by presentation, film, pamphlets, placard, pop up screens, etc.), focusing on how employees gather knowledge

about cyber threats that they may face, and ensuring safety of the organization while employees fulfil their responsibilities.

5.1.6 Cybersecurity training

This topic concerns ICT professionals. It entails providing regular training to develop ICT employees' skills related to dealing with potential cyber threats, as well as to protect the organization's ICT assets.

5.1.7 Cybersecurity risk management

The organization must define, document, approve, implement, and communicate its risk management process. Cybersecurity risks are primarily focused on protecting the confidentiality, integrity, and availability of data stored, transmitted, or processed by ICT systems. It also ensures integration with the organization's enterprise risk management (ERM) and other relevant business processes.

5.1.8 Cybersecurity compliance and audit

The organization must establish a cybersecurity compliance process to periodically perform audits based on cybersecurity policies, legal and regulatory requirements, and industry standards or certifications (ISO/IEC 27001, PCI DSS, HIPAA, etc.) in order to identify, evaluate, manage, communicate, and monitor cybersecurity compliance. It should report its findings and any cybersecurity critical compliance violations needing the attention of the top management.

5.2 IT network, cybersecurity operations, and technology

5.2.1 Architecture security

Security requirements, standards, and controls are developed and defined based on the business requirements and criticality. The architecture is defined and documented according to the functionalities and security requirements described above. It is periodically reviewed and amended.

5.2.2 Infrastructure security

The organization must define, document, approve, and implement infrastructure security requirements and controls while periodically adjusting, monitoring, and evaluating its compliance to meet all security requirements, standards, and controls for infrastructure (external communication, endpoint devices, operating systems, database, servers, virtual devices, network devices, security devices, and file sharing).

5.2.3 Application security

Security controls should be applied for the development, procurement, and/or acquisition of an application, including mobile applications, Web services, and APIs, to ensure the secure closure of known gaps in the application, which can be exploited. These gaps might increase the risk of the organization.

A special application that should be monitored carefully is the email server; emails should also be monitored because they are used by everyone in the organization.

5.2.4 Physical security

Physical and environmental security measures are essential to prevent unauthorized physical access, damage and interference to the information assets, and information processing facilities.

5.2.5 Change and project management

The organization must define, document, approve, and implement change and project management requirements and controls that are incorporated and integrated into all stages of a project lifecycle' it should accordingly change management processes. Cybersecurity personnel are monitoring the project phases to ensure compliance. All the changes/projects must be tested in a test environment before implementation in the production environment.

5.2.6 Assets management

The asset management process starts at acquisition, followed by recording (register) with ownership, and finally classification and labelling. The management process then follows the stewardship of the asset throughout its lifecycle until disposal.

5.2.7 Business continuity and disaster recovery (DR) management

This framework is based on a business impact analysis study and the availability of a DR centre and an alternative work area. It is generally guided by the ISO 22301.

5.2.8 Identify access management

The organization must establish security controls to ensure the protection of information systems and data against unauthorized access, while still providing access to users who have a business need. It allows for various methods to authenticate genuine users and adopt stringent authentication processes for privileged users.

5.2.9 Threat management

The organization must define, prepare, approve, implement, and document a threat management process to describe a factual background for understanding the risks, early warnings, breach prevention, detection, and remediation efforts.

5.2.10 Vulnerability management

The organization must ensure timely identification and effective mitigation of application and infrastructure vulnerabilities to reduce the likelihood of incidents and business impact on the organization.

5.2.11 Incident management

The organization must define, approve, and implement a cybersecurity incident management to ensure the timely identification and handling of cybersecurity incidents, in order to reduce the (potential) business impact and to allow timely recovery from cybersecurity incidents. Disaster recovery plans include different scenarios of cybersecurity incidents.

5.2.12 Event and log management

The organization should define, approve, and implement a security event management process to analyse operational and security loggings as well as respond to security events. It should ensure timely identification of threats and respond to anomalies or suspicious events, particularly with regard to information assets.

5.2.13 Cryptography

Whenever needed, a cryptographic security standard should be defined, approved, and implemented. It usually covers data at rest, data transmitted over third-party communication lines, and the encryption of sensitive data when transmitted inside the infrastructure.

5.2.14 Bring your own device (BYOD)

Whenever the organization allows the use of personal devices (e.g., smartphones, tablets, laptops, etc.) for business purposes, their use should be supported by a defined, approved, and implemented cybersecurity standard in addition to staff agreements and cybersecurity user-awareness training.

5.2.15 Safe disposal security

The organization must dispose of its information assets safely when they are no longer needed. Thus, it can ensure the safety of its business, customers, and sensitive information from unauthorized disclosure upon disposal.

5.2.16 Human resources security

Human resources security is the weakest link in cybersecurity. Organizations should incorporate cybersecurity requirements into human resources processes. This will help ensure that staff cybersecurity responsibilities are included in staff agreements and that the staff are screened before their employment and made aware during their employment.

5.3 Third-party cybersecurity

5.3.1 Contract and vendor management

The organization must set up, define, approve, implement, and monitor the required cybersecurity controls within the contract and vendor management processes. This will help ensure that its approved cybersecurity requirements are appropriately addressed before signing contracts and that the compliance with the established cybersecurity requirements is being monitored and evaluated during the contract duration.

5.3.2 Outsourcing

The organization must define, set up, implement, and monitor the required cybersecurity controls within outsourcing policy and outsourcing process to ensure that its cybersecurity requirements are appropriately addressed before, during, and after exiting outsourcing contracts.

5.3.3 Cloud computing security

This framework is used to address the organization's use of cloud hosting services whenever applicable. It includes setting up policies and procedures to address the cloud service provider's cybersecurity setup, data privacy, and data recovery at termination.

5.4 Quick comparison between the methodology and current standards

Most standards address the specific issues of the ICT governance coverage and then recommend securing each section of the ICT installation (like securing a network or a web-service). The merits of these standards is that they provide the necessary guidelines in any implementation; however an implementation needs specific roles and responsibilities that are not considered in NIST and CIS, for example, but are required by ISO 27001 These standards imply the role and responsibility of organizations to implement the guidelines and recommendations as they see fit. The methodology covers these points and others, such as cyber-strategy, so that the standards can be fully implemented.

6 FINAL WORD IN 2021

The information in this paper will be relevant for only sometime (3–4 years); the methodology may be applicable for a longer "sometime" but eventually will become obsolete in 6–8 years. The standards are being regularly reviewed, and cybersecurity software are evolving. Nevertheless, the cybersecurity framework described in the methodology above will be applicable for some time. It could be augmented, developed, and amended until the introduction of new technologies related to ICT, which might have their own set of cybersecurity problems.

REFERENCES

[1] Meharchandani, D., 10 major cyber attacks witnessed globally in Q1 2021. https://securityboulevard.com/2021/04/10-major-cyber-attacks-witnessed-globally-in-q1-2021/.

[2] Selfkey, All data breaches in 2019–2021: An alarming timeline. https://selfkey.org/data-breaches-in-2019/.

[3] Winterburn, T., Pentagon blames 'Chinese hackers' for global Microsoft attacks. https://www.euroweeklynews.com/2021/03/06/pentagon-blames-chinese-hackers-for-global-microsoft-attacks/.

[4] Global HR, The role of hr in mitigating cyber security threats. https://www.ghrr.com/the-role-of-hr-in-mitigating-cyber-security-threats/.

[5] PA Knowledge, Putting human resources at the heart of cyber security. https://www.paconsulting.com/insights/putting-human-resources-at-the-heart-of-cyber-security/.

[6] Zielinski, D., Five top cybersecurity concerns for HR in 2019. https://www.shrm.org/resourcesandtools/hr-topics/technology/pages/top-cybersecurity-concerns-hr-2019.aspx.

[7] Williams, L., 22 BEST cyber security software tools in (2021 List). https://www.guru99.com/cybersecurity-software-tools.html.

[8] Cassetto, O., 21 top cyber security threats and how threat intelligence can help. https://www.exabeam.com/information-security/cyber-security-threat/.

[9] Centers for Disease Control and Prevention (CDC). Health Insurance Portability and Accountability Act of 1996 (HIPAA). https://www.cdc.gov/phlp/publications/topic/hipaa.html.

[10] PCI Security Standards Council, https://www.pcisecuritystandards.org/ and PCI Compliance Guide, https://www.pcicomplianceguide.org/faq/.

[11] The American Institute of CPAs (AICPA)'s, SOC 2® – SOC for Service Organizations: Trust Services Criteria. https://www.aicpa.org/interestareas/frc/assuranceadvisoryservices/aicpasoc2report.html/.

[12] National Organization for Standards and Technology (NIST), NIST Cybersecurity Framework (CSF) Reference Tool. https://www.nist.gov/cyberframework/nist-cybersecurity-framework-csf-reference-tool.

[13] ISO, ISO/IEC 27001 Information Security Management. https://www.iso.org/isoiec-27001-information-security.html.

[14] Wikipedia, The CIS critical security controls for effective cyber defense. https://en.wikipedia.org/wiki/The_CIS_Critical_Security_Controls_for_Effective_Cyber_Defense.

DEVELOPMENT OF A HYBRID EXERCISE FOR ORGANIZATIONAL CYBER RESILIENCE

YUITAKA OTA, ERIKA MIZUNO, KOKI WATARAI, TOMOMI AOYAMA,
TAKASHI HAMAGUCHI, YOSHIHIRO HASHIMOTO & ICHIRO KOSHIJIMA
Nagoya Institute of Technology, System Management and Engineering, Japan

ABSTRACT

In recent years, COTS (commercial off the shelf, such as Windows OS, Intel PC, and open source applications) have been proposed to reduce the cost of deploying operational technology (OT) systems. Also, DX efforts are being made to shift from physical operation to virtual operation by using virtualization with IoT, AI, and cloud servers. Current ransomwares, therefore, infect without distinguishing between IT systems and OT systems. For example, in May 2021, the Colonial Pipeline Company, a major oil pipeline company in the United States, was infected with ransomware and shut down its pipeline operation. As a countermeasure against cyberattacks, many companies focus on creating a less vulnerable environment. However, attackers exist worldwide, and they are constantly searching for new attack surfaces and developing new attack methods. It is also difficult for defenders to prevent all attacks, no matter what measures they take. Therefore, companies need to educate employees to ensure the safety of their factories in the event of a cyberattack. The authors developed a series of table-top BCP (business continuity plan) exercises to acquire the meta-knowledge necessary to respond to cyberattacks targeting the OT system for the above reasons. However, we found that the learning effect of these exercises depended on how the participants imagined cyberattacks. Therefore, in this paper, we propose a hybrid learning system that combines cyberattack simulations and table-top BCP exercises to increase the cyber resilience of participants.
Keywords: cyber-incident, cyber resilience, business continuity plan, exercise.

1 INTRODUCTION

It is essential to recognize that cyberattacks are conducted against information technology (IT) systems, such as credit card fraud and information leakage, and against the operational technology (OT) systems used to operate plants. Since the cyberattack on the controller of a uranium enrichment plant not connected to the external net was discovered in 2010 [1], the number of cyberattacks on OT systems has increased every year [2]. Moreover, the HatMan malware [3], which targets controllers and safety instrumentation, was discovered in Saudi Arabia in 2019. In addition, in May 2021, a pipeline in the United States [4] was damaged by an attack not on the plant but on the management system that controls the plant, making it impossible to provide service for an extended period of time. Accordingly, companies that own plants must not only respond to cyberattacks on IT systems to continue their businesses but must also take measures against cyberattacks on OT systems in a way that considers their plants and the series of systems used to operate these plants.

1.1 Approaches to security measures targeting OT systems

To minimize damage from cyberattacks, many companies have now focused on building environments that are less vulnerable to cyberattacks, for example, introducing mail checkers, whitelists, and information security training for employees.

However, there are attackers worldwide and new vulnerabilities are discovered and reported every day [5], [6]. In addition, the reported vulnerabilities are only a fraction of the total, and there remains a high possibility of attacks (zero-day attacks) using unreported vulnerabilities. In other words, no matter how much a company takes measures to build an

www.witpress.com, ISSN 1743-3509 (on-line)
doi:10.2495/SAFE210051

environment that is resistant to cyberattacks, it is impossible to reduce the possibility of cyberattacks to zero. We believe that in order for companies to respond to cyberattacks targeting OT systems, the following two points are required, based on the assumption that cyberattacks will occur, to build an environment that is resistant to cyberattacks.

- **Formulation of an industrial control system (ICS)-business continuity plan (BCP) considering cyberattacks**
  Usually, in companies that own plants, the production department prepares a safety-BCP for physical problems such as fires, spills of hazardous materials, and natural disasters and conducts drills based on this plan. In addition, the information system department develops an IT-BCP to deal with cyber-incidents on the business network, including information leakage. A cyberattack on an OT system uses IT systems to attack the target OT system to damage the plant or stop the services of a company. Therefore, it is essential to analyze the risks considering cyberattacks that can cause damage across OT and IT systems and to create a plan in advance to minimize the damage of such risks.
- **Human resource development**
  Even if an ICS-BCP is developed, if the employees who respond to incidents cannot do so, the damage from cyberattacks cannot be minimized. Therefore, companies need to educate their employees to respond based on the plan that they have developed.

However, many Japanese companies may have experience dealing with cyber-incidents involving IT systems but have little experience with cyber-incidents involving OT systems. Without sufficient experience with cyber-incidents targeting OT systems, it is not easy to understand the impact of cyberattacks on OT systems and imagine the actions that need to be taken in the event of such an incident. This makes it difficult for companies to formulate ICS-BCPs for OT systems and develop human resources.

1.2 Approaches to gaining experience with cyber-incidents

To get an accurate picture of cyberattacks, companies need to experience responding to cyber-incidents. However, as mentioned above, there are few opportunities to experience cyberattacks.

Thus, to address this issue, we developed an exercise to experience a simulated control system cyber-incident to help company personnel increase their experience with cyber-incidents and acquire the meta-knowledge necessary to respond to cyberattacks targeting OT systems. We developed a hybrid exercise that combines a table-top format with an operational simulation format using a simulated plant. The effectiveness of the hybrid exercise is enhanced because the exercise participants gain a better image of a cyberattack on an OT system.

2 EXERCISES TO UNDERSTAND SAFETY RESPONSES

The purpose of this exercise is to understand how a cyberattack targeting OT systems affects a plant and to learn the actions (safety responses) and concepts required to make the plant safe after it has been rendered insecure by the cyberattack.

2.1 Structure of the exercises

This exercise is a combination of simulation exercises (cyberattack demo and operation testbed) based on our testbed and table-top exercises (card-based incident response exercise and ICS-BCP exercise). In this combination, we examine how a company would respond to a cyberattack as a member of a virtual company (card-based incident response exercise and ICS-BCP exercise).

Fig. 1 shows the objectives of each exercise and the flow for conducting the exercises. The exercise participants have different backgrounds in terms of work experience and knowledge. Prior to the exercise, classroom lectures are conducted in order to prepare the learning base of the exercise participants. Then, the participants operate the testbed to perform start-up, steady-state operations, and emergency operations. We deepen their understanding of plants and OT systems using these simulation exercises. Next, they engage in organizational collaboration during cyber-incidents via table-top-style exercises. By experiencing this sequence of events, the exercise participants deepen their knowledge of cyberattacks on OT systems and understand the safety responses required to make a plant safe.

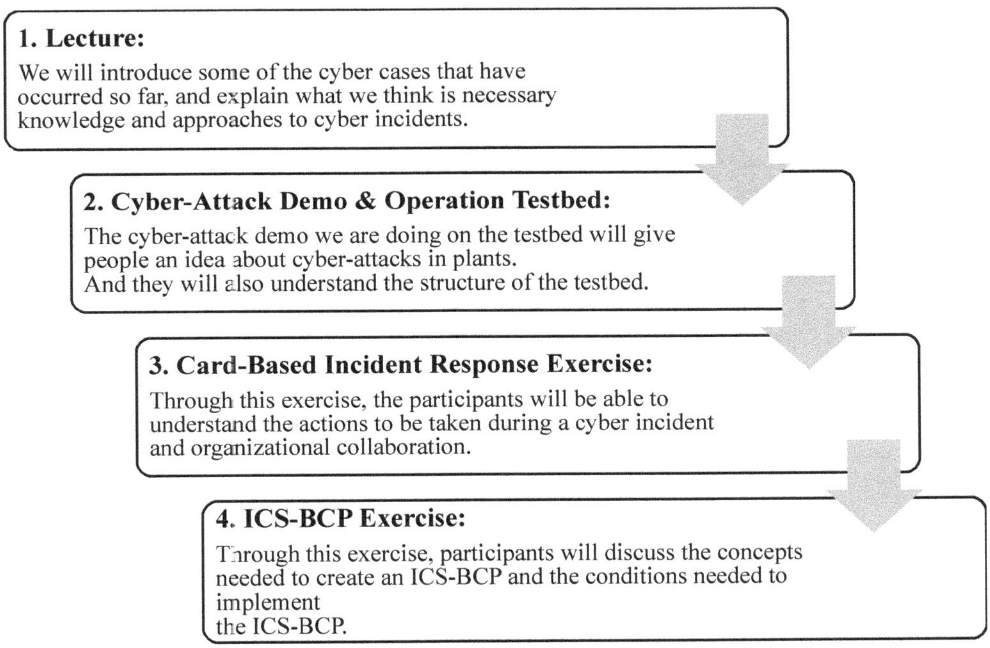

Figure 1: Relationship between the exercises.

2.2 Simulation exercises

This exercise aims to help participants understand the steps involved in a cyberattack on an OT system and how it can affect the plant via the network. The exercise participants observe a cyberattack demonstration on the testbed (Fig. 2) in our laboratory. In this cyberattack demonstration, the OT system operating the plant is attacked via the Internet through the IT system, and the control equipment of the plant is damaged. Then, the exercise participants experience the start-up, shutdown, and emergency operations of the testbed by themselves. In this experience, participants deepen their understanding of the relationship between the plant and the controller and the communication flow between the OT systems (e.g., the data management server, human–machine interface, and data-historian server) built to operate the plant. This simulation gives the exercise participants an idea of the actions that need to be taken to keep the plant safe and operational during a cyber-incident and the locations from which the effects of the cyber-incident need to be removed.

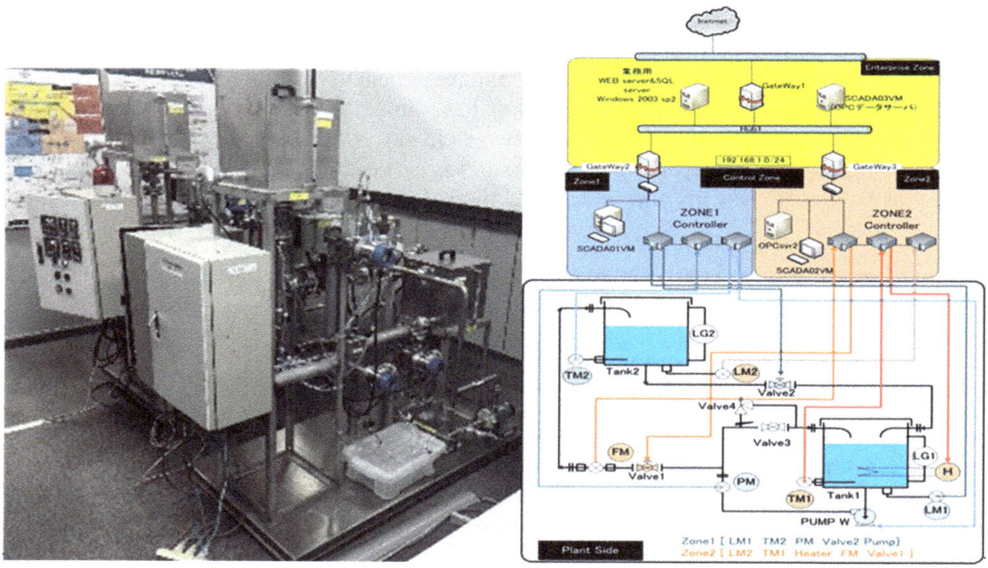

Figure 2: Cyber security testbed and industrial control system (ICS) network [7].

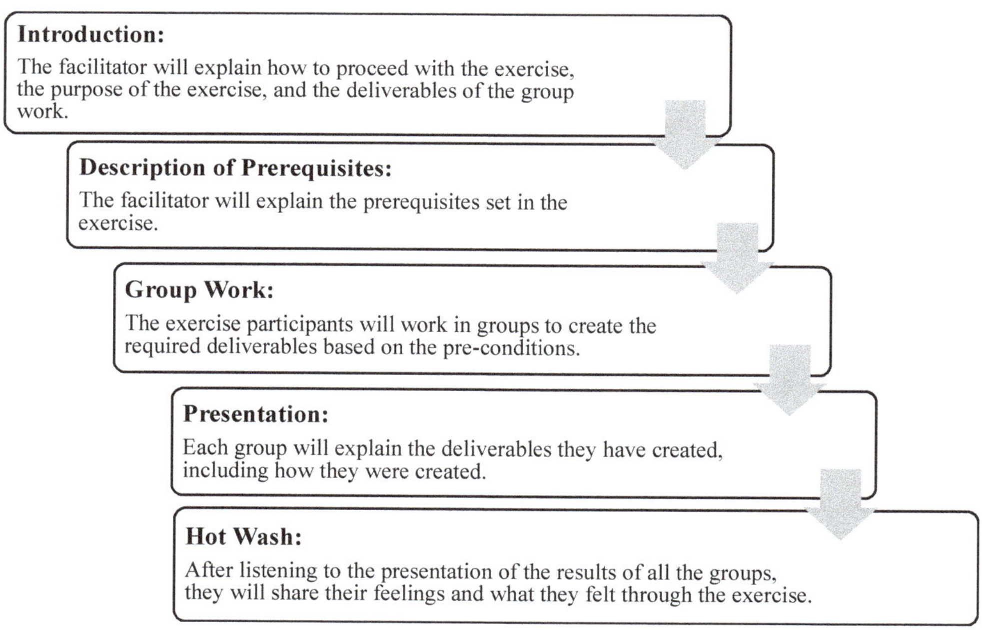

Figure 3: Human resource development training flow.

2.3 Table-top exercises

In this exercise, based on the knowledge and experience gained in the simulation exercise, the response to transition the plant to a specific state after a cyberattack on the plant and the response to remove the effects of the cyberattack that made the plant insecure are discussed.

In the table-top-style exercise, the exercise participants progress by role-playing as members of a virtual company. Each exercise consists of five steps: introduction, prerequisite explanation, group work, presentation, and hot wash (Fig. 3).

2.3.1 Card-based incident response exercise

This exercise mainly aims to increase the participants' understanding of the impact of cyberattacks on a plant and to discuss what actions are needed to protect the plant from cyberattacks. In this exercise, the participants need to understand the situation of the virtual company from the conditions provided by the facilitator and to examine and create a workflow to minimize the damage that a cyberattack can inflict on the plant. The group is given a total of 38 normal cards of three types, i.e., "Field-side action", "Head office action", and "Shut down/restart plant", and two special cards, "Suspect/confirm cyberattack" and "Information sharing". The normal cards contain the action details, the department-taken action, and the result after the taken action; the exercise participants then consider their subsequent action based on this information. The information sharing cards are used to share the information obtained by each department with other departments. In the group work, the participants need to create a workflow (Fig. 4) using these given cards.

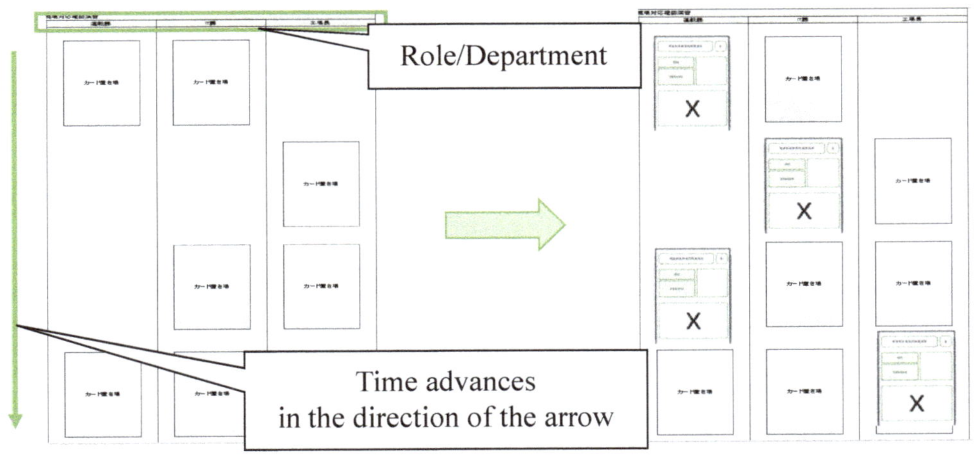

Work-flow(before group-work) Work-flow(after group-work)

Figure 4: Workflow development (image).

2.3.2 ICS-BCP exercise

This exercise guides participants in discussing the perspectives needed to create an ICS-BCP. The exercise requires creating a scenario, action cards, and worksheets to conduct a card-based incident response exercise. Group work is conducted as follows.

1. Participants analyze the possible risks from the hypothetical company and plant scenarios given by the facilitator.
2. Participants create a scenario of a possible cyberattack that could cause the analyzed risk and analyze the impact on a hypothetical company or plant caused by that scenario.
3. Participants examine how to respond to the cyberattack scenario to minimize the damage and create an ideal response workflow. Then, they create an action card based on these scenarios for a card-based incident response exercise.
4. Finally, participants review their exercise or workflow and improve the accuracy of the workflow by conducting their own card-based incident response exercise on other groups.

Figure 5: Team discussion in group work.

In this exercise, exercise participants experience the sequence of creating an ICS-BCP by (1) analyzing the risks to the company, (2) examining the possibility that these risks may occur, and (3) examining countermeasures by imagining the company's situation when the risks occur. This exercise helps exercise participants understand what perspectives are needed and what they should imagine to create an ICS-BCP.

2.4 Findings from these exercises

We have been conducting this exercise in workshops since 2015. We have received excellent evaluations from the participants indicating that the table-top exercise was easy to understand and that they were able to understand the safety measures necessary to make a plant safe, which was the primary purpose of the exercise, by experiencing the cyberattack demonstration and plant operation simulation using the actual equipment (testbed). At the same time, the results of the exercises revealed the following issues:

- **Lack of discussion on how to respond to cyberattacks before signs appear in a plant**
  These exercises focus on responses to ensure the safety of a plant after a cyberattack when the plant shows signs of abnormalities. Many of the exercise participants confirmed that they were able to gain an understanding of this stage. However, to

minimize the impact of cyberattacks, proactive measures should be taken prior to a cyberattack, and it is necessary to develop exercises that can discuss these security responses.

- **Space for setting up a simulated plant is required to conduct the exercise**
  One of the features of this exercise is that it was developed using a simulated plant that can be operated manually. The exercise participants can easily understand the impact of cyberattacks on the plant by touching the simulated plant and watching the demonstration of a cyberattack on it. Then, participating in the table-top exercises with an understanding of the subject of the exercise, the educational effect of this hybrid exercise is enhanced. However, it is essential to have a physical space for the simulated plant to conduct the exercises. As a result, the place where the exercise is provided is limited to the physical location of the simulated plant.
- **Structure makes it challenging to check the effects of the exercises**
  This exercise was developed to acquire the meta-knowledge necessary to respond to cyberattacks targeting OT systems. Feedback from the exercise participants has indicated that they feel that they have gained such meta-knowledge. However, we have failed to confirm whether the capabilities of the participants improved. Even if the exercise is conducted under a different scenario to measure the effect of the exercise, if the subject of the exercise (e.g., a simulated plant or a scenario of a disguised company) is the same, it may not be an exercise but rather a drill to confirm the procedures.
- **Structure makes it challenging to understand the point of view of the cyberattacker**
  To protect a plant from cyberattacks, it is essential to understand how cyberattackers think and conduct their attacks. These exercises focus on the perspective of the defender. A cyberattack scenario, including the point of view of the cyberattackers, was created during the scenario development phase of these exercises, and these exercises were developed based on this cyberattack scenario. However, these exercises were not designed to make the exercise participants intensely aware of the cyberattacker's perspective.

3 EXERCISE TO UNDERSTAND INCIDENT MITIGATION AND RESPONSE

The focus of the exercises developed so far has been on the aftermath of a cyberattack on a plant that results in its operational failure. However, to mitigate the effects of cyberattacks, it is necessary to develop an exercise that allows the user to understand proactive measures that should be taken before any abnormalities occur in the plant. The purpose of this new hybrid exercise is to understand the flow of cyberattacks on plants, the proactive measures that can be taken to protect plants from any cyberattack, and the responses that can be taken to eliminate the effects of cyberattacks when they occur. Accordingly, a new exercise was conducted by adopting a game called "Red vs. Blue gamification" [8] developed and provided by ThreatGEN for simulations (Fig. 6) using a simulated system.

This game from ThreatGEN meets our requirements as shown in Table 1.

3.1 About the Red vs. Blue gamification portal

ThreatGEN Red vs. Blue is the industry's first online multiplayer strategy computer game designed to teach real-world cybersecurity. The game consists of a turn-based system (3 min/turn) for a total of 75 turns, where the player chooses a given action (Fig. 7) under the constraints of time, money, and human resources. Players can play as either the attacker (Red) or the defender (Blue).

Figure 6: Red vs. Blue gamification portal.

Table 1: Reasons for adopting the Red vs. Blue gamification portal.

Our demands	Features of Red vs Blue gamification portal
The structure should make the exercise participants aware of the point of view of a cyberattacker.	Gain experience on both the cyberattacker and defender sides.
There should be a mechanism to check the effects obtained by the exercises.	Because the results of the game are expressed as a score, the learning effect can be measured.
The structure should be able to discuss actions to be taken after an operational abnormality occurs in the plant and the actions to be taken before an abnormality occurs in the plant.	Experience the sequence of a cyberattack. Experience in pre-action to build an environment that is less susceptible to cyberattacks and in incident response to control devices that have been subjected to cyberattacks.
Exercises can be conducted at any location without the need for a space to conduct them in.	Because it can be conducted online, there is no need to consider the location of the exercise as long as a network environment is available.

The Red side chooses actions to shut down the plant using various attack methods, whereas the Blue side chooses actions to protect the plant from attackers and continue their business. The method of play can be player vs. computer or player vs. player. Using ThreatGEN Red vs. Blue, players can enjoy learning about the actions and mindset of an attacker (Red side) during a cyberattack and the actions and mindset of a defender acting to protect (Blue side) the plant from that attack.

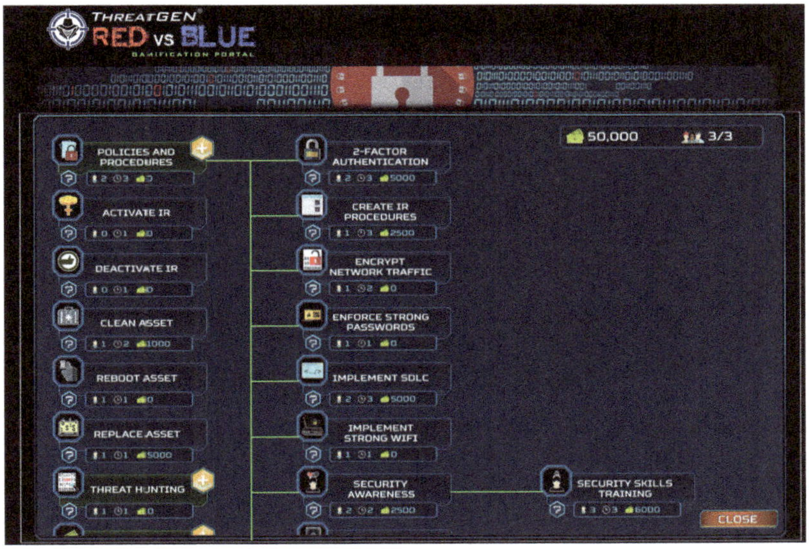

Figure 7: Play screen of the Blue side (select action).

3.2 Exercise method using ThreatGEN Red vs. Blue

In this game, exercise participants only have to select an action to proceed through the scenario. In other words, if they play the game without thinking about it, the benefits they obtain from the game are reduced. Thus, to increase the effect of the game, we combined ThreatGEN (the simulation exercise) with the following additional actions.

3.2.1 Lecture prior the game

As in the previous exercises, a classroom lecture is conducted before playing the game. This classroom lecture focuses on the NIST Framework [9] and Cyber Kill Chain [10]. The NIST framework can prioritize actions to reduce cybersecurity risks by classifying them into five categories, that is, identity, protect, detect, respond, and recover. Cyber kill chain, as proposed by the Lockheed Corporation, models the sequence of actions in a targeted attack and divides the attack into seven phases. The NIST framework is explained for reference when thinking about Blue side behavior, while the cyber kill chain is explained for reference when thinking about Red side behavior.

3.2.2 Play the game

After the classroom lecture, the exercise participants then play the game. The objective of this phase is to understand the primary usage of the game. After the game, exercise participants understand how to play the game, and the game play helps them understand their abilities. Then, they record their scores.

3.2.3 Analyze the action

In this phase, the exercise participants analyze the actions that can be selected in the game. At the same time, they analyze the actions selected in the game, the type of content they have, and their impact. Table 2 shows the phases of the cyber kill chain in which the Red side can choose actions.

Table 2: Categorize attacker's actions (Red-side).

RED ACTION	RECONNAISSANCE	WEAPONIZATION	DELIVERY	EXPLOTATION	INSTALLATION	COMMAND & CONTROL	ACTION ON OBJECTIVES
HOST SCAN	✓						
PORT SCAN	✓						
SERVICE ENUMERATION	✓						
FIND PUBLIC VULNERABILITIES	✓						
ATTACK				✓			
MANIPULATION				✓			
DENIAL				✓			
FUZZING	✓						

Next, exercise participants examine which categories of the NIST framework the Blue side can be chosen from for its actions. Table 3 is used to organize which phase of the cyber kill chain the selected actions are effective against.

Table 3: Categorize defender's actions (Blue-side).

BLUE ACTION	Category (NIST FRAMEWORK)	RECONNAISSANCE	WEAPONIZATION	DELIVERY	EXPLOTATION	INSTALLATION	COMMAND & CONTROL	ACTION ON OBJECTIVES
POLICIES AND PROCEDURES	Respond & Recover	✓		✓	✓			
2-FACTORS AUTHENTICATION	Protect	✓			✓	✓	✓	
CREATE IR PROCEDURES	Respond				✓			
ENCRYPT NETWORK TRAFFIC	Protect	✓						
ENFORCE STRONG PASSWORDS	Protect	✓			✓	✓	✓	
IMPLEMENT SDLC		✓			✓		✓	
IMPLEMENT STRONG Wi-Fi		✓		✓	✓			
SECURITY AWARENESS		✓	✓	✓	✓	✓	✓	✓
SECURITY SKILLS TRAINING		✓	✓	✓	✓	✓	✓	✓

By organizing the actions in this way, the attacker can have better a grasp on the phases required to accomplish the attack and selects the actions to be taken for each phase. On the other hand, the defenders select actions to prevent cyberattacks by understanding the attacker's attack flow and organizing the actions to be taken to stop this flow.

3.2.4 Play the game strategically

The participants then play the game again based on the actions and strategies they analyzed and discussed in Phase 3.2.3. They compare their new scores with their previous scores to see if the strategy they considered is appropriate or needs to be revised.

3.2.5 Hot wash

Participants in the exercise share their insights and impressions throughout the game-based exercise. They also share the strategies they developed and the results of the exercise. They then discuss what constraints exist in applying the actions in the game and the strategies they formulated to a real-world company to use what they learned in the game.

3.3 Exercise results

Using this exercise method, the scores of the exercise participants generally improved. We believe that the fact that they played the game a couple of times has an effect; however, we also believe that, now that they understand the perspective of the attacker/defender, they can strategically choose their actions. The exercise participants gave positive feedback, saying that they were able to gain various things through the game; at the same time, they were able to measure their growth by being evaluated by a score. However, the following areas for improvement were identified.

- **Participants need to be made very aware of the intent of the exercise**
  Because of the structure of the game, where selecting an action advances the scenario, the game can be carried out even if choices are made without thinking. When playing on the Red side, it is essential to understand the attack methods. It is also necessary to understand that, as the attack phase progresses, the target of the attack changes. When playing on the Blue side, participants can choose actions under the managing constraints, such as budget and human resources. Participants, however, may not be aware of the inter-organizational cooperation required to perform counter-actions. Therefore, we believe that a mechanism to make participants aware of the purpose and intent of the exercise (e.g., adding content to the classroom lecture) is necessary.

4 CONCLUDING REMARKS

In this paper, we presented elements required for companies to respond to cyberattacks targeting OT systems and described a hybrid exercise, which we developed and implemented, combining a table-top exercise with a simulation exercise to enhance resilience to cyberattacks on corporate plants. The hybrid exercise combines table-top and simulation exercises to enhance the resilience of corporate plants against cyber threats. A newly developed exercise was also described. In the future, we will further develop these exercises and consider and construct new hybrid exercises that combine video and other media to make it easier for participants to gain insights from the exercises. We will also create a system to measure the effectiveness of the exercises.

REFERENCES

[1] Kaspersky, Stuxnet: Zero victims. https://securelist.com/stuxnet-zero-victims/67483/. Accessed on: 7 Sep. 2021.

[2] Cybersecurity and Infrastructure Security Agency, 2021 Alerts. https://us-cert.cisa.gov/ncas/alerts/2021. Accessed on: 7 Sep. 2021

[3] MAR-17-452 HatMan – Safety System Targeted Malware (Update B). https://us-cert.cisa.gov/ics/MAR-17-352-01-HatMan-Safety-System-Targeted-Malware-Update-B. Accessed on: 7 Sep. 2021.

[4] Pipeline Cybersecurity Initiative. https://www.cisa.gov/pipeline-cybersecurity-initiative. Accessed on: 20 Aug. 2021.

[5] Cybersecurity and Infrastructure Security Agency, National Cyber Awareness System, Bulletins. https://us-cert.cisa.gov/ncas/bulletins. Accessed on: 7 Sep. 2021.

[6] Software Engineering Institute. https://www.kb.cert.org/vuls/. Accessed on: 7 Sep. 2021.

[7] Aoyama, T., Koike, M., Koshijima, I. & Hashimoto, Y., A unified framework for safety and security assessment in critical infrastructures. *Safety and Security Engineering V*. WIT Press: Southampton and Boston, 2013.

[8] Red vs. Blue Gamification and Training. https://threatgen.com/red-vs-blue/. Accessed on: 7 Sep. 2021.

[9] NIST Cybersecurity Framework. https://www.nist.gov/cyberframework/framework. Accessed on: 7 Sep. 2021.

[10] The Cyber Kill Chain. https://www.lockheedmartin.com/en-us/capabilities/cyber/cyber-kill-chain.html. Accessed on: 7 Sep. 2021.

SECTION 3
DISASTER AND EMERGENCY MANAGEMENT

DEVELOPMENT AND IMPLEMENTATION OF A REAL-TIME SEDIMENT DISASTER ALERT SYSTEM

TAKEYASU SUZUKI[1], TAKUMI ITO[2] & SATOSHI GOTO[1]
[1]Disaster and Environmentally Sustainable Administration Research Center, University of Yamanashi, Japan
[2]Civil and Environmental Engineering Course, Integrated Graduate School of Medicine, Engineering, and Agricultural Sciences, University of Yamanashi, Japan

ABSTRACT
For flooding, visual information of rising water levels can signal the timing of residents' evacuation. For impending sediment disasters, however, the state of the groundwater level or soil moisture content in a slope's soil layer cannot be visually confirmed. Therefore, the authors developed a sediment disaster alert system, in part using a color-coded chart that signals evacuation urgency – an "evacuation switch". The system uses a slope's real-time rainfall observation as input data and performs real-time stability analysis based on rainwater runoff and infiltration simulations. In Nishikatsura Town, Yamanashi Prefecture, Japan, a slope designated as a mountain stream from which debris flow is likely to occur was selected as the target. The simulation model was constructed based on results of soil exploration and soil testing. The website for a sediment disaster alert system was developed by combining the rainfall observation system and the simulation. To provide the evacuation switch for residents, this website expresses risk of sediment disasters not by a safety factor but by a chart showing the current groundwater level with respect to the groundwater level at which the slope becomes unstable and collapses.
Keywords: sediment disaster, real-time rainfall observation, simulation, sediment disaster alert system, switch for evacuation.

1 INTRODUCTION

Abnormal weather caused by global warming has brought about unprecedented torrential rain, causing rain-related disasters in various parts of Japan. The 2017 Northern Kyushu heavy-rain disaster caused 42 casualties (40 dead and two missing), and the 2018 Western Japan heavy-rain resulted in 232 casualties (224 dead and eight missing). In 2019, Typhoon No. 19 killed 104, and in July 2020, a heavy-rain disaster killed 84. Many of these casualties were victims of sediment disasters, which, among heavy-rain-related disasters, are most likely to be fatal.

The most effective means of protecting lives against sediment disasters is early evacuation, and for early evacuation, municipalities have to anticipate sediment disasters, issue evacuation information, and establish a reliable evacuation system. In Japan, in line with the 2001 Sediment Disaster Prevention Act, municipalities issue evacuation information based on sediment disaster warnings announced by prefectures and the Japan Meteorological Agency. For short-duration heavy rainfall, however, sediment disaster warning information might not be announced in a timely fashion. But even if the information is announced, a sediment disaster often occurs before the announcement. And unfortunately, residents might not evacuate even if evacuation information is issued.

Indeed, evacuation due to a sediment disaster warning involves several issues. First, information is based on the critical line of a two-dimensional plot – with the soil–water index on the horizontal axis and 60 minutes' accumulated rainfall on the vertical axis. When, on this two-dimensional plane, a moving point is judged to have exceeded the critical line (after 2 hours), sediment disaster alert information is issued.

WIT Transactions on The Built Environment, Vol 206, © 2021 WIT Press
www.witpress.com, ISSN 1743-3509 (on-line)
doi:10.2495/SAFE210061

The soil water index, calculated with 60 minutes' accumulated rainfall as input, is the most important parameter for sediment disaster warning information [1]. Sixty minutes' accumulated rainfall is analysed by the Automated Meteorological Data Acquisition System precipitation radar and corrected by rainfall observation results on the ground. Although the Japan Meteorological Agency announces the amount of precipitation at a resolution of 1 km, the accuracy of rainfall observation points on the ground is only ~10 km. Thus, it is unsuitable for local rainfall estimation.

Therefore, we placed real-time weather sensors in sediment disaster hazard areas and, to issue evacuation information, predicted disasters based on local observational rainfall data. Furthermore, we demonstrated the possibility of forecasting rainfall and sediment disaster by evaluating auto-correlation and cross-correlation of rainfall intensity between rainfall observation stations in real-time. In adjacent municipalities' sediment disaster hazard areas, when rainfall observation is conducted at specific points in real-time, rainfall of ~1 hour can be forecast using cross-correlation between observation points, regardless of municipal boundaries [1].

This paper reports a new simulation method for evaluating risk of sediment disasters in real-time by combining rainwater infiltration into the soil, groundwater level generation and rise, and unsaturated surface slope stability by inputting real-time rainfall observation data. Additionally, this simulation is incorporated into a website providing a signal to evacuate, or an "evacuation switch", for community residents [2].

2 PREDICTION OF SEDIMENT DISASTERS BASED ON SLOPE STABILITY ANALYSIS

To improve the accuracy of predicting sediment disasters, improving the accuracy of predicting heavy rainfall and of local slopes' stability is important. Since slopes' inclination angle, surface thickness, and physical properties, including strength constants and infiltration characteristics of the soil's surface layer, differ, their stability differs depending on location even under the same rainfall conditions.

Many methods have been proposed to estimate beginning landslides and debris flows due to heavy-rain via stable analysis of a soil mass on a slope's base rock. Sliding resistance of a soil mass can be expressed by Mohr–Coulomb's theory of shear strength with parameters of cohesion, normal stress, and internal friction angle. On the other hand, sliding external force is a downward component of the weight of the soil mass on base rock. That is, if F_s expressed by eqn (1), which is the ratio of slip resistance to slip external force, is greater than 1, the soil mass is stable; if less than 1, the soil mass is unstable; and at exactly 1, the ratio is balanced, and the soil mass is expected to slide.

$$F_s = \frac{\tau_g}{\tau_r}. \tag{1}$$

As a result of rainwater infiltration, when a groundwater layer forms on the surface layer covering the slope's base rock, the weight of the soil mass increases, and the numerator of F_s formula increases. On the other hand, since effective stress decreases due to buoyancy, the denominator corresponding to shear resistance decreases, and consequently, it can be expected that when the F_s falls below 1, the surface layer starts to slide. Here three methods are used for handling rainwater infiltration into the surface layer. The first is a method in which all rainwater permeates; in the second, only part of rainwater determined by the runoff coefficient permeates; and, in the third, an inflow amount is calculated by runoff analysis.

To calculate F_s accurately, evaluation of vertical force by saturation of unsaturated ground formed by rainwater infiltration is necessary. Determining conditions under which the groundwater level begins to form is also necessary. When some heavy-rain permeates from the ground surface to the surface layer and permeates downward in an unsaturated state to reach base rock, saturation begins from the soil layer's lower part, and the groundwater level is formed. Groundwater flows downstream along the slope's surface layer, supplied from upstream and from the ground surface. By analyzing groundwater seepage in this way, the inclination angle β of the slope shown in Fig. 1, the groundwater level h_w, the saturated unit weight γ_{sat} of surface soil below groundwater level, and the wet unit weight γ_t of surface soil above groundwater level can be obtained. Then, the safety factor F_s can be expressed as follows:

$$F_s = \frac{c+\sigma' \tan\varphi}{(\gamma_t(H-h_w)+\gamma_{sat}\cdot h_w)\sin\beta}, \tag{2}$$

where, c is the soil's cohesion, φ is the soil's internal friction angle and σ' is an effective stress on base rock and can be expressed by the following equation:

$$\sigma' = (\gamma_t(H-h_w)+(\gamma_{sat}-\gamma_w)h_w)\cos\beta. \tag{3}$$

Thus, slope stability due to straight sliding on an infinite slope with inclination angle β is evaluated by the following safety factor, F_s:

$$F_s = \frac{c+\{\gamma_t(H-h_w)+(\gamma_{sat}-\gamma_w)h_w\}\cos\beta\tan\varphi}{\{\gamma_t(H-h_w)+\gamma_{sat}h_w\}\sin\beta}, \tag{4}$$

where, c is cohesion (kN/m^2); γ_t is wet unit weight soil (kN/m^3); γ_{sat} is saturated unit weight of soil (kN/m^3); γ_w is unit weight of water (kN/m^3); H is the surface layer thickness of (m); β is the target slope's inclination angle; φ is the soil's internal friction angle; and h_w is the groundwater level (m).

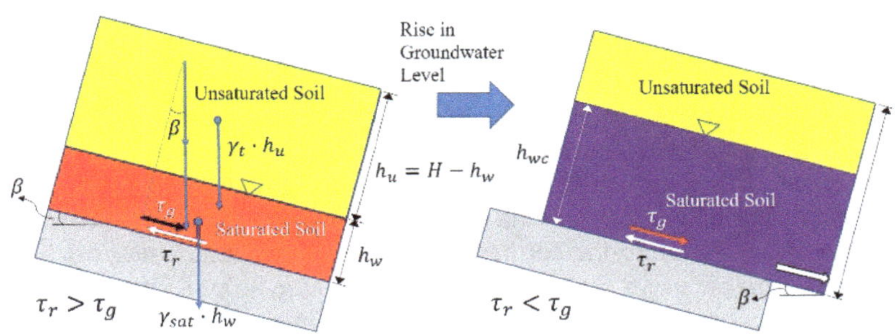

Figure 1: Conceptual diagram of slope stability analysis.

For slopes of constant gradient, Montrasio et al. proposed that groundwater level be a function of time related to precipitation intensity [3]. Kawabata et al. conducted rainfall experiments with rainfall intensities of 5 mm/hour, 25 mm/hour, and 50 mm/hour, finding that a high-water content zone with a degree of saturation corresponding to these rainfall intensities was formed from the ground surface's vicinity [4]. By groundwater seepage analysis, they proposed to determine the process of groundwater level formation when the high-water content zone moves downward, reaches the rock surface, and the groundwater level is formed and rises. Furthermore, to analyze F_s in each soil layer cross section when the slope and surface thickness are given, Kawabata et al. proposed a method of analyzing two-dimensional groundwater seepage along the slope [4].

As described above, for improving the accuracy of slope stability analysis, accurate evaluation of the soil's shear resistance and the soil layer's weight is necessary. To calculate the soil layer's weight accurately, tracking the soil's wet density and groundwater level (saturated soil height), which change according to rainfall intensity, is necessary. On the other hand, these parameters are also necessary for correct evaluation of effective stress in shear resistance.

The method of evaluating occurrence of sediment disasters was proposed from the ratio of the groundwater level to the collapsed groundwater level, in which the safety factor F_s reaches 1, by observing pore water pressure on the local slope and estimating the groundwater level [5]. Furthermore, a system for evaluating a slope's safety factor by estimating the soil's cohesion, the internal friction angle, the soil's weight, and the pore water pressure based on field observation of the soil layer's water content has also been adopted for practical use [6].

3 CONSTRUCTION OF SEDIMENT DISASTER RISK ANALYSIS AND VISUAL REPRESENTATION OF RISK

3.1 Runoff analysis using the tank model with observed real-time rainfall as input

This study developed a simulation method that combines real-time rainfall observation and slope stability analysis. With rainfall (mm) observed every minute as input, runoff analysis is performed using the tank model proposed by Ishihara and Kobatake (also used by the Japan Meteorological Agency for sediment disaster warning information) [7], and the amount of rainwater stored in the surface layer $S_1 + S_2$ is calculated every minute. $S_1 + S_2$ is rainwater (mm) that actually permeates the surface layer, calculated by subtracting surface flow and outflow from the outermost layer from the precipitation. In the tank model, $S_1 + S_2$ can be calculated as the storage capacity of the first and second tanks.

3.2 Seepage analysis for groundwater generation and groundwater level determination

When $S_1 + S_2$ reaches a certain threshold, the high-water content zone proposed by Kawabata et al. [4] is assumed to have has reached base rock, and calculation of groundwater level generation begins. This threshold needs to be set according to the local slope's soil layer characteristics. In the study by Kawabata et al. [4], the volumetric water content ratio θ_h of the high-water content ratio zone formed under constant rainfall intensity was defined. Obtaining the volumetric water content ratio θ_h according to rainfall intensity r that changes every minute is necessary. Therefore, we decided to obtain an approximate expression of the volumetric water content ratio θ_h' that fits the experimental results of Kawabata et al. and to estimate the volumetric water content ratio by extrapolation even at rainfall intensity over 50 mm/hour.

$$\theta_h' = 0.0388 \ln r + 0.2064. \tag{5}$$

However, observations have clarified that rainfall intensity changes every minute, and the amount of change is great. Especially in short-time heavy-rain, the peak changes in a cycle of about 10 minutes, so in calculation of volumetric water content using the approximate eqn (1), we decided to use the average value of rainfall intensity for 10 minutes up to the current time. In addition, because the experiment by Kawabata et al. [4] was conducted on decomposed granite soil, it cannot be applied to soil with geotechnical properties significantly different from those of decomposed granite soil.

When $S_1 + S$ exceeds a certain threshold, the increased groundwater level due to rainwater that permeates the soil from the ground surface is calculated. Since the volumetric water content ratio corresponds to the pore space in the saturated state, it can be expressed by the following equation using the void ratio, *e*.

$$\theta_{sat} = \frac{e}{1+e}. \tag{6}$$

The difference between θ_{sat} and the volumetric water content ratio θ of the unsaturated soil in the high-water content zone before saturation, θ_r, means the volumetric water content ratio at which rainwater permeating from the ground surface fills the unsaturated soil's voids.

$$\theta_r = \theta_{sat} - \theta_h. \tag{7}$$

Therefore, the increase in groundwater level, ΔH (m), due to rain water, r (mm) can be calculated by the following equation:

$$\Delta H = \frac{r}{\theta_r}. \tag{8}$$

In this study, however, determining soil parameters by a general soil test was important. Thus, the volumetric moisture content ratio, θ_h according to the rainfall intensity used in this analysis is estimated using eqn (5), obtained by experiments using materials differing from that of the local soil sample. Therefore, in this study, θ_h in eqn (7) is replaced with θ_h' in eqn (5), and the amount of groundwater level rise is estimated by eqn (9):

$$\Delta H = \frac{r}{\theta_r} = \frac{r}{\theta_{sat}-\theta_h'}. \tag{9}$$

After the groundwater level is generated, $r = S_1 + S_2$ is the increment of groundwater supplied from the ground surface to the surface layer in 1 minute. In addition, calculating groundwater's inflow and outflow from the slope's adjacent surface blocks every minute is necessary [4].

The density of saturated soil ρ_{sat} can be expressed using the density of water ρ_w and the void ratio *e* as follows:

$$\rho_{sat} = \frac{G_s+e}{1+e}\rho_w. \tag{10}$$

On the contrary, the density of unsaturated soil ρ_t with a volumetric water content ratio θ_h can be given as follows:

$$\rho_t = \frac{G_s+\theta_h e \rho_w}{1+e}. \tag{11}$$

However, since $\theta_h \neq \theta_h'$, the unsaturated soil's density ρ_t is treated as a constant value as the soil's wet density as sampled in the field analysis. Slope stability analysis is performed in real-time based on fluctuating groundwater levels.

4 REAL-TIME REPRESENTATION OF THE RISK OF SEDIMENT DISASTERS USING A CHART BASED ON THE GROUNDWATER LEVEL

Slope stability analysis shows that when safety factor F_s falls below 1, the slope becomes unstable and collapses. Although engineers easily understand the safety factor, residents are not very familiar with it. Therefore, we thought that residents would find it difficult to use F_s for evacuation decisions. Thus, the groundwater level when Fs becomes one is defined as the critical groundwater level h_{wc}, and α, the ratio of the current groundwater level h_w to the critical groundwater level, is used as an index indicating the degree of urgency of sediment disaster.

$$\alpha = \frac{h_w}{h_{wc}}. \tag{12}$$

Since 2020, evacuation information has been released in Japan by classifying alerts into five levels. Table 1 shows the relationship between alert level and evacuation information. Alert levels 1 and 2 are announced by the Japan Meteorological Agency as weather information, and warning levels 3 and above are issued by municipalities as evacuation information. Alert level 3 is for evacuation of the elderly, that is, the timing when elderly people and people with disabilities start evacuation with support; alert level 4 is the timing when healthy people start and complete evacuation; and alert level 5 calls for ensuring emergency safety where a disaster has already occurred. Among these, α in eqn (10) corresponds to alert level 4, considered effective in communicating the degree of sediment disaster urgency as the groundwater level rises.

Fig. 2 shows this study's proposed chart. The green chart on the far left shows stages to alert level 2. The yellow chart shows the stage before the groundwater level is formed, but if rainfall continues, the groundwater level will be generated after approximately 1 to 2 hours. In this analysis, a threshold value is set for $S_1 + S_2$, and when this threshold value is reached, the alert level is 3, and a yellow chart is displayed. When $S_1 + S_2$ further increases and reaches

Table 1: Alert levels and evacuation information.

Alert level	Name of information	Required behavior	Administrative agency to issue
5	Ensuring emergency safety	Move to the upper floors or a safe building	Municipality
Complete evacuation up to alert level 4			
4	Evacuation order	Residents living in areas at high risk of disaster begin to evacuate	
3	Evacuation of elderly people, etc.	Information to urge elderly people to evacuate	
2	Heavy rain/flood/storm surge warning	Check evacuation behavior in advance	Japan Meteorological Agency
1	Early warning information	Early warning information	

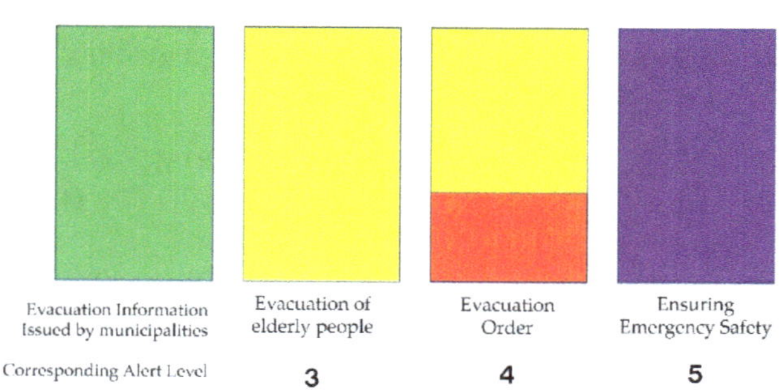

Figure 2: Charts to represent the state of slope stability corresponding to alert levels.

the next threshold value, calculation of groundwater begins, and the chart turns to yellow and red with different ratios depending on the value of α. That is, the red chart's height is shown as the ratio of the current groundwater level to the critical groundwater level. Finally, when the groundwater level reaches the critical groundwater level and the safety factor falls below 1, the alert level changes to 5, and a purple chart on the far right in the figure is displayed.

5 LOCAL APPLICATION

5.1 Overview of the target area

The authors are conducting real rainfall observations in Nishikatsura Town, Yamanashi Prefecture, Japan. The Shimokurechi district of Nishikatsura Town is an alluvial fan located upstream of the Shakunagare River in the town's north-western part. The area has 330 households and a population of 900. Almost all districts are designated as sediment disaster warning areas, and some are designated as sediment disaster special warning areas. Fig. 3, along with the Shimogurechi area's hazard map, shows the location of the rainfall observation point and of the slope that is this study's field.

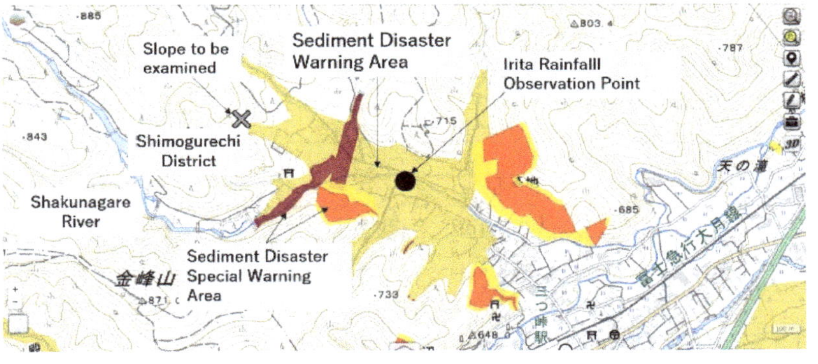

Figure 3: Locations of the rainfall observation point and the slope to be examined.

5.2 Modelling the slopes based on a field survey

Fig. 4(a) shows a photograph taken from the local slope of a residential area. This slope has had several slides, and after rainfall, its surface layer was muddy, with a lot of water, and rainwater flowed from the ground surface toward the residential area. A simple, standard penetration test was conducted along the slope of the valley to estimate surface thickness and slope angle of base rock. Results confirmed that the surface layer was almost uniform, of sandy soil, and with weathered conglomerate with an N-value of 1 to 2; Fig. 4(b) shows its profile. Using this cross section, seepage flow analysis was performed, and the slope safety factor was calculated for the No. 1 to No. 3 cross sections.

5.3 Input parameters for surface layer

Undisturbed soil samples were taken from the slope, and tests were conducted for density of soil particle, permeability, and direct shear resistance. Based on these results and the surface profile in Fig. 4(b), input parameters were summarized, as shown in Table 2.

(a)

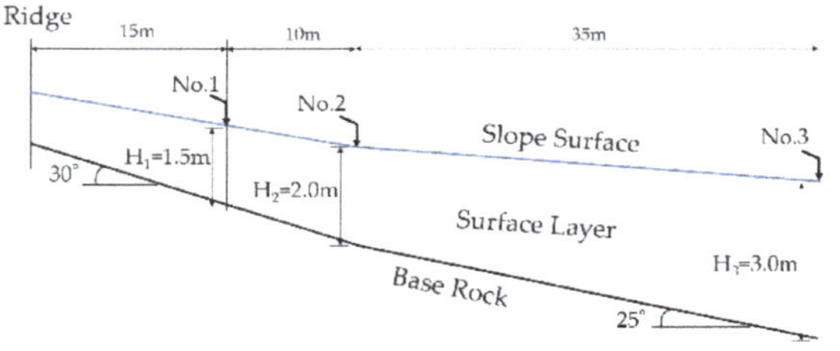

(b)

Figure 4: Local slope and residential area (sediment disaster warning area). (a) View of the residential area; and (b) Subsurface profile of the modelled local slope.

Table 2: Input data used in the simulation.

	No. 1	No. 2	No. 3
Angle of slope, β	30.0	30.0	25.0
Thickness of surface layer, H (m)	1.5	2.0	3.0
Density of soil particle, ρ_s (g/cm^3)		2.585	
Saturated density, ρ_{sat} (g/cm^3)		1.577	
Wet density, ρ_t (g/cm^3)		1.336	
Void ratio, e		1.75	
Volumetric moisture content ratio, θ		0.241	
Saturated volumetric moisture content ratio, θ_{sat}		0.636	
Cohesion, c (kN/m^3)		0	
Internal friction angle, φ (degree)		40.03°	
Coefficient of permeability, k (m/s) $\times 10^{-4}$	3.80	3.84	3.84

6 SETTING THE THRESHOLD OF $S_1 + S_2$ FOR SIMULATION

To apply this simulation to the actual slope, indices must be established for alert levels 3 and 4, respectively, to begin calculating groundwater level. Using the tank model, these indices were established according to the threshold value of the output value $S_1 + S_2$ of the runoff analysis. Therefore, we conducted parametric studies in which $S_1 + S_2$ was changed for the slope analysis model in the Shimogurechi district of Nishikatsura Town.

In 2019, Typhoon No. 19 brought heavy rainfall with accumulation of about 500 mm in Nishikatsura Town (Fig. 5), which issued an early evacuation advisory, based on rainfall information from the sediment disaster alert system developed by the author, to residents of sediment disaster warning areas, including the Shimogurechi district. According to a questionnaire survey conducted by Suzuki et al. [8], only 30% of residents evacuated even within the area designated as a sediment disaster warning area, and voluntary evacuation was conducted in addition to issuance of evacuation advisories. Therefore, in addition to Nishikatsura Town's issuance of evacuation advisories, we decided to create new information – "an evacuation switch" – to encourage residents to evacuate voluntarily. Real-time rainfall observation was conducted at the Irita observation point at the confluence of small rivers, almost at the Shimogurechi district's center. Fortunately, no large debris flow occurred on the town's slopes, but violent surface flow was observed in the area's swamps, indicating a state just previous to a sediment disaster.

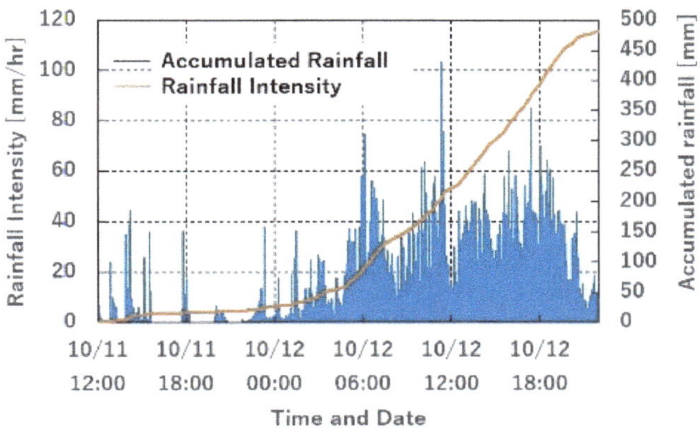

Figure 5: Rainfall observation record at Irita during Typhoon No. 19.

Fig. 6 shows results of the tank model's runoff analysis of observed rainfall data recorded every minute. The peak value of $S_1 + S_2$ was about 210 mm, and the soil–water index, which is a sum of $S_1 + S_2$ and S_3, infiltration into the basement, was about 250 mm at the maximum. When $S_1 + S_2$ reaches 140, 150, 160, 170, or 180 mm, respectively, assuming that the high-water content ratio zone reaches the basement, the subsequent inflow of rainfall infiltration ($r = S_1 + S_2$) and seepage along the slope's surface layer was analyzed. Then, according to the generated groundwater level, the slip safety factor was analyzed every minute. As a result of these parametric studies, Fig. 7 shows the time transition of the slip safety factor in cross sections No. 1 to No. 3. Results in which threshold values are 140, 150, 160, 170, and 180 mm, respectively, are compared in the figures.

For this rainfall event, the threshold value of $S_1 + S_2$ was set so that the slope's slip safety factor was approximately 1, that is, the verge of a sediment disaster. From these figures, in the case the threshold value of $S_1 + S_2$ is 170 mm, the safety factor is not less than one in the No. 1 and No. 3 cross sections, but is definitely less than one in No. 2. Therefore, we decided that the threshold value of $S_1 + S_2$ for starting calculation of groundwater level was 170 mm, so evaluation of alert level 4 was made at this stage. In contrast, the timing for displaying the alert level 3 chart was set to 130 mm, about 75% of 170 mm.

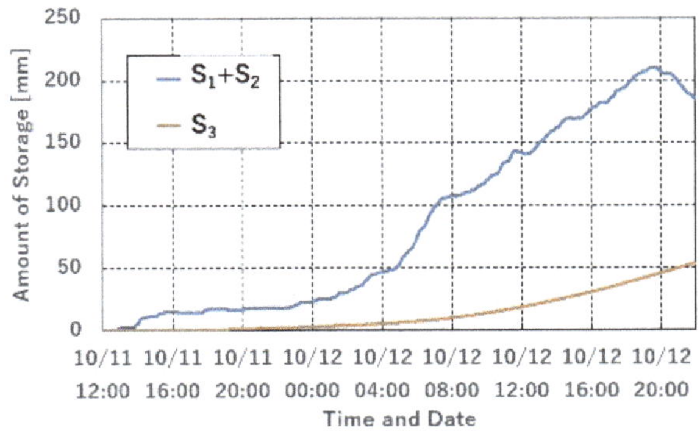

Figure 6: Results of runoff analysis by the tank model.

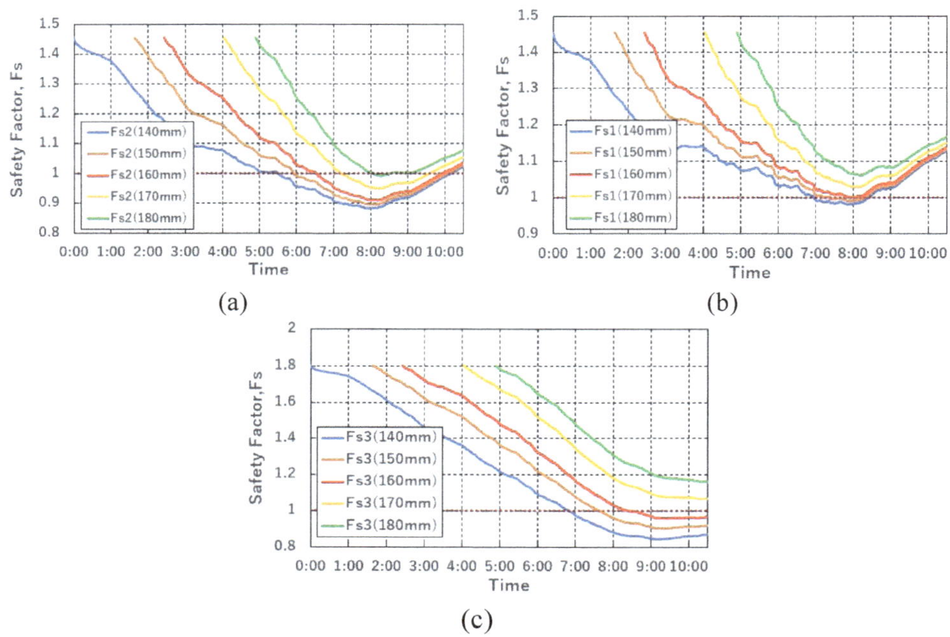

Figure 7: Time histories of the slip safety factor in the cross sections. (a) No. 1; (b) No. 2; and (c) No. 3

7 DEVELOPMENT OF SEDIMENT DISASTER ALERT SYSTEM

The authors are conducting real-time meteorological observations in Nishikatsura Town, Yamanashi Prefecture, Japan. An observation system that graphs the amount of rainfall observed every minute over a time span of 1 hour, 24 hours, and 72 hours is being used for disaster response. Furthermore, the sediment disaster risk web system that displays these on a map by color-coding rainfall intensity (mm/h) and dividing 72-hour cumulative rainfall into six levels from 0 to 5 is used by Nishikatsura Town staff and fire brigade members during heavy rainfall.

Yamori calls the deciding level for evacuation of residents an "evacuation switch," a signal when a river's flooding reaches a certain level or water begins to accumulate in the lowlands [2]. However, in evacuation from debris flow disasters, groundwater level and soil–water content cannot be grasped without on-site observation, so it is difficult and dangerous for local residents to monitor a visual on-site evacuation switch. Therefore, in our sediment disaster alert system, a chart showing degree of urgency is displayed, so local residents can view it as an evacuation switch.

Fig. 8 shows the sediment disaster alert system's configuration. The proposed simulation is incorporated as a Fortran Module receiving rainfall data every minute in cooperation with a PHP application and analyzing the flag, groundwater level ratio, safety factor and so on that indicate the alert level. In addition, the groundwater level ratio is displayed on the sediment disaster alert system using an html application. Fig. 9 is a screen representation of this system.

The system was introduced by one of the authors at a gathering of Nishikatsura townspeople, to ward mayors and disaster prevention leaders who are participants in the

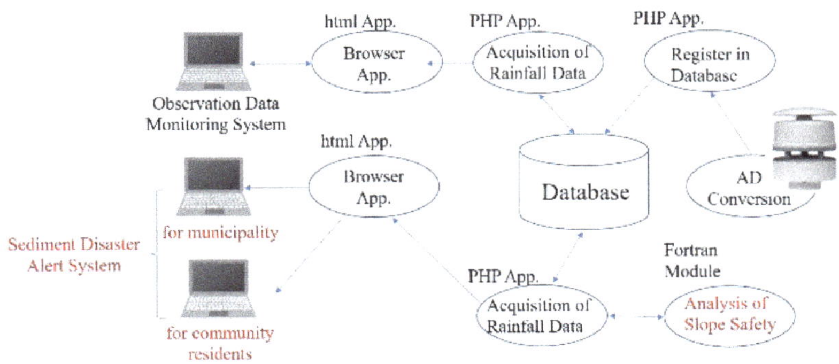

Figure 8: Configuration of the developed sediment disaster alert system.

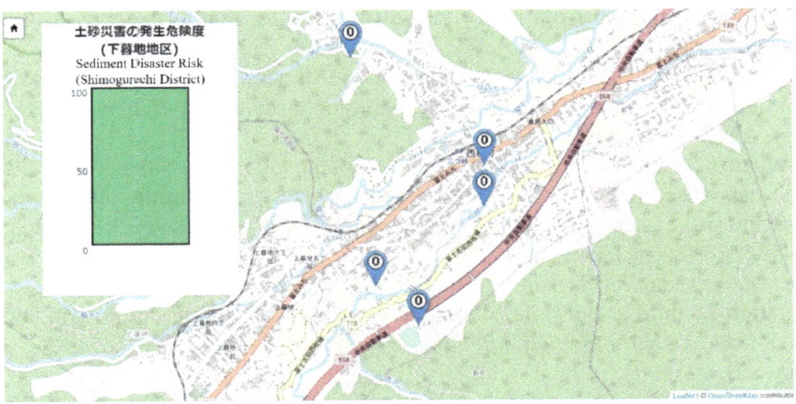

Figure 9: Sediment disaster alert system for residents.

gathering, and was evaluated as easy for residents to understand. In the future, we would like to add a supplementary explanation of the chart, so it will be even easier for residents to understand, and hold workshops with residents in the Shimogurechi district.

8 CONCLUDING REMARKS

For Nishikatsura Town, Yamanashi Prefecture, Japan, the authors developed a sediment disaster alert system with color-coded charts serving as an evacuation switch indicating sediment disasters' urgency. The system uses real-time rainfall observation as input data and performs real-time stability analysis based on a slope's rainwater runoff and infiltration simulations. A slope designated as a mountain stream from which debris is likely to flow was selected as the target. The simulation model was constructed based on-site exploration and soil testing results. For the sediment disaster alert system, a website was developed by combining the rainfall observation system and the simulation. To provide an evacuation switch for residents, the website expresses sediment disaster risk not by a safety factor but by a chart showing current groundwater compared to the groundwater level at which the slope becomes unstable and collapses.

This system was constructed for the Shimogurechi area of Nishikatsura Town and has been experimentally operated as a switch for local residents' voluntary evacuation since the spring of 2021. We hope this system will be accepted by community residents, be operated in other districts of the town, and then be expanded to other municipalities.

ACKNOWLEDGEMENTS

The authors would like to express our gratitude to the General Incorporated Foundation Kyousou, Koden Electronics Co., Ltd., and the Yahoo! Fund for their research funding. The authors would like to thank Enago (www.enego.jp) for English language review.

REFERENCES

[1] Suzuki, T. & Ito, T., Prediction of sediment disasters due to short-duration heavy rain based on real-time rainfall observation. *WIT Transaction on The Built Environment*, vol. 189, WIT Press: Southampton and Boston, pp. 91–99, 2019.

[2] Yamori, K., Missing of evacuation Information/FACP model/Evacuation switch: Rethinking evacuation from heavy rain disasters. *Firefighting Science*, **134**, pp. 7–11, 2018. (In Japanese.)

[3] Montrasio, L., Valentino, R. & Losi, G.L., Towards a real-time susceptibility assessment of rainfall-induced shallow landslides on a regional scale. *Nat. Hazards Earth Syst. Sci.*, **11**, pp. 1927–1947, 2011.

[4] Kawabata, S., Tsuchida, T., Kano, S., Yuri, A., Hanaoka, T. & Nakagawa, S., Evaluation of the risk of sediment-related disasters during heavy rains in mountain streams based on ground surveys and slope stability analysis. *Japanese Geotechnical Journal*, **8**(1), pp. 119–131, 2013. (In Japanese.)

[5] Yabe, M., Terada, Y., Nonami, K., Tobita, K., Ueno S. & Sone, Y., Pore water pressure estimation method at the time of surface collapse using observation data and the tank model. *Proc. Japan Society of Civil Engineers*, **C73**(2), pp. 141–156, 2017. (In Japanese.)

[6] NEC, Sediment disaster sign detection system. https://jpn.nec.com/press/201708/20170830_01.html. Accessed on: 2 May 2021.

[7] Ishihara, Y. & Kobatake, S., Runoff model for flood forecasting. *Bulletin of the Disaster Prevention Research Institute*, **29**(1), pp. 27–43, 1979.

[8] Suzuki, T., Lyu, J. & Ito T., Efforts for community disaster management to promote autonomous evacuation in sediment disasters: Efforts conducted at Shimogurechi district, Nishikatsuwra Town. *Community Bosai*, **21**, 2021. (In Japanese.)

RESILIENCE AND EMERGENCY MANAGEMENT OF ROAD TUNNELS: THE CASE STUDY OF THE SAN ROCCO AND STONIO TUNNELS IN ITALY

FABIO BORGHETTI[1], ALESSIO FRASSOLDATI[2], MARCO DERUDI[2], IGINO LAI[3] & CRISTIAN TRINCHINI[3]
[1]Design Department, Mobility and Transport Laboratory, Politecnico di Milano, Italy
[2]Department of Chemistry, Materials and Chemical Engineering, Politecnico di Milano, Italy
[3]Strada dei Parchi S.p.A, Italy

ABSTRACT

In Italy, mainly due to its complex orography, there are more than 50% of road tunnels longer than 500 m belonging to the Trans European Road Network. Moreover, many road tunnels were designed and built several years ago, where both traffic flows and vehicle sizes were different. Generally speaking, it is possible to state that if on one hand road tunnels allow the improvement of the plano-altimetric coordination of road layouts, on the other hand they can represent a safety problem for users and for the infrastructure in case of a major event. With this in mind, it is interesting to study the resilience of a road tunnel following an event: how can the resilience of a tunnel be increased by considering emergency management? How can negative consequences be reduced by adopting specific safety measures, such as emergency teams? Starting from these two questions, the main goal of this work was to analyze two fire events that occurred in two tunnels located on the A24 highway in Italy between 2019 and 2021 and managed by Strada dei Parchi S.p.A. Both fires involved a light vehicle and there were no injuries or fatalities. For each of the two events, the response time of the emergency teams and the resolution time of the event were analyzed. In the first case it is possible to evaluate the effectiveness of the emergency teams in controlling and/or extinguishing a fire, while in the second case the impact on road traffic caused by the tunnel closure is analyzed. These analyses can be used for two important purposes: (i) risk assessment of a tunnel considering the emergency team as a safety measure; and (ii) implementation of traffic management plans to define the best strategy to be adopted according to the estimated time of closure of a tunnel.

Keywords: road tunnel, tunnel safety, emergency management, traffic management, transport resilience, tunnel resilience, tunnel operations, emergency teams effectiveness, tunnel fire, tunnel accident.

1 INTRODUCTION

In Italy, mainly due to its complex orography, there are more than 50% of road tunnels longer than 500 m belonging to the Trans European Road Network (TERN) [1], [2]. Moreover, many road tunnels, were designed and built several years ago, where both traffic flows and vehicle sizes were different. The use of road tunnels makes it possible to guarantee the mobility of people and goods, limiting, for example, the tortuosity, the slope and length of roadways, and fuel consumption; this leads to a reduction of potential danger factors. Generally speaking, it is possible to state that if on one hand road tunnels allow the improvement of the plano-altimetric coordination of road layouts, on the other hand they can represent a safety problem for users and for the infrastructure in case of a major event. This is true not only for tunnels located in suburban areas but also for urban ones: today tunnels play an important role, as they allow the reorganization of traffic in the city and therefore the rethinking of city spaces, making them more human-scale [3]. After the fire that occurred in the Mont Blanc tunnel in 1999, which caused 39 deaths and extensive damage to the infrastructure (almost 3 years of closure), the European Community issued Directive 2004/54/EC on minimum safety requirements for tunnels belonging to the TERN of more

WIT Transactions on The Built Environment, Vol 206, © 2021 WIT Press
www.witpress.com, ISSN 1743-3509 (on-line)
doi:10.2495/SAFE210071

than 500 m in length. Thanks to this Directive, a normative framework has been implemented that today includes technical regulations and guidelines on good standards for the design, construction and operation of road tunnels [4]. Due to the serious consequences and deaths caused also by the events that occurred in other European tunnels (Gotthard in 2001, Tauri in 2002 and Fréjus in 2005), public and political awareness of the problem has increased: specifically, the importance of this type of infrastructure on a human, cultural, economic and social level has been highlighted. Specifically, the 2004 European Directive regulates tunnel safety by establishing minimum safety requirements in terms of equipment, infrastructure and management procedures, and by identifying risk analysis as an analytical and well-defined methodology for estimating the risk level of each tunnel. In practice, the Directive regulates safety:

- identifying the tasks of the entities and administrative authorities responsible for the management and control of tunnels;
- establishing minimum requirements for equipment and infrastructure based on certain parameters (e.g., tunnel length and traffic volumes);
- identifying and recommending Risk Analysis as the appropriate analytical methodology to determine the level of risk of each tunnel.

With reference to this last issue, the main goal of the quantitative risk analysis (QRA) for road tunnels is to assess the risk for the specific "tunnel system" as illustrated in Fig. 1.

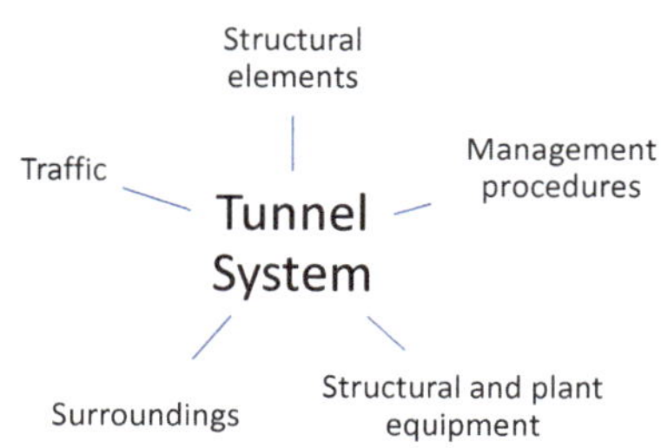

Figure 1: Tunnel system components.

The analysis includes consideration of several parameters such as accident rate, traffic, tunnel geometry, structures/equipment, infrastructure and management measures. Among the different steps of the risk analysis there is the estimation of the frequencies of possible accidental events, which has to consider: (i) the type of vehicles involved (light and/or heavy); (ii) the location of the accident inside the tunnel; and (iii) the number of injuries and/or fatalities. One possible technique, based on event tree analysis, is to define top event, and then assess the frequencies and consequences of each accidental scenario [5], [6]. In this regard, emergency teams can influence the evolution of an accidental scenario, both in terms of the frequency of occurrence and the consequences of the event itself. In the first case, emergency teams can control or extinguish a fire while preventing the fire from involving other vehicles; in the second case, they can help and coordinate the evacuation of users from the tunnel [7], [8]. In general, the probability of occurrence of an accident and the probability of being injured is lower in tunnels than in open road sections. It should be noted, however, that if an accident (e.g., fire) occurs in a tunnel, the potential consequences inside this confined environment are significantly greater than in open sections [9], [10]. Vehicle drivers are more careful driving in tunnels than in open sections: as a result, the crash risk is estimated to be about half that of open sections [11]–[13]. Because of this aspect, safety studies and risk analyses for road tunnels typically consider as reference accidental scenarios vehicle fires and dangerous goods releases, which can cause major consequences [9], [14]–[16]. Prolonged closure of a tunnel following an event can also result in socio-economic impacts that can be observed across a large area (region or country): such a situation can cause increased costs to the community. More specifically, these impacts can be related to the loss (or reduction) of freight and passenger traffic (including tourism) and increased travel time caused by the detour of traffic to alternative routes. In some studies, the risk assessment is done by also considering the decision-making process from an economic perspective [17]. With this in mind, it is interesting to study the resilience of a road tunnel following an event: how can the resilience of a tunnel be evaluated and increased by considering emergency management? How can negative consequences and effects be reduced by adopting specific safety measures, such as emergency teams? Starting from these two questions, the main goal of this work was to analyze two fire events that occurred in two tunnels located on the A24 highway in Italy between 2019 and 2021 and managed by Strada dei Parchi S.p.A. Both fires involved a light vehicle and there were no injuries or fatalities. For each of the two events, the response time of the emergency teams and the resolution time of the event were analyzed. In the first case it was possible to evaluate the effectiveness of the emergency teams in controlling and/or extinguishing a fire, while in the second case the impact on road traffic caused by the tunnel closure (resilience) was analyzed. These analyses can be used for two important purposes: (i) risk assessment of a tunnel considering the emergency team as a safety measure; and (ii) implementation of traffic management plans to define the best strategy to be adopted according to the estimated time of closure of a tunnel.

2 RESILIENCE AND EMERGENCY MANAGEMENT

Highway accident and emergency management professionals are continually working to improve strategies to ensure safe and rapid incident clearance [18]. When an event occurs, several activities need to be executed including protecting the crash site by realizing detours or implementing other traffic management measures when necessary, such as rescuing stopped traffic, and rescuing and evacuating injured people. In the event of long road closures, domino effects can occur that result in high levels of congestion in the surrounding road network, so it is important to alert drivers about the road/tunnel closures as soon as possible so that they can take alternative routes and avoid being trapped and increasing congestion in the area where the event occurred. When an accident occurs, it is difficult to estimate the cost of delays due to accident consequences, because generally speaking, not only the loss of time must be taken into account, but also, for example, late freight deliveries, missed appointments, missed flights at airports, personal inconvenience, and frustration. From an analytical point of view, the various factors that can most affect the response time due to an accident are: location, time of day, direction of travel, accident type, weather conditions, number and type of vehicles involved, number and location of lanes involved, number and type of responders required on scene, traffic queues (delay). In the scientific literature, several studies have addressed the issue of resilience of a transportation infrastructure following a major event [19], [20]. To date, there are several definitions of resilience; one of which can be "the ability of an entity, e.g., asset, organization, community, region – to anticipate, withstand, absorb, respond to, adapt to, and recover from a disruption" [21]. When a roadway accident occurs, including those in tunnels, the recovery time should be considered, which could be longer than that of a common accident due to the confined space. Fig. 2 shows the trend and components of resilience with reference to a transportation system following a major event. The performance of a transportation system could be evaluated by considering the satisfied transportation demand (e.g., percentage of traffic satisfied).

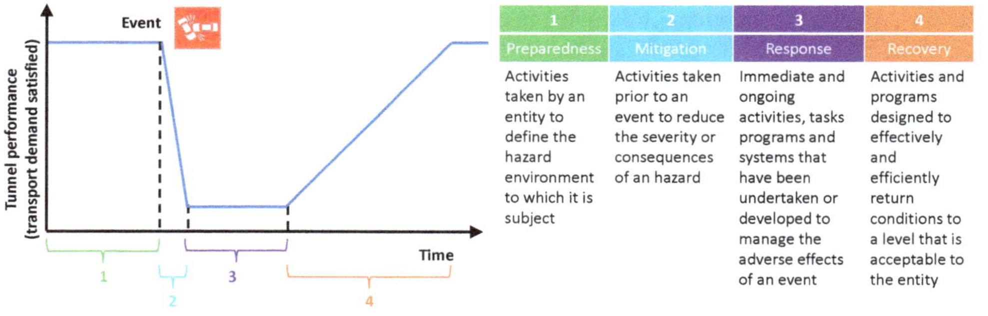

Figure 2: Resilience components of a transport system. *(Source: Adapted from [21]–[23].)*

A number of international studies have confirmed that the presence of firefighting crews or emergency teams located near tunnel entrances (capable of rapid response) brings significant benefits in terms of effectiveness in managing a tunnel fire [5], [24]. A study performed on some Japanese and European tunnels states that firefighting team intervention should occur within 7 min [25]. In the report "Fire Safe Design, Technical Report – Part 3" [26], it is stated that the intervention of emergency teams is effective if it occurs within the first 10 min of the accidental event. An additional aspect relates to the emergency teams' situational awareness to effectively execute their operations. To ensure this, it is necessary that teams are dedicated to a specific tunnel and specific communication systems are used with the control center, which oversees the evolution of the accidental scenario. The added value in using emergency teams, in addition to their speed of intervention, is their detailed knowledge of the tunnel system with particular reference to infrastructure measures, equipment and management procedures. In this way, emergency teams are able to operate to control or extinguish fire principles before they evolve into larger fires [5]. Moreover, as abovementioned, emergency teams can support tunnel users in their evacuation process by quickly providing useful information about the direction of the closest emergency exit to be reached [26]. In case of relevant events, thanks to the specific knowledge of the tunnel system, these teams can eventually support the intervention of other rescue teams (e.g., fire brigade) by providing detailed information about the evolution of the scenario and the infrastructure also through a continuous communication with the control center [5], [27], [28]. In another research, several tunnel fires were analyzed to estimate the response time and effectiveness of teams in controlling or extinguishing the fire. The study also analyzed the intervention strategies of fire teams for different types of tunnels (e.g., one-way or two-way, with or without vehicle queues in the smoke zone, and different ventilation strategies) [24]. When a major event occurs and it requires the prolonged closure of a tunnel, alternative routes can reduce traffic demand on the roadway and prevent users from approaching the accident area. Identifying practicable alternative routes is not always easy; this task requires the implementation of specific traffic management plans. However, not all alternative routes may be suitable for all types of vehicles. As an example, for heavy vehicles, there may be restrictions related to mass and shape limits that could restrict passage on certain routes (e.g., presence of a bridge or bottleneck). In these cases, separate alternative routes may be identified for light and heavy vehicles. The analysis of alternative routes and more general traffic management measures also depend on the traffic flows affected by an event. Assuming that the average daily traffic (ADT) of a road is known, it is possible to represent its trend as a function of time during the 24 hours as shown in Fig. 3. The duration of an accident event may lead to a reduction or even an interruption of traffic depending on the scenario and the type of infrastructure. In this way, it is possible to estimate how many vehicles will potentially be involved and will need to be managed (e.g., on alternative routes or stopped at suitable service areas) depending on the period in which the event occurs (e.g., peak hour). Fig. 3 also shows the timeline for the different management activities from the time an accident occurs until vehicle traffic is recovered.

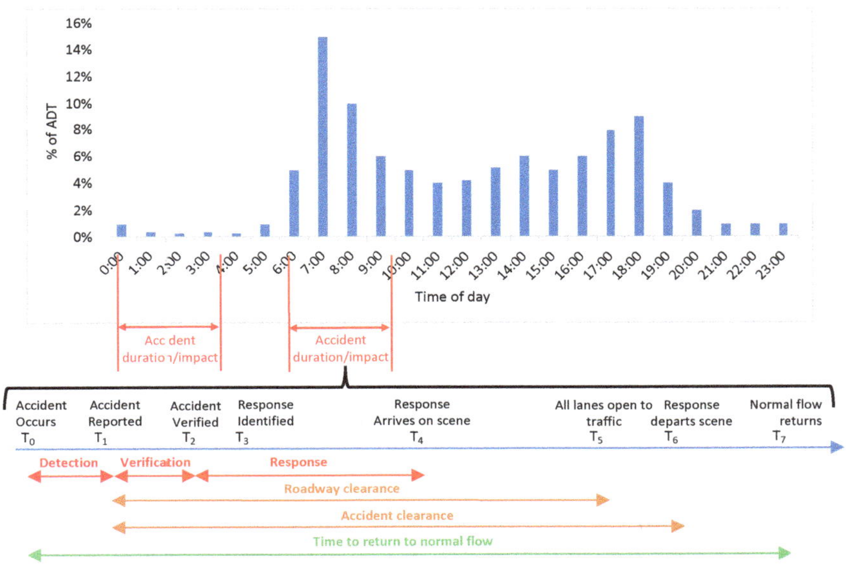

Figure 3: Qualitative trend of average daily traffic and timeline of traffic accident elements. *(Source: Adapted from [29].)*

3 CASE STUDY: THE TUNNELS OF THE A24 AND A25 HIGHWAYS IN ITALY

The A24 and A25 highways form a freeway system in Italy currently managed by Strada dei Parchi S.p.A, which starts in Rome and ends in Teramo (A24). About halfway along the route, at Torano, there is a junction that allows to go towards Pescara (A25). The two highways have a total length of about 280 km (~165 km of A24 from Rome to Teramo and ~115 km of A25 from Torano to Pescara). The route is characterized by the presence of Apennines: it follows that there are numerous bridges, viaducts and tunnels. Specifically, on the two highways there are 153 long bridges and viaducts for a total length of 118.8 km. In addition, there are 54 tunnels for a total length of 70.8 km, which corresponds to 12.6% of the entire route. 14 tunnels have a length greater than 500 m and belong to the TERN network: eight of these tunnels exceed 2,000 m and one exceeds 10,000 m (Gran Sasso tunnel). Fig. 4 shows the layout of the A24 and A25 highways and the location of the 14 tunnels longer than 500 m. Twelve tunnels are located on the A24 highway (highlighted using blue dashed line) while two tunnels are located on the A25 highway (green dashed line). The two highways consist of two carriageways and each carriageway has two lanes in each direction plus an emergency lane (not present inside the tunnels). Table 1 summarizes the main features of the two tunnels analyzed: San Rocco and Stonio.

Figure 4: Location of the 14 tunnels managed by Strada dei Parchi S.p.A. *(Source: Strada dei Parchi S.p.A.)*

Table 1: Main features of the San Rocco tunnel and Stonio tunnel. *(Source: Strada dei Parchi S.p.A.)*

	San Rocco tunnel		Stonio tunnel	
Feature	Right tube	Left tube	Right tube	Left tube
Year of opening to traffic	1969	1969	1969	1969
One-way tunnel (yes/no)	Yes	Yes	Yes	Yes
Length (m)	4181	4176	1242	1192
Cross-section area (m^2)	70	70	54	54
Central height (m)	8.5	8.5	6.5	6.5
Available height (m)	5	5	5	5
Lane width (m)	3.75	3.75	3.75	3.75
Overtaking lane width (m)	3.75	3.75	3.75	3.75
Number of lanes	2	2	2	2
Emergency lane (yes/no)	No	No	No	No
Average longitudinal slope (%)	2.2%	–2.2%	2.6%	–2.6%
Minimum bending radius (m)	1200	1200	1010	1010
Average daily traffic (vehicle/day)	7122	6992	15398	15176
Light vehicles (LV) (%)	86	87	88	88
Heavy goods vehicles (HGV) (%)	14	13	12	12
Speed limit (km/h)	110	110	110	110
Overtaking heavy vehicles (> 3.5 t) (yes/no)	No	No	No	No
Dangerous goods transport	Yes	Yes	Yes	Yes

3.1 Emergency teams description

One of the safety measures adopted by Strada dei Parchi S.p.A. concerns the emergency teams located along the A24 and A25 highways near the tunnels. The firefighting service consists of emergency teams with special vehicles and equipment to control and extinguish fires. The personnel have specific qualifications such as, for example, that of "Emergency and first aid workers in tunnels/confined areas". The proximity of the emergency teams to the tunnels allows them to quickly reach the scene of the accidental event and to control/suppress a possible fire. In addition, team personnel can support tunnel users during the evacuation process. Both vehicles are equipped with an emergency firefighting system using micronized water. Fig. 5 shows the main technical features and Fig. 6 shows an example of location of emergency teams near a tunnel.

Description
1 stainless steel tank with a capacity of about 60 L of extinguishing liquid
2 compressed air cylinders of 9 L each at 300 bar
1 fire hose with a length of about 40 m
1 spray gun of 1 L

Description
1 stainless steel tank with a capacity of about 400 L of extinguishing liquid
5 compressed air cylinders of 9 L each at 300 bar
1 fire hose with a length of about 40 m
1 spray gun of 1 L
1 high pressure pump
1 fractional Water Lance

Figure 5: Main technical characteristics of vehicles used by firefighting teams. *(Source: Strada dei Parchi S.p.A.)*

Figure 6: Example of location of emergency teams near a tunnel. *(Source: https://www.google.it/maps/.)*

3.2 San Rocco tunnel event

On 23 February 2021, a light vehicle fire occurred within the right tube of the San Rocco tunnel, which is 4,181 m long. The possible cause of the fire is associated with an engine failure. Fortunately, there were no injuries or fatalities. Damage to infrastructure and equipment was also minor. Fig. 7 shows the representation of the event within the tunnel and the location of the emergency teams about 450 m away from the tunnel entrance. Fig. 8 shows two pictures of the involved vehicle after the fire.

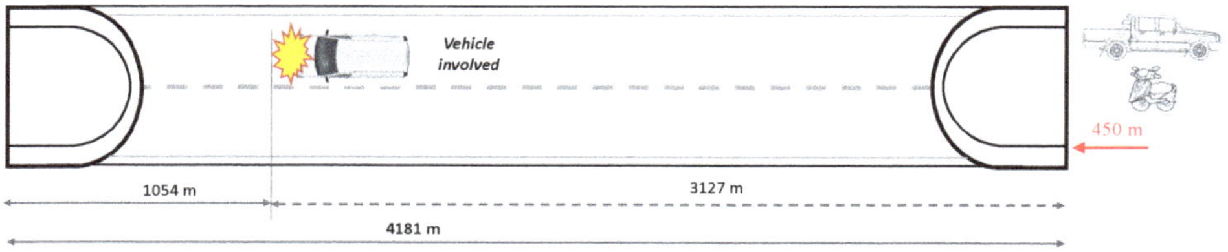

Figure 7: Graphical representation of the event inside the San Rocco tube. *(Source: Strada dei Parchi S.p.A.)*

Figure 8: Vehicle involved in the event inside the San Rocco tunnel. *(Source: Strada dei Parchi S.p.A.)*

The event occurred at 12:23 and as soon as it was detected the emergency team was activated. The tunnel is equipped with AID (Automatic Incident Detection) cameras that allow to quickly detect an event and communicate it to the control center and emergency teams. The emergency team arrived at the location of the event in approximately 2 min (12:25) and took approximately 15 min to extinguish the fire and secure the vehicle (12:40). It then took approximately 34 min to remove the vehicle, perform authority surveys, clean the roadway, and reopen the two lanes to traffic (13:14). Fig. 9 shows the timeline of this accident and highlights the transport demand not satisfied (TDNS) area (in gray).

Regarding the impact of the event on road traffic, it is necessary to consider the total duration of the event (about 51 min) and the ADT (7,122 vehicles per hour). Considering eqn (1) it is possible to estimate the unsatisfied transport demand which results to be of 6054 vehicles:

$$\text{TDNS} = \text{ADT} \times \text{event duration} = 7{,}122\ \frac{\text{vehicles}}{\text{hours}} \times 0.85\ \text{h} = 6{,}054\ \text{vehicles}. \quad (1)$$

It follows that this value (6,054) represents the number of vehicles that will be stopped in the queue and need to be managed during the tunnel closure (impact).

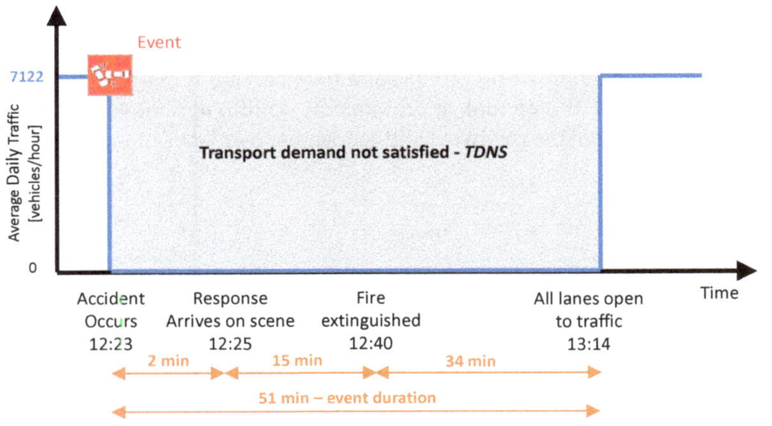

Figure 9: Timeline of traffic accident elements inside the San Rocco tunnel.

3.3 Stonio tunnel event

On 22 June 2019, a light vehicle fire occurred within the right arch of the Stonio tunnel, which is 1,242 m long. Again, the possible cause of the fire is associated with an engine failure. Damage to infrastructure and equipment was also minor. Fig. 10 shows the representation of the event within the tunnel and the location of the emergency teams about 400 m away from the tunnel entrance. Fig. 11 shows two pictures of the involved vehicle after the fire.

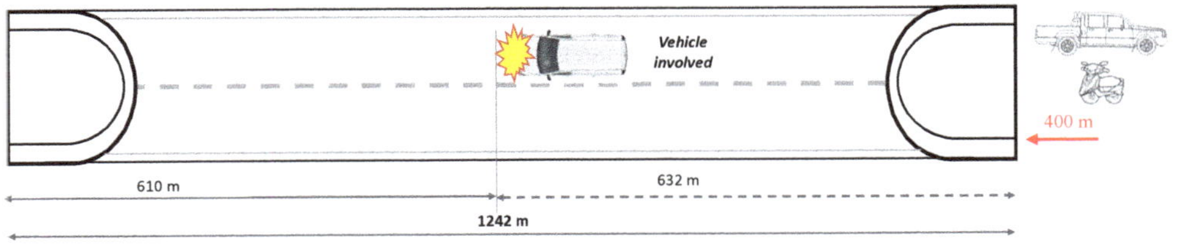

Figure 10: Graphical representation of the event inside the Stonio tube. *(Source: Strada dei Parchi S.p.A.)*

Figure 11: Vehicle involved in the event inside the Stonio tunnel. *(Source: Strada dei Parchi S.p.A.)*

A timeline of the Stonio tunnel event is shown in Fig. 12. The event occurred at 20:43 and after about 3 min (20:46) the emergency team was activated. The emergency team arrived at the scene of the event in about 2 min (20:48) and took about 6 min to extinguish the fire and secure the vehicle (20:54). It then took approximately 45 min to remove the vehicle, conduct surveys by authorities, clean the roadway, and reopen the two lanes to traffic (at about 21:39).

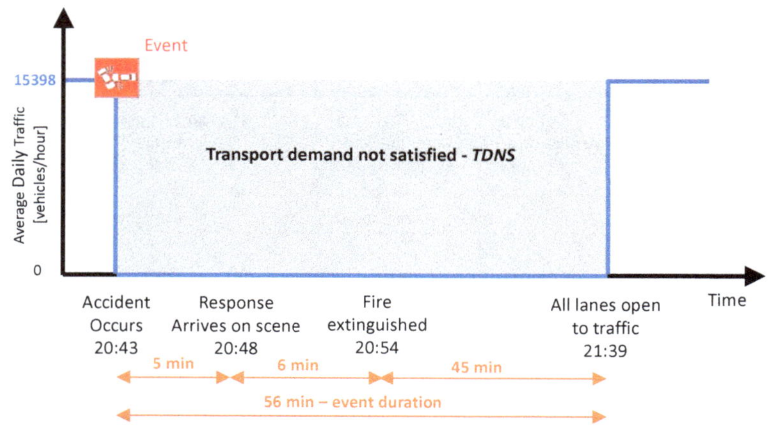

Figure 12: Timeline of traffic accident elements inside the Stonio tunnel.

Also in this second event, as far as the impact on vehicle traffic is concerned, it is necessary to consider the total duration of the event equal to about 56 min and the ADT (15,398 vehicles per hour). By means of eqn (2) it is possible to estimate the unsatisfied transport demand which is 14371 vehicles:

$$\text{TDNS} = \text{ADT} \times \text{event duration} = 15{,}398\ \frac{\text{vehicles}}{\text{hours}} \times 0.93\ \text{h} = 14{,}371\ \text{vehicles}. \quad (2)$$

The light vehicle fire in the Stonio tunnel resulted in a traffic disruption: approximately 14371 vehicles were stopped in the queue and had to be handled during the tunnel closure (impact).

4 CONCLUSIONS

The main goal of this work was to analyze two fire events that occurred in two tunnels located on the A24 highway in Italy between 2019 and 2021 and managed by Strada dei Parchi S.p.A. Both fires involved a light vehicle and there were no injuries or fatalities. For each of the two events, the response time of the emergency teams and the resolution time of the event were analyzed. The effectiveness of the emergency teams in controlling and/or extinguishing a fire, and impact on road traffic caused by the tunnel closure were analysed for the two cases. These analyses can be used for two important purposes: (i) risk assessment of a tunnel considering the emergency team as a safety measure; and (ii) implementation of traffic management plans to define the best strategy to be adopted according to the estimated time of closure of a tunnel. Both events analyzed involved a light vehicle. From a quantitative point of view in the event that occurred in the San Rocco tunnel, fire crews intervened in 2 min and the total event lasted about 51 min. The closure of the tunnel resulted in an impact that can be estimated at about 6,054 vehicles to be managed. With reference to the Stonio tunnel, the emergency teams intervened in 2 min from the moment of activation while the total event lasted about 56 min. In this case closure of the tunnel resulted in 14,371 vehicles to be managed (impact on the road system). In both cases the prompt intervention of the emergency teams allowed to effectively control and extinguish the fire.

REFERENCES

[1] Senato della Repubblica Italiana, Relazione concernente lo stato di attuazione degli interventi relative all'adeguamento delle gallerie stradali della Rete Transeuropea, Ministero delle Infrastrutture e dei Trasporti, 2013.

[2] Borghetti, F., Derudi, M., Gandini, P., Frassoldati, A. & Tavelli, S., Safety in road tunnels. DOI: 10.1007/978-3-319-49517-0_1, 2017.

[3] Król, A. & Król, M., Numerical investigation on fire accident and evacuation in a urban tunnel for different traffic conditions. *Tunnelling and Underground Space Technology*. DOI: 10.1016/j.tust.2020.103751, 2021.

[4] European Commission, Directive 2004/54/EC of the European Parliament and of the Council on Minimum Safety Requirements for Tunnels in the Trans-European Road Network 2004, Brussels, 2004.

[5] Beard, A. & Carvel, R., *The Handbook of Tunnel Fire Safety*, Thomas Telford Publishing, 2005.

[6] PIARC, Technical Committee 3.3 Road Tunnel Operation. Experience with significant incidents in road tunnels, 2016R35EN, ISBN 978-2-84060-444-0, 2017.

[7] Bjørnsen, G. & Njå, O., Competence constraints for fire and rescue personnel involved in tunnel fire safety as part of the tunnels' risk acceptability. *Proceedings of the 30th European Safety and Reliability Conference and the 15th Probabilistic Safety Assessment and Management Conference*, ISBN: 978-981-14-8593-0, 2020.

[8] Bosco, D., Lovreglio, R., Frassoldati, A., Derudi, M. & Borghetti, F., Queue formation and evacuation modelling in road tunnels during fires. *Chemical Engineering Transactions*, **67**, pp. 805–810, 2018. DOI: 10.3303/CET1867135.

[9] Nævestad, T. & Meyer, S., A survey of vehicle fires in Norwegian road tunnels 2008–2011. *Tunnelling and Underground Space Technology*, **41**(1), pp. 104–112, 2014. DOI: 10.1016/j.tust.2013.12.001.

[10] Borghetti, F., Cerean, P., Derudi, M. & Frassoldati, A., Road tunnels risk analysis, 2019. DOI: 10.1007/978-3-030-00569-6_1.

[11] Nussbaumer, C., Comparative analysis of safety in tunnels. Austrian Road Safety Board, Young Researchers Seminar 2007, Brno, 2007.

[12] Caliendo, C. & De Guglielmo, M.L., Accident rates in road tunnels and social costs evaluation. SIIV – 5 International Congress – Sustainability of Road Infrastructures. *Procedia-Social and Behavioral Sciences*, **53**, pp. 166–177, 2012.

[13] Bassan, S., Overview of traffic safety aspects and design in road tunnels. *IATSS Research*, **40**(1), pp. 35–46, 2016. DOI: 10.1016/j.iatssr.2016.02.002.

[14] Mashimo, H., State of the road tunnel safety technology in Japan. *Tunnelling and Underground Space Technology*, **17**(2), pp. 145–152, 2002. DOI: 10.1016/S0886-7798(02)00017-2.

[15] Vuilleumier, F., Weatherill, A. & Crausaz, B., Safety aspects of railway and road tunnel: Example of the Lötschberg railway tunnel and Mont-Blanc road tunnel. *Tunnelling and Underground Space Technology*, **17**(2), pp. 153–158 2002. DOI: 10.1016/S0886-7798(02)00018-4.

[16] Caliendo, C., Ciambelli, P., De Guglielmo, M.L., Meo, M.G., Russo, P., Simulation of fire scenarios due to different vehicle types with and without traffic in a bi-directional road tunnel. *Tunnelling and Underground Space Technology*, **37**, pp. 22–36, 2013. DOI: 10.1016/j.tust.2013.03.004.
[17] Gehandler, J., Road tunnel fire safety and risk: A review. *Fire Sci. Rev.*, **4**(1), pp. 1–27, 2015. DOI: 10.1186/s40038-015-0006-6.
[18] Haule, H.J., Sando, T., Lentz, R., Chuan, C. & Alluri, P., Evaluating the impact and clearance duration of freeway incidents. *International Journal of Transportation Science and Technology*, **8**(1), pp. 13–24, 2019. DOI: 10.1016/j.ijtst.2018.06.005.
[19] Reggiani, A., Network resilience for transport security: Some methodological considerations. *Transport Policy*, pp. 63–68, 2013.
[20] Mattsson, L. & Jenelius, E., Vulnerability and resilience of transport systems: A discussion of recent research. *Transportation Research Part A: Policy and Practice*, pp. 16–34, 2015.
[21] Carlson, J.L., Haffenden, R.A., Bassett, G.W., Buehring, W.A., Collins III, M.J., Folga, S.M., Petit, F.D., Phillips, J.A., Verner, D.R. & Whitfield, R.G., Resilience: Theory and application, 2012. DOI: 10.2172/1044521.
[22] Ayyub, B.M., Systems resilience for multi-hazard environments: Definition, metrics, and valuation for decision making. *Risk Analysis*, **34**(2)2, pp. 340–355, 2013. DOI: 10.1111/risa.12093.
[23] Borghetti, F., Petrenj, B., Trucco, P., Calabrese, V., Ponti, M. & Marchionni, G., Multi-level approach to assessing the resilience of road network infrastructure. *International Journal of Critical Infrastructures*, in press.
[24] Kim, H.K., Lönnermark, A. & Ingason, H., Effective firefighting operations in road tunnels. SP Technical Research Institute of Sweden: Sweden, 2010.
[25] Chen, F., Chien, S., Lee, Y., Lin, C. & Sie, H., The integrated strategies for fire safety of long road tunnels in Taiwan. *Paper presented at the Procedia Engineering*, **62**, pp. 36–45, 2013. DOI: 10.1016/j.proeng.2013.08.042.
[26] Rhodes, N. & MacDonald, M., FIT-European thematic network fire in tunnels, Fire response management (Technical Report Part 3), Brussels, Belgium, 2005.
[27] Borghetti, F., Derudi, M., Frassoldati, A., Lai, I. & Trinchini, C., Tunnel risk analysis: A quantitative evaluation of the effectiveness of emergency teams inside the A24 and A25 motorway tunnels. *Chemical Engineering Transactions*, **82**, pp. 277–282, 2020. DOI: 10.3303/CET2082047.
[28] Caliendo, C., Genovese, G. & Russo, I., A numerical study for assessing the risk reduction using an emergency vehicle equipped with a micronized water system for contrasting the fire growth phase in road tunnels. *Applied Sciences (Switzerland)*, **11**(11), 2021. DOI: 10.3390/app11115248.
[29] Amer, A., Roberts, E., Mangar, U., Kraft, W.H., Wanat, J.T., Cusolito, P.C., Hogan, J.R. & Zhao, X., *Traffic Incident Management: Gap Analysis Primer*. Federal Highway Administration, 2015.

ITERATIVE OUTLIER ANALYSIS HEURISTIC TO STUDY HIGH IMPACT TERROR ATTACKS OF THE MENA REGION AND EUROPE

PRABAL PRATAP SINGH* & DEEPU PHILIP†
Department of Industrial and Management Engineering, IIT Kanpur, India

ABSTRACT
Terrorism perpetrated by both internal and external actors destabilizes the economy, hampers social harmony, and jeopardizes any nation's internal security. While various forms of terror attacks occur globally, most terrorist organizations focus on high impact attacks (HIA) that maximize human fatalities and associated damages to precipitate fear within the community. Realistic modeling of terrorism is essential to identify critical trends and patterns to devise appropriate countermeasures. global terrorism database (GTD) is the most comprehensive terrorism-related database available for research purposes. Comparing terrorist attacks require specific benchmarks. A specific attribute from the GTD (nkill) and a derived composite parameter from GTD (global terrorism index (GTI)) are compared to establish their efficacy in successfully segregating HIAs from other terrorist attacks. This research focuses on the Middle East and North Africa (MENA) region and Europe. A sequential filtering heuristic based on local point outlier analysis utilizes both measures to extract HIAs of the MENA region. The proposed heuristic using nkill reports 1,055 unique HIAs in comparison with GTI and 522 vice versa. A simple geospatial analysis of these HIAs indicates the most fertile regions for terror organizations. Terrorism proliferation plots are developed to visually identify the migration timelines of major terrorist organizations that are capable of HIAs from the MENA region to Europe. The study concludes that virulent terrorist groups of the MENA region like Al-Qaida in the Arabian Peninsula (AQAP), Islamic State of Iraq and the Levant (ISIL), Muslim Extremists (ME), etc. migrated to Europe successfully.
Keywords: high impact attack, global terrorism database (GTD), global terrorism index (GTI), outlier analysis, sequential filtering heuristic, MENA region, terrorist migration, geospatial analysis.

1 INTRODUCTION

Terrorism is a global threat that jeopardizes the peace, development, and economic progress of a nation. From 2000 to 2018, terrorism cost the world economy an estimated $US 855 billion [1]. In these 18 years, countries in the Middle East and North Africa (MENA) region have witnessed the highest level of terrorism due to various armed conflicts. MENA region is the cradle of many notorious terrorist groups like Islamic State (IS), Al-Qaida in the Arabian Peninsula (AQAP), etc., which are responsible for many devastating domestic and transnational terror attacks [2]. For example, ISIS (Islamic State of Iraq and Syria), also known as ISIL (Islamic State of Iraq and the Levant), is a Sunni jihadist group with extreme violent ideology. Al-Qaida inspired ISIS origins. Agencies are actively analyzing ISIS's financing, management, and organization to overcome various threats posed by it. The European Union's expenditure to strengthen law enforcement agencies has increased over the years to tackle this menace from its immediate south [3].

A database like global terrorism database (GTD) [4] can foster studies on terrorism trends to develop various insights that can help to devise appropriate countermeasures. GTD follows an annual database release schedule and reports an ensemble of terror incidents from 1970

* *ORCID: https://orcid.org/0000-0002-0738-7629*
† *ORCID: http://orcid.org/0000-0002-4607-9020*

doi:10.2495/SAFE210081

using various sources, as shown in Fig. 1. GTD contains details of 181,691 terrorism incidents with 135 attributes for each terror incident collected by different organizations such as Pinkerton Global Intelligence Services (PGIS), Centre for Terrorism and Intelligence Studies (CETIS), Institute for the Study of Violent Groups (ISVG), and the National Consortium for the Study of Terrorism and Responses to Terrorism (START) [5].

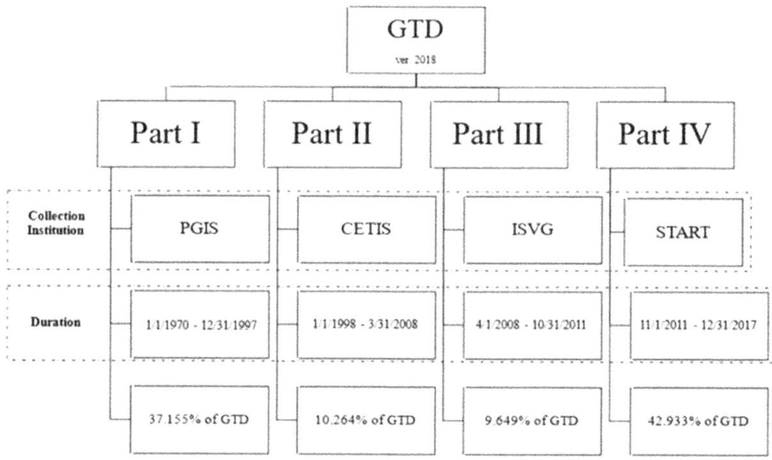

Figure 1: Constituents of global terrorism database (GTD).

Terrorism is a contested concept, and hence, its definition changes from one region to another [6]. GTD utilizes its own definition to classify and include an event as a terrorist attack. It defines terrorism as the threatened or actual use of illegal force and violence by a non-state actor to attain a political, economic, religious, or social goal through fear, coercion, or intimidation [5].

Global terrorism index (GTI) [7] utilizes the latest 5-year incident information from the GTD to rank nations based on the level of terrorism faced by them. GTI is a composite score synthesized using GTD's three main attributes: (i) number of fatalities (nkill); (ii) number of people wounded (nwound); and (iii) the property damage reported in the attack (propextent) as shown in eqn (1):

$$GTI - IS = 3 \times nkill + 1 \times nwound + 2 \times propextent + 1 \times incidence. \quad (1)$$

Next, a weighted average of the GTI impact score (GTI-IS) for the last 5 years is calculated to obtain a weighted impact raw score (WIRS). Then, the WIRS is scaled to a 1–10 interval to rank a nation, thereby creating GTI index.

The main aim of any terrorist attack is to instill fear in the target population by inflicting pain and misery on its victims. An event that results in a large magnitude of pain and misery, known as a high impact attack (HIA), requires coordination, planning, sophisticated weapons, trained and motivated individuals. Such attacks are infrequent and hence usually considered as anomalies or outliers in usual statistical analysis.

Statistically, an outlier is an observation that deviates significantly from other observations in the dataset, thereby arousing suspicion that it was generated by a different mechanism [8]. A survey on outlier detection methods by Blázquez-Garcia et al. [9] reports that these techniques vary based on input data type, the outlier extraction process, and the technique's nature.

Box-and-Whisker plot is a statistical tool that utilizes five number diagrammatic summary: (i) minimum; (ii) first quartile (25th percentile); (iii) median (50th percentile); (iv) third quartile (75th percentile); and (v) maximum on any univariate data series, which can also retrieve outliers [10]. Box-and-Whisker plot provides two sets (lower and upper) of fences: (i) inner fences; and (ii) outer fences constructed using the interquartile range (IQR). The data points beyond the lower and upper outer fences are considered as outliers, while those in between the inner and outer fences are considered potential outliers. However, this study considers data points beyond the upper inner fence (UIF) as outliers, representing the HIAs. Parrinello et al. [11] proposed a similar iterative approach to detect outliers in the laboratory recalibration for removing the drift in Uric acid measurements from the atherosclerosis risk in communities (ARIC) study. However, they considered data points greater than the three standard deviations from the mean as outliers. Their procedure iteratively determines anomalies until no further outliers are reported during a particular iteration. A similar yet heavily modified iterative approach using the UIF value of the Box-and-Whisker plot was developed for this work.

2 METHODOLOGY

Each record in the GTD is a terrorism incident that contains information about casualties, perpetrator information, and details of the attack type, weapon type, target type, etc. Terrorist organization that aims to perpetrate maximum fear tends to maximize fatalities associated with HIAs, which is captured by the nkill attribute. The primary aim of this work is to extract HIAs from GTD for a particular region of interest. For this, an iterative outlier analysis (IOA) heuristic that utilizes the UIF value of the Box-and-Whisker plot is developed. With a univariate data series as input, the heuristics extract local point outliers for each year, which are considered as HIAs. The iterative extraction stops when the recalculated UIF values converge for two consecutive iterations, resulting in no event labeled as HIA.

It should be noted that the nkill attribute has the highest weight in the composite index GTI-IS as per eqn (1), which inspires an efficacy analysis of both nkill and GTI-IS in extracting HIAs. Thus, the proposed IOA heuristic is implemented separately using Python ecosystem, as shown in Fig. 2 on GTI-IS and nkill univariate data series to extract HIAs.

The incident records for the MENA region from GTD part IV were used in this study. Using either nkill or GTI-IS univariate data series, IOA iteratively computes the UIF values for each year until convergence while extracting local outliers simultaneously. Finally, the algorithm combines these annual outliers to obtain the final HIA dataset. A simple geospatial analytics of HIAs was conducted using the location information of the attack to visualize the fertility of a particular region towards terrorism. Additionally, terrorist attacks by the most active organizations reported in the Europe region are compared with the HIAs of the MENA region to elicit probable migration patterns of such organizations.

3 RESULTS

GTD 2018 part IV contains data from 2011 to 2017 that amounts to 42.93% of the total database. As mentioned earlier, this study focuses on attacks reported in the MENA and Europe region. GTD part IV reports 78,005 total incidents, out of which the MENA region contributes 30,246 incidents. Table 1 summarizes the countries of the MENA region along with their respective number of reported terror incidents. Iraq tops the list with 56.68% of all reported attacks of the region, followed by Yemen, Libya, and Syria reporting 9.69%, 7.38%, and 6.71% of the attacks, respectively.

Algorithm 1 Iterative Outlier Analysis

Input: GTD part IV Dataframe - MENA region
Output: Two datasets containing outliers based on nkill and GTI Impact score
Require: nkill, GTI score

```
procedure IOA(Dataframe with nkill and gti score)
    for each attribute attrin{nkill, GTIscore} do
        Divide the dataframe annually
        repeat
            for each year i with attribute attr do
                uif_val_i = Q3 + 1.5 * (Q3 - Q1)
                uof_val_i = Q3 + 3 * (Q3 - Q1)
                uif_val_steady_i ← uif_val_i
            end for
        until uif_val becomes steady at n iteration
        repeat
            for each year i with attribute attr do
                Initialize Set1_i & Set2_i
                for each incidence j of current dataset do
                    if attr_j < uif_val_steady then
                        Set1_i ← Set1_i + incidence_j
                    else
                        Set2_i ← Set2_i + incidence_j
                    end if
                end for
            end for
        until done
        OutlierDataframe = Σ_i Set2_i
    end for
end procedure
```

Figure 2: Iterative outlier analysis algorithm.

Table 1: Number of terrorist incidents in MENA countries.

Country	Incident count	Percentage
Iraq	17,144	56.682
Yemen	2,930	9.687
Libya	2,232	7.379
Syria	2,028	6.705
Egypt	1,975	6.530
Turkey	1,475	4.877
West Bank and Gaza Strip	710	2.347
Israel	535	1.769
Lebanon	453	1.498
Saudi Arabia	307	1.015
Bahrain	163	0.539
Algeria	116	0.384
Tunisia	88	0.291
Iran	55	0.182
Jordan	27	0.089
United Arab Emirates	3	0.010
Kuwait	3	0.010
Qatar	1	0.003
Morocco	1	0.003

Also, GTD reports attacks in Eastern and Western Europe separately, accounting for 2,249 and 1,576 events, respectively. This study combined both and then compared them with the HIAs extracted from the MENA region to obtain possible insights into the influence of terrorism from one neighboring region to another.

Additionally, the nkill, nwound, and propextent attributes of the GTD were analyzed due to their usage in the GTI-IS equation. Both nkill and nwound attributes exhibit a positive skew of 54.07 and 50.40, respectively, for the MENA region. An empirical cumulative distribution (ECDF) [12] plot of nkill and nwound is given in Fig. 3, indicating that around 80% of the reported fatalities/wounds are in single digits.

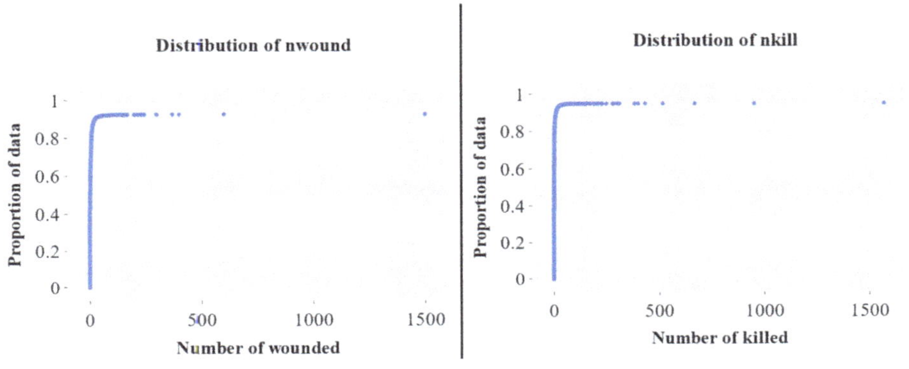

Figure 3: ECDF of nkill and nwound.

Table 2: Category levels for extant property damage in the MENA region.

Damage code		Description	Incident counts
Codebook	GTI		
1	3	Catastrophic	0
2	2	Major	8
3	1	Minor	7,550
4	0	Unknown	22,688

Table 2 summarizes the levels of the propextent attribute with category codes from the GTD Codebook [5] and GTI [7]. No catastrophic property damage from the MENA region is reported for the duration of this HIA analysis. Also, 75.01% of all incident records from the MENA region have an unknown amount of property damage.

The proposed IOA heuristic extracted local point outliers continuously until the UIF value converged between two successive iterations. Typical convergence occurred around the fourth iteration for every annual dataset. The confirmatory fifth iteration ensures convergence of fence values, thereby terminating the heuristic. The five-statistic data summary of the Box-and-Whisker plot for the first and fourth iteration of the heuristic using both nkill and GTI-IS datasets is provided in Table 3.

A comparative analysis of HIAs extracted using nkill and GTI-IS data series is depicted in Fig. 4, which indicates that the proposed IOA can find more HIAs using the nkill attribute than the composite score GTI-IS. About 1,055 HIAs were exclusively captured by the IOA

Table 3: Iteration details of IOA.

Iteration details	nkill outlier iter 1	nkill outlier iter 4	GTI-IS outlier iter 1	GTI-IS outlier iter 4
Count	3,849	4,162	3,029	3,629
Mean	15.67	14.74	73.96	65.66
Standard deviation	39.61	38.23	139.57	128.88
Min	3	3	13	14
25th percentile	6	5	34	30
50th percentile	9	8	47	41
75th percentile	15	14	73	65
Max	1,570	1,570	4,711	4,711

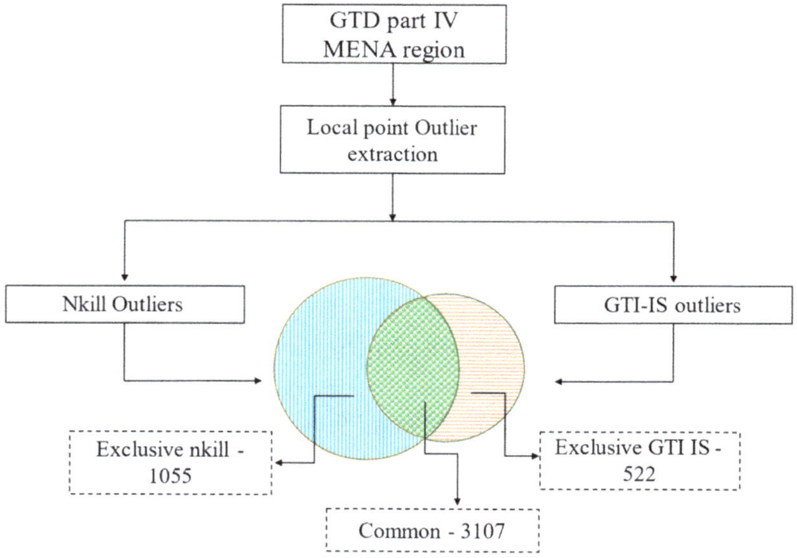

Figure 4: Overview of comparative analysis of outliers.

when using the nkill attribute as input, whereas only 522 HIAs were exclusively captured by the same heuristics while using GTI-IS as the input data. Also, 3,107 HIAs were commonly identified from both data series by the heuristic.

Table 4 compares the Pearson and Spearman's rank correlation analysis between the nkill and nwound attributes within the three HIAs datasets generated by the IOA: (i) exclusive nkill dataset; (ii) exclusive GTI-IS dataset; and (iii) common HIAs dataset along with the input data. It is evident that the nkill and nwound attributes in the input data are positively correlated, but the exclusive GTI-IS HIAs indicate a negative correlation w.r.t other HIAs.

Geospatial visualization of the extracted HIAs is provided in Fig. 5 for exclusive nkill, Fig. 6 for exclusive GTI-IS, and Fig. 7 for common HIAs to facilitate the visual identification of fertility patterns. It is evident that the exclusive HIAs from nkill (Fig. 5) have similar patterns as common HIAs (Fig. 7).

Table 4: Correlation analysis of three HIAs datasets against input data.

Correlation analysis	Pearson	Spearman
Input data	0.284	0.425
Exclusive GTI	–0.041	**–0.365**
Exclusive nkill	–0.132	–0.163
Common	0.063	–0.162

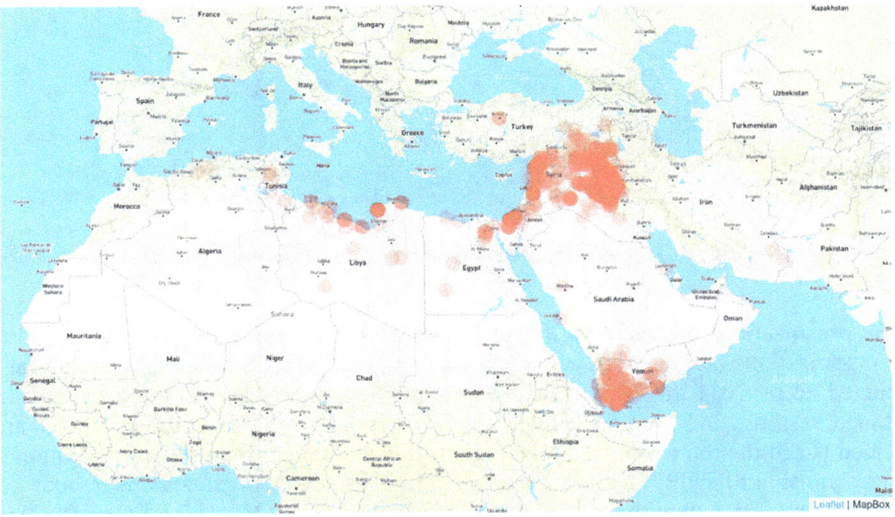

Figure 5: Geographical visualization of exclusive nkill HIAs.

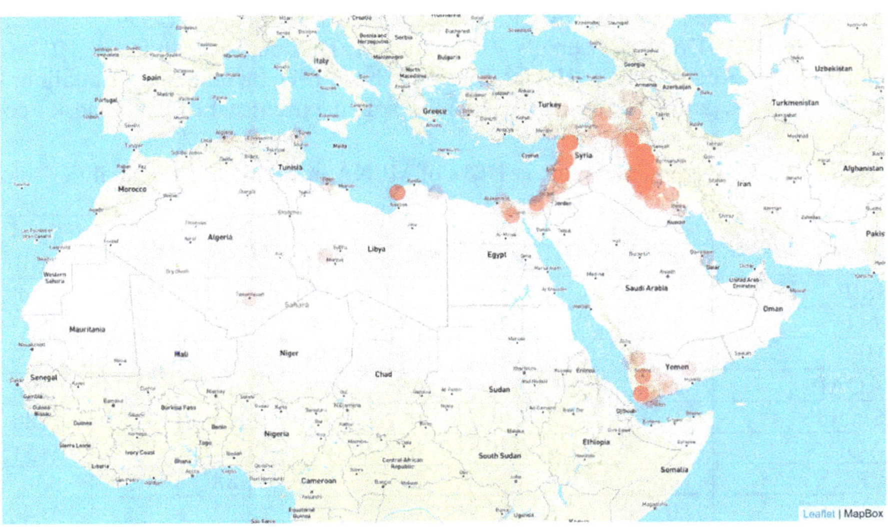

Figure 6: Geographical visualization of exclusive GTI-IS HIAs.

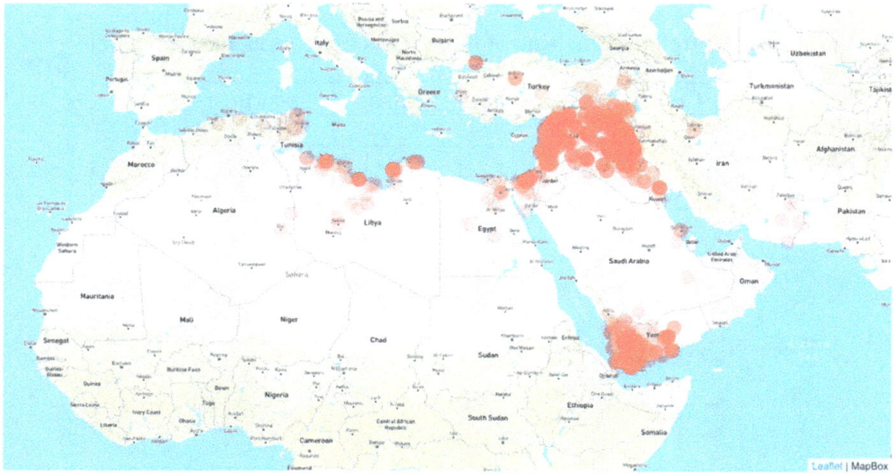

Figure 7: Geographical visualization of common HIAs.

The organizations orchestrating HIAs in the MENA region are of interest. To identify such organizations, aggregating and sorting of the nkill/GTI-IS values for each year was conducted from 2011 to 2017. Each year's top ten organizations were extracted and combined, resulting in 37 unique active organizations in both datasets. Tables 5 and 6 are the truncated list of the top ten most active organizations obtained from the aforementioned 37 groups against nkill and GTI-IS, respectively. The most active organizations in both Tables 5 and 6 are quite similar, with the top four groups: (i) ISIL; (ii) ANF; (iii) HE; and (iv) AQAP, remaining the same.

Aliases of all the organizations tabulated in Tables 5–8 are as follows: ANF, Al-Nusrah Front; AQAP, Al-Qaida in the Arabian Peninsula; AQI, Al-Qaida in Iraq; Adan IS, Adan-Abyan Province of the Islamic State; CE, Caucasus Emirate; CPIS, Caucasus Province of the Islamic State; DPR, Donetsk People's Republic; DR, Dissident Republicans; HE, Houthi extremists (Ansar Allah); HTAS, Hay'at Tahrir al-Sham; ISB, Imam Shamil Battalion; ISIL, Islamic State of Iraq and the Levant; JIE, Jihadi-inspired extremists; LPR, Luhansk People's

Table 5: Most active organizations in HIAs of MENA inflicting highest fatalities (nkill).

Organizations	Year							Total nkill
	2011	2012	2013	2014	2015	2016	2017	
ISIL	0	0	1,045	9,532	7,379	10,770	6,521	35,247
ANF	0	298	676	751	919	193	0	2,837
HE	0	0	54	186	1091	711	417	2,459
AQAP	15	814	258	588	279	115	157	2,226
Sinai IS	0	0	0	0	487	205	588	1,280
AQI	103	729	319	0	0	0	0	1,151
PKK	3	173	0	0	0	288	72	536
Adan IS	0	0	0	0	0	323	71	394
Tripoli IS	0	0	0	0	117	205	0	322
HTAS	0	0	0	0	0	0	313	313

Table 6: Most active organizations in HIAs of MENA reporting highest aggregated GTI-IS.

Organizations	Year							Total GTI-IS
	2011	2012	2013	2014	2015	2016	2017	
ISIL	0	0	5,682	34,214	28,169	39,837	23,194	131,096
ANF	0	1,889	2,284	2,744	3,078	718	0	10,713
HE	0	0	254	757	4,928	2,555	1,682	10,176
AQAP	51	3,067	1,138	2,380	1,095	483	591	8,805
AQI	605	4,175	1,730	0	0	0	0	6,510
Sinai IS	0	0	0	0	1,816	738	2,097	4,651
PKK	11	836	0	0	0	1,857	287	2,991
Adan IS	0	0	0	0	0	1,272	295	1,567
Sanaa IS	0	0	0	0	1,534	0	0	1,534
ME	0	492	0	349	617	0	0	1,458

Table 7: Most active organizations in Europe region inflicting highest fatalities (nkill).

Organizations	Year							Total nkill
	2011	2012	2013	2014	2015	2016	2017	
DPR	0	0	0	1,006	597	30	24	1,657
LPR	0	0	0	173	77	0	0	250
ISIL	0	0	0	4	137	40	27	208
JIE	0	0	0	0	10	112	25	147
CE	0	23	42	28	0	3	0	96
ME	0	28	3	10	9	4	29	83
RS	0	0	0	44	7	0	0	51
CPIS	0	0	0	0	0	12	30	42
ISB	0	0	0	0	0	0	16	16
PRM	0	0	0	11	0	0	2	13

Table 8: Most active organizations in Europe region reporting highest aggregated GTI-IS.

Organizations	Year							Total GTI-IS
	2011	2012	2013	2014	2015	2016	2017	
DPR	0	0	0	3,998	3,079	122	105	7,304
ISIL	0	0	0	0	838	397	243	1,478
LPR	0	0	0	980	431	0	0	1,411
JIE	0	0	0	0	59	878	185	1,122
ME	0	109	39	56	51	39	268	562
CE	0	188	248	122	0	0	0	558
RS	0	0	0	155	38	0	0	193
CPIS	0	0	0	0	0	70	113	183
UN	0	0	0	0	156	0	0	156
DR	0	0	38	22	48	31	16	155

Republic; ME, Muslim extremists; PKK, Kurdistan Workers' Party; PRM, Pro-Russia Militia; RS, Right Sector; Sanaa IS, Sanaa Province of the Islamic State; Sinai IS, Sinai Province of the Islamic State; Tripoli IS, Tripoli Province of the Islamic State; UN, Ukrainian nationalists.

Tables 7 and 8 summarize similar data as that of Tables 5 and 6, respectively, but for the European region. It should be noted that the data for Europe did not undergo the HIA extraction process using the IOA heuristic due to the sparsity in data for Europe in comparison with the MENA region. Usually, a terrorist attack in the European region attracts global attention, resulting in a larger impact compared to the MENA region.

Organizations common in the complete list of the most active organizations for both regions are virulent. ISIL, AQAP, ME, SME, and Gunmen are active in both regions, while the GTI-IS dataset additionally identifies PKK. Fig. 8 depicts the proliferation plot of the activities of these virulent organizations from their host to the migrated region against the nkill attribute. It can be seen from Fig. 8 that ISIL took an additional year before becoming active in Europe.

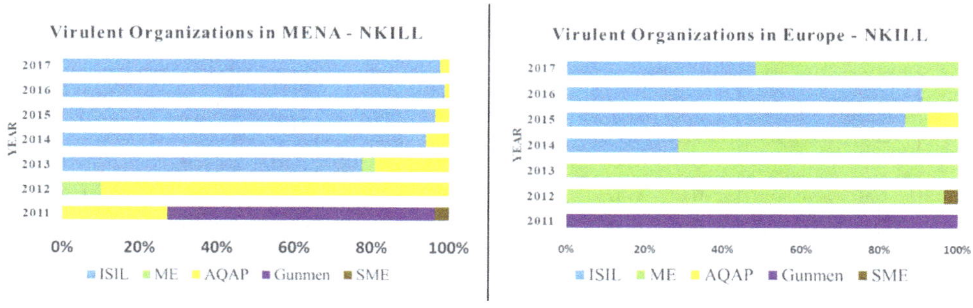

Figure 8: Virulent organization in MENA and Europe based on nkill.

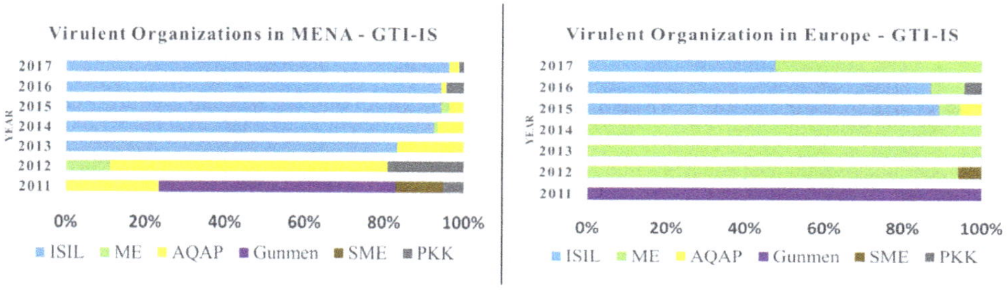

Figure 9: Virulent organization in MENA and Europe based on GTI-IS.

Fig. 9 is a similar proliferation plot as of Fig. 8 but by utilizing GTI-IS as the input. Contrary to the analysis using nkill, Fig. 9 indicates that ISIL took two additional years to emerge in the Europe region.

4 CONCLUSION

This study suggests that using a pure database attribute (nkill) provided a richer set of HIAs than the composite derived GTI-IS value. In totality, the IOA heuristic extracted 14.69% more HIAs using nkill than GTI-IS. We attribute this to the averaging effect of GTI-IS value along with the choice of the coefficients in the GTI-IS calculation in eqn (1). The exclusive nkill HIAs and the common HIAs maintained the same fertility patterns whereas, exclusive GTI-IS HIAs displayed somewhat deviant patterns. Proliferation plots of most active organizations is an effective tool to identify migrational patterns. Further, time-series analysis of HIAs for the MENA region indicates the global threat of ISIS and its potential to migrate to neighboring regions.

REFERENCES

[1] Bardwell, H. & Iqbal, M., The economic impact of terrorism from 2000 to 2018. *Peace Econ. Peace Sci. Public Policy*, **27**, pp. 227–261, 2021. https://doi.org/10.1515/peps-2020-0031.

[2] Kim, W. & Sandler, T., Middle East and North Africa: Terrorism and conflicts. *Glob. Policy*, **11**, pp. 424–438, 2020. https://doi.org/10.1111/1758-5899.12829.

[3] Sofija, V., *The Fight Against Terrorism*, p. 12.

[4] LaFree, G. & Dugan, L., Introducing the global terrorism database. *Terror Polit. Violence*, **19**, pp. 181–204, 2007. https://doi.org/10.1080/09546550701246817.

[5] GTD Codebook, 2018. https://www.start.umd.edu/gtd/downloads/Codebook.pdf. Accessed on: 2 Aug. 2021.

[6] Schmid, A.P. ed., *The Routledge Handbook of Terrorism Research*, Routledge: London and New York, 2013.

[7] Hyslop, D. & Morgan, T., Measuring terrorism with the global terrorism index. *Contrib. Confl. Manag. Peace Econ. Dev.*, **22**, pp. 97–114, 2014. https://doi.org/10.1108/S1572-8323(2014)0000022010.

[8] Hawkins, D.M., *Identification of Outliers*, Springer Netherlands: Dordrecht, 1980. https://doi.org/10.1007/978-94-015-3994-4.

[9] Blázquez-García, A., Conde, A., Mori, U. & Lozano, J.A., A review on outlier/anomaly detection in time series data. ArXiv200204236, Cs Stat, 2020.

[10] Ramachandran, K.M. & Tsokos, C.P., *Mathematical Statistics with Applications in R*, 3rd ed., Elsevier: Philadelphia, 2020.

[11] Parrinello, C.M. et al., Iterative Outlier Removal: A method for identifying outliers in laboratory recalibration studies. *Clin. Chem.*, **62**, pp. 966–972, 2016. https://doi.org/10.1373/clinchem.2016.255216.

[12] University of Virginia Library Research Data Services and Sciences, Understanding Empirical Cumulative Distribution Functions. https://data.library.virginia.edu/understanding-empirical-cumulative-distribution-functions/. Accessed on: 2 Aug. 2021.

SECTION 4
EARTHQUAKES AND NATURAL HAZARDS

EVALUATION OF SAFETY WHEN A SEISMIC ISOLATED BUILDING COLLIDES WITH RETAINING WALLS CONSIDERING DIFFERENT NUMBERS OF FLOORS

CARLOS CUADRA & KENTA KIKUCHI
Faculty of System Science and Technology, Akita Prefectural University, Japan

ABSTRACT

Large dynamic responses of long-period buildings were observed during strong earthquakes, especially when these input motions are of large amplitude or long duration and low frequency content. A base isolated building is basically a structure of long period and therefore large displacement response could be observed during mentioned earthquakes. This large displacement of the isolation layer could cause the collision of the isolated structure against the surrounding retaining walls. This research uses 5-storey and 10-storey three-dimensional building models to investigate the safety of base isolated structures in the event of impact against retaining walls. The analytical models consider gap link elements to study the collision against the retaining wall. As input acceleration, a record of the Kobe earthquake was used, scaling this record by 1, 1.7, 1.8 and 2. For both models, maximum acceleration in upper structure increases in case of collision. Since this acceleration could cause furniture to tip over, a safety analysis is performed based on predominant frequency of vibration and maximum acceleration observed on each floor. In case of no collisions, there is less than a 30% probability that the furniture will tip over in all cases of seismic waves. When a collision occurs, the acceleration at each floor increases significantly and the probability that the furniture will fall is 50% or more.

Keywords: rubber isolator, collision, seismic response analysis, finite element method, gap element.

1 INTRODUCTION

In Japan, located in an active seismic region, major earthquakes such as the Great East Japan earthquake (2011) and the Kumamoto earthquake (2016) have occurred, and many people lost their homes due to the collapse of buildings. Seismic isolation is a technology that protects building structures from shaking due to earthquake motions [1]. According to the data collection of the Japan Society of Seismic Isolation, the stock of isolated buildings reaches approximately 5,000 structures, and the number of isolated housing structures (residential houses) reaches approximately 4,700 units [2].

During the 2011 East Japan earthquake, low frequency (or long period) content acceleration records were observed in many points of Japan [3] and these signals induced large displacement responses or long duration responses of isolated buildings, since these structures have a long period of vibration and therefore resonance phenomenon could occur. On the other hand, in 2016, during the Kumamoto earthquake (near field earthquake) high amplitude signals were observed and therefore large amplitude response of the isolated buildings was also observed (displacement of the order of 40 cm). Then, it is necessary to investigate the case of probable collision of an isolated structure against surrounding retaining walls [4], [5]. When the collision occurs, the input acceleration is amplified on the upper floors and the accelerations on each floor could cause damage to the structure itself or could cause the furniture to tip over, affecting the safety of the structure and its content. Fig. 1 shows, schematically, the effect of the impact of isolated building against the retaining walls. In addition to damage to building and overturning of furniture, the collision could lead to crushing of retaining walls and buckling of isolation devices [6], [7].

WIT Transactions on The Built Environment, Vol 206, © 2021 WIT Press
www.witpress.com, ISSN 1743-3509 (on-line)
doi:10.2495/SAFE210091

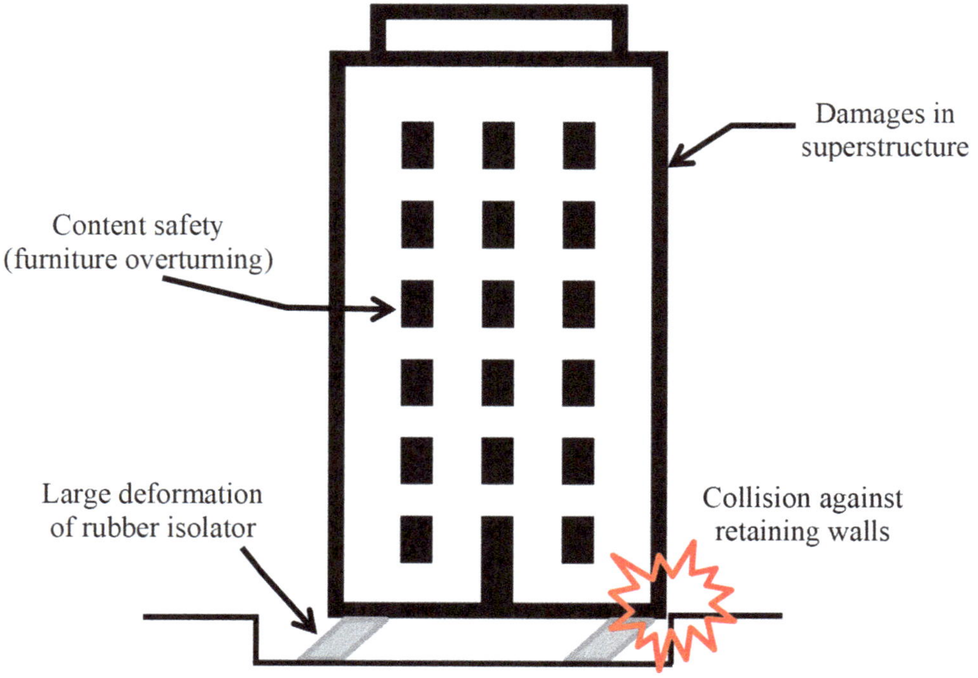

Figure 1: Collision of a seismic isolated building against retaining walls.

In this research, two building models are chosen to investigate the effect of the collision of isolated structures. The 5-storey building and the 10-storey building are modelled using structural analysis software called Etabs. Isolation devices are modelled using rubber isolator link element and the impact against retaining wall is modelled with a spring gap link element [8]. Kobe earthquake record (Japan Meteorological Agency (JMA), 1995) is used as the input motion. From this time–history response analysis, maximum accelerations on each floor are obtained and with the analysis of the predominant frequency a verification of the safety of the furniture is carried out [9].

2 TARGET ISOLATED STRUCTURES AND FEM MODELS

In the present analysis, a 5-storey mid-rise building and a 10-storey tall building are considered to investigate the effect of the difference in the number of stories on the response and safety of these target buildings. In both cases the buildings correspond to reinforced concrete frame structures. If they are considered as non-isolated buildings, they have different periods of vibration. However, for isolated buildings, in both structures the isolated devices are designed in such a way that they have the same period of vibration.

The design criteria of target buildings consider that the period of vibration of the 5-storey structure should be less than 0.5 s and for 10-storey structure the period should be less than 1 s. In both cases the period of vibration for isolated building must be between 3 s to 4 s.

Currently, the design clearance is 60 cm; however, for the first isolated buildings, the design considered a clearance of 40 cm. Therefore, in this study, this clearance of 40 cm will be considered for the impact analysis.

2.1 Characteristics of 5-storey building

Fig. 2 shows the characteristics of the 5-storey building such as the dimensions of the building and its properties of mass and clearance of the isolation floor.

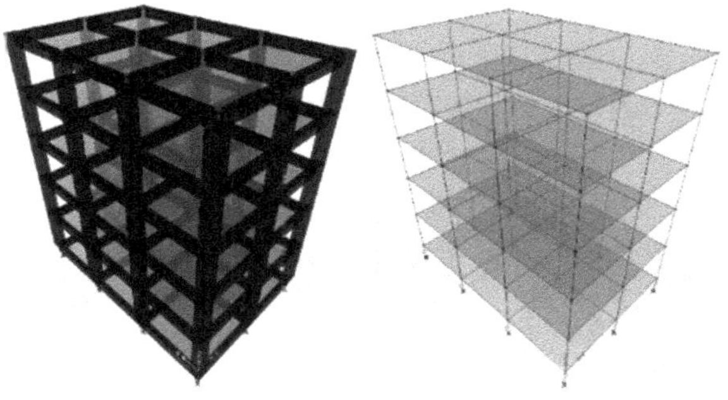

Height: 20 m
Storey height: 4 m
Plan dimension: 16 m x 18 m
Floor area: 1,300 m^2
Structure: Reinforced concrete
Column cross section: 800 x 800
Beam cross section: 750 x 400
Floor slab: 350 mm
Total weight: 1483 t
Clearance: 400 mm

Figure 2: 5-storey building model.

2.2 Characteristics of 10-storey building

Fig. 3 shows the characteristics of the 10-storey building such as the dimensions of the building and its properties of mass and clearance of the isolation floor.

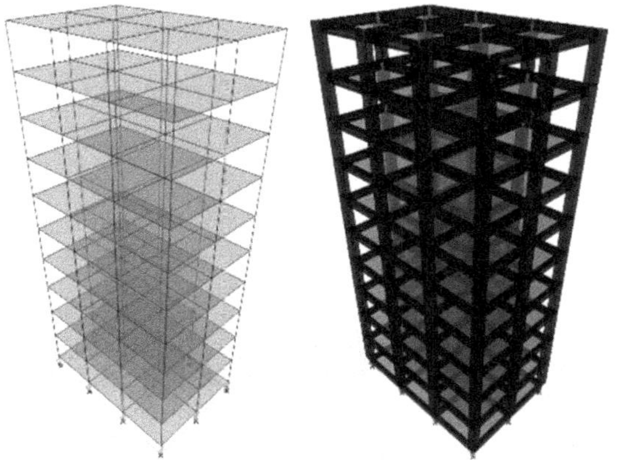

Height: 40 m
Storey height: 4 m
Plan dimension: 16 m x 18 m
Floor area: 2,400 m^2
Structure: Reinforced concrete
Column cross section: 800 x 800
Beam cross section: 750 x 400
Floor slab: 350 mm
Total weight: 2759 t
Clearance: 400 mm

Figure 3: 10-storey building model.

2.3 Base isolated models

For base-isolated structures, the equivalent stiffness of the isolation layer is calculated considering that the target period of vibration is 3.5 s. If the entire structure above the isolated layer is considered as a lumped mass (m) and if the stiffness of the isolation devices is K, then the period of vibration of the isolated structure is given by the following equation:

$$T = 2\pi\sqrt{\frac{m}{K}}. \quad (1)$$

Then, considering $T = 3.5$ s, the equivalent stiffness of the isolation layer is obtained from the following equation:

$$K = \frac{4\pi^2}{3.5^2}m. \quad (2)$$

If K is the total stiffness of the isolated layer and considering that 10 devices are used, then the stiffens of a device (rubber isolator) is $k = K/10$. The stiffnesses required for the isolation layer and the devices for each model are as follows:

5-storey building: $K = 4779$ kN/m, $k = 478$ kN/m
10-storey building: $K = 8891$ kN/m, $k = 889$ kN/m

High damping rubber isolators are selected for the target buildings. The equivalent lateral stiffness of each isolator must be closer to the calculated stiffness k, however not greater than this value in order to obtain the estimated period. Table 1 shows the properties of the selected isolation devices for both target buildings.

Table 1: Properties of rubber isolators.

Device property	Target building	
	5-storey building	10-storey building
Name or code	HN060X3R	HL065X4S
Type	High damping rubber	High damping rubber
Diameter	600 mm	650 mm
Weight	5 kN	6.3 kN
Initial stiffness	3,060 kN/m	4,730 kN/m
Second stiffness	306 kN/m	473 kN/m
Yield load	23.7 kN	53.0 kN
Equivalent lateral stiffness	424 kN/m	799 kN/m
Equivalent damping factor	0.17	0.24

A modal analysis was carried out for non-isolated and isolated buildings to obtain their predominant periods of vibration. Table 2 shows results of this modal analysis. For no-isolated buildings, the target periods were shorter that 0.5 s for the 5-storey building and shorter that 1 s for the 10-storey building. The periods obtained are 0.44 s and 0.86 s for the 5-storey building and the 10-storey building respectively. For isolated buildings was assumed that the period of vibration must be between 3 s to 4 s. The actual periods of vibration obtained for isolated buildings are 3.32 s and 3.30 s for the 5-storey building and the 10-storey building, respectively.

Table 2: Modal analysis results.

Period of vibration	5-storey building		10-storey building	
	Non-isolated	Isolated	Non-isolated	Isolated
Target period	$T < 0.5$ s	$3\text{ s} < T < 4$ s	$T < 1.0$ s	$3\text{ s} < T < 4$ s
Analysis result	0.44 s	3.32 s	0.86 s	3.30 s

3 SEISMIC RESPONSE ANALYSIS

The target isolated buildings were subjected to a time–history analysis. As input motion, a record obtained during the Kobe earthquake (JMA, 1995) was used. Then the displacement responses and acceleration responses in the event of a collision with the retaining wall are studied.

3.1 Input motion

Fig. 4 shows the original input motion (kobe-1) and the amplified signal (kobe-1.7, kobe-1.8, kobe-2). The amplification factors of 1.7, 1.8 and 2 were chosen to produce the collision or to obtain a displacement larger than the clearance, which is 40 cm in this study.

The original Kobe signal with a maximum acceleration of 817 cm/s^2 does not cause collisions of isolated buildings. As a verification of the proper behaviour of the rubber insulators, the responses of these devices were obtained in the case of the original Kobe input movement as shown in Fig. 5. These responses are shown for the 5-storey building (5F) and the 10-storey building (10F).

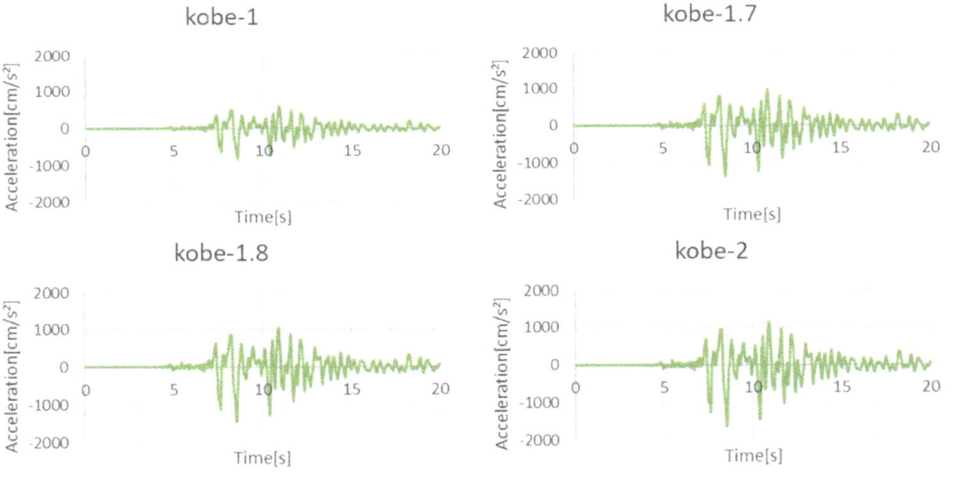

Figure 4: Input motion for time–history analysis.

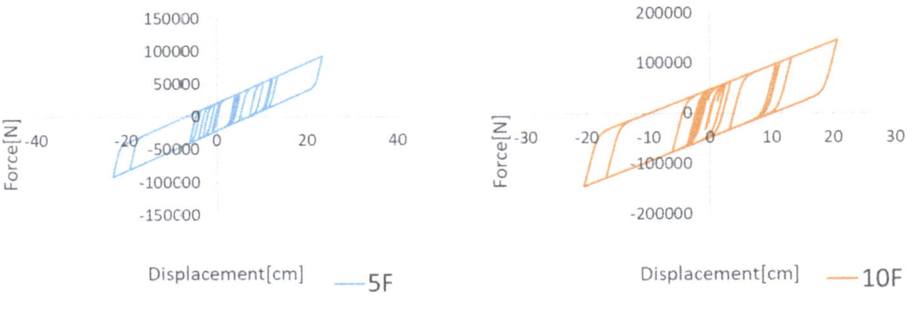

Figure 5: Responses of rubber isolators for isolated buildings.

3.2 Response analysis

Fig. 6 shows the displacement responses of the isolated 5-storey building during various levels of Kobe input motion. For the original Kobe record (kobe-1) the maximum displacement is 23.4 cm, which is less than 40 cm of clearance. If the input signal is amplified by 1.7, 1.8 and 2, the maximum response is 43.8 cm, 48.9 cm, and 62.6 cm, respectively. In these cases, the maximum responses exceed the clearance of 40 cm and, therefore, it is assumed that the structure collides with the retaining walls.

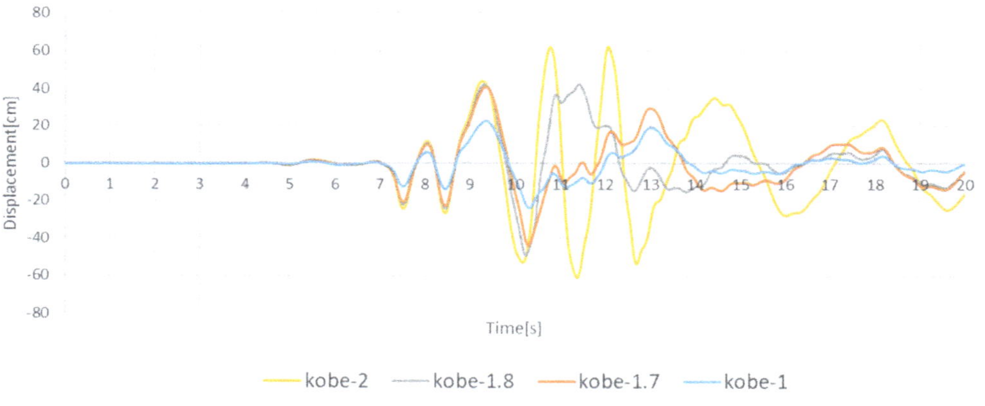

Figure 6: Displacement response for the isolated 5-storey building.

Fig. 7 shows the displacement responses of isolated 10-storey building when subjected to various level of Kobe input motion. For the original Kobe record (kobe-1) and amplification factor of 1.7, the maximum displacements are 20.8 cm and 39.1 cm, respectively, which are less than 40 cm of clearance. When the input signal is amplified by 1.8 and 2; the maximum responses are 41.3 cm, and 48.6 cm respectively. In these cases, the maximum responses exceed the clearance of 40 cm and, therefore, is assumed that the structure collides with the retaining walls.

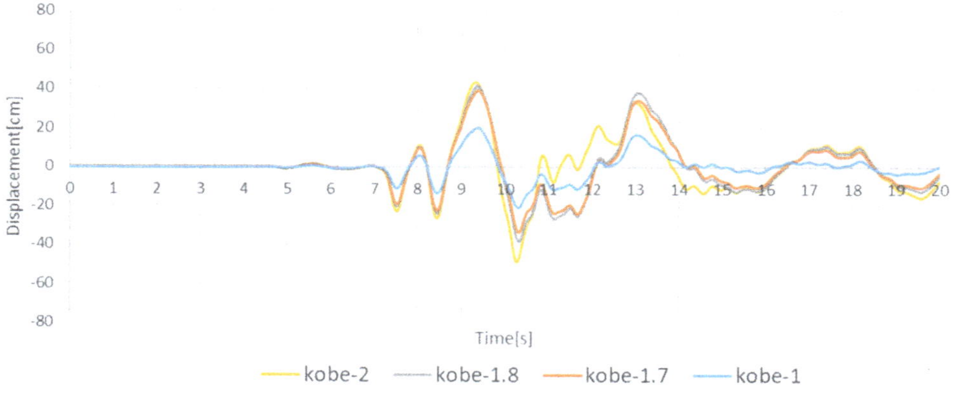

Figure 7: Displacement response for the isolated 10-storey building.

Comparing the acceleration responses, it can be seen that when the displacement response exceeds 40 cm of clearance, the maximum acceleration increases significantly due to the collision with the retaining wall. However, since the maximum displacement differs for each building, the displacement crush of the retaining walls will affect the reaction force due to impact and the maximum acceleration will be different for each building and for each amplification factor. The reaction force in case of collision is modelled by a linear spring of high stiffness that reacts when the clearance is exceeded.

Fig. 8 shows the acceleration response of the 5-storey building. For original Kobe earthquake (kobe-1) the maximum acceleration at the top of the building is 135 cm/s^2. Since the maximum input acceleration is 817 cm/s^2, in this case the isolation effect is confirmed by observing this decrease in acceleration. The acceleration amplification factor is less than 1 (135/817 = 0.17). In the case of amplified input motions, the maximum accelerations are 1,283 cm/s^2, 2,151 cm/s^2, and 3,594 cm/s^2 for amplification factors of 1.7, 1.8 and 2 respectively. For this 5-storey building, it is clear that when the maximum displacement response exceeds clearance, a collision occurs and the acceleration is amplified.

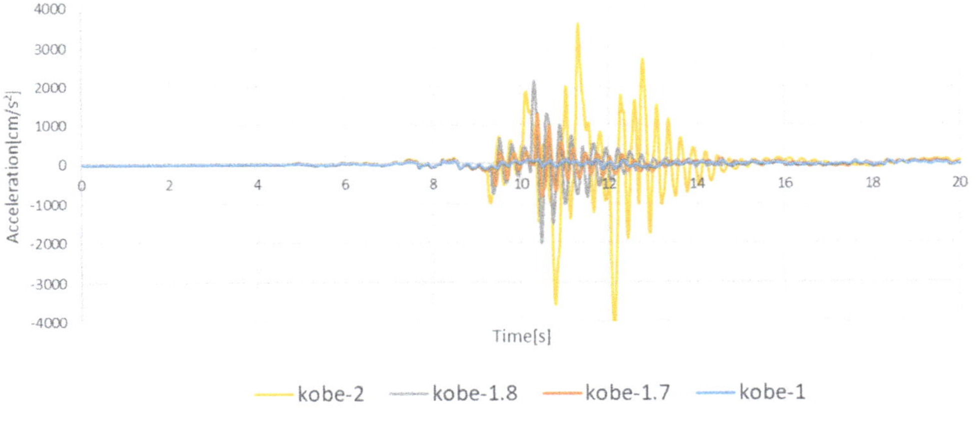

Figure 8: Acceleration response for the isolated 5-storey building.

Fig. 9 shows the acceleration response of the isolated 10-storey building. For the original Kobe earthquake (kobe-1), the maximum acceleration at the top of the building is 257 cm/s^2. In this case, the isolation effect is confirmed with the acceleration amplification factor less than 1 (257/817 = 0.32). When the input is amplified by 1.7, the maximum displacement is still less than the clearance and therefore there is no collision, and the isolation effect is still confirmed with a maximum acceleration of 326 cm/s^2. For 1.8 of amplification factor, the clearance is exceeded by a small amount and the maximum acceleration increases to 370 cm/s^2. Finally, for the amplification factor of 2, the maximum acceleration is 1514 cm/s^2.

Although the 5-storey isolated building and the 10-storey isolated building have almost the same predominant period of vibration, the corresponding maximum displacement and acceleration responses differ. It was expected that in the event of an impact from the same earthquake, the 10-storey building could produce a greater acceleration response due to greater mass. However, this larger mass causes smaller displacement responses and therefore the number of impacts or penetration against the retaining walls is also less. This results in a lower reaction force due to impact, also resulting in a lower acceleration response.

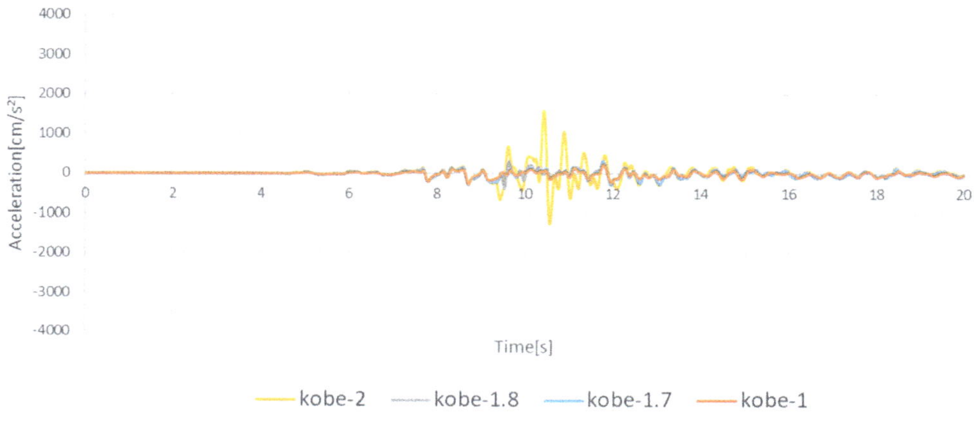

Figure 9: Acceleration response for the isolated 10-storey building.

4 ANALYSIS OF BUILDING CONTENT SAFETY

The safety of the building content is evaluated based on the failure flow and sliding of furniture proposed by Kaneko et al. [9]. An overturning limit acceleration and a slipping start acceleration are defined considering the predominant frequency of vibration of each floor. The predominant frequency of each floor is obtained from Fourier spectrum of signal obtained at the corresponding floor. The acceleration parameters also depend on the type of furniture, the dimensions, and the friction coefficient between the furniture and the floor slab. In this analysis, a representative piece of furniture (bookshelf) is selected. Dimensions of this piece of furniture is 200 cm high, 40 cm wide and 0.3 is assumed as the friction coefficient.

Fig. 10 shows the results of the furniture overturning probability analysis for the 5-storey building when is subjected to the original Kobe earthquake. In this case, the building does not collide against retaining walls and acceleration response of each floor is below the overturning limit. In other words, the contents of the building are safe.

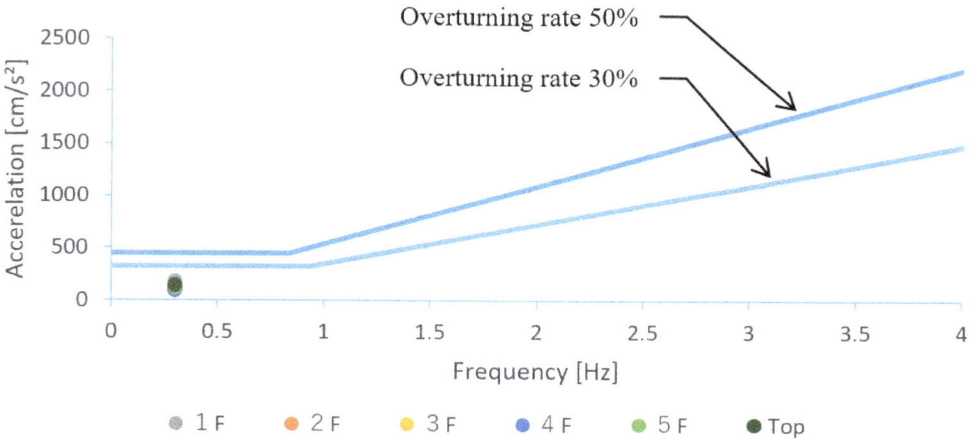

Figure 10: Acceleration and floor response for the isolated 5-storey building (no collision).

Fig. 11 shows the results of the analysis of the probability of furniture overturning for a 10-storey building when subjected to the original Kobe earthquake. In this case, the building also does not collide with the retaining walls and the acceleration response of each floor is below the overturning limit. In this case also the contents of the building are safe.

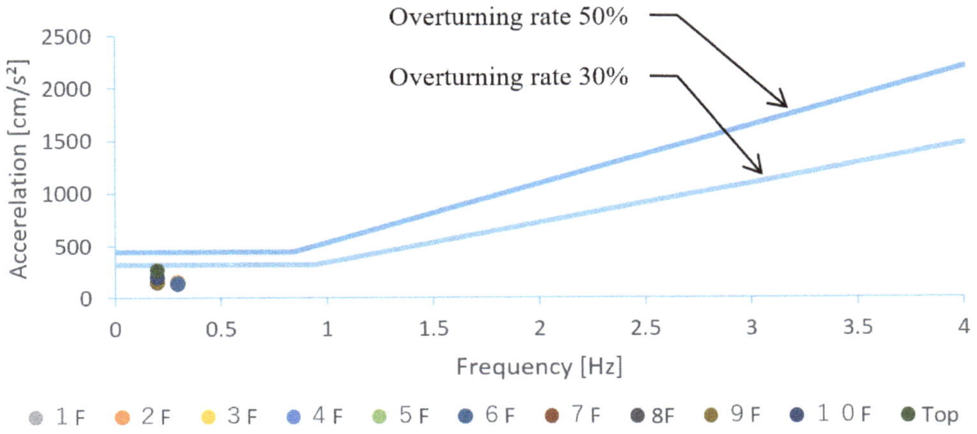

Figure 11: Acceleration and floor response for the isolated 10-storey building (no collision).

To study collision cases, for the 5-storey building the results corresponding to the input motion with an amplification factor of 1.8 were selected and for the 10-storey building the results of the input motion with an amplification factor of 2 were chosen. In these both cases the maximum displacements are similar (48.9 cm for the 5-storey building and 48.6 cm for the 10-storey building) and the collision produces higher accelerations.

Fig. 12 shows the safety analysis of the 5-storey building in the event of a collision. Only the first floor is below the 50% overturning probability curve. The rest of the floors exceed the 50% probability of overturning.

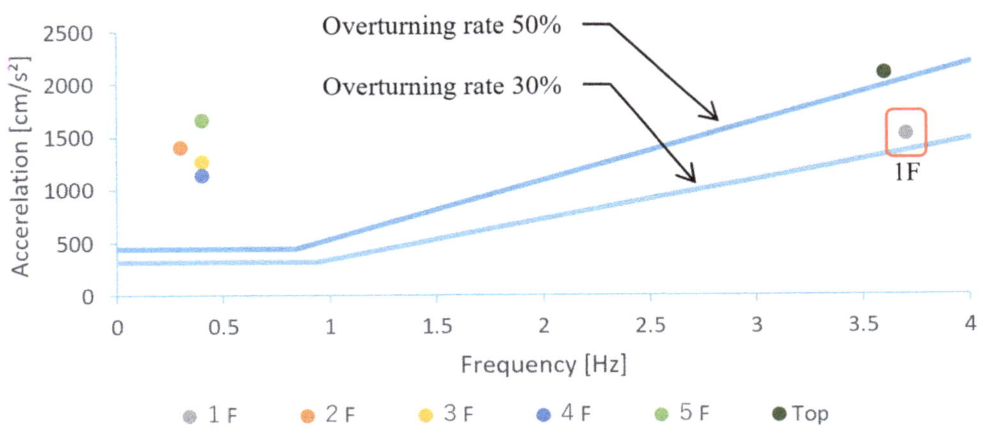

Figure 12: Safety analysis of the isolated 5-storey building in case of collision.

Fig. 13 shows the safety analysis of the 10-storey building in the event of a collision. In this case, the acceleration of the first floor, the second floor and the third floor are below the 50% overturning probability curve. The rest of floors exceed the 50% probability of overturning.

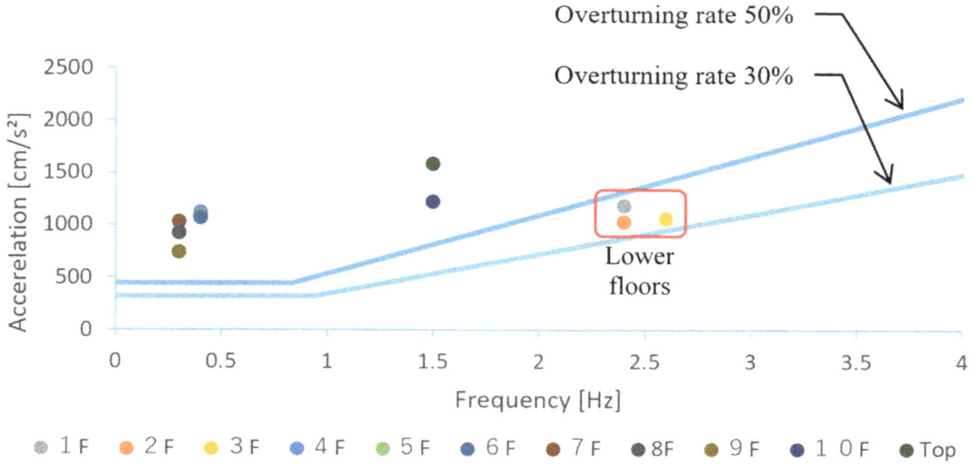

Figure 13: Safety analysis of the isolated 10-storey building in case of collision.

5 CONCLUSIONS

The 5-storey and 10-storey base-isolated buildings selected in this study have similar periods of vibration. However, the collision was confirmed in the first building when the Kobe seismic wave was amplified 1.7 times, and in the second building an amplification factor of 1.8 was required to produce the collision.

It was found that the behaviour of the isolated building at the time of collision with the retaining wall is greatly affected by the amount of displacement in which the response displacement exceeds the clearance.

If the isolated building does not collide with the retaining wall, the probability of the interior furniture will fall is less than 30%.

In cases of collision with the retaining wall, the probability of furniture overturning is greater than 50% in most floors. The equivalent frequency of the floor response tends to increase near the first floor and the vicinity of the upper floor, reducing the probability of overturning on these floors. It was also observed that the probability of furniture overturning is higher on the upper floors compared to the lower floor.

REFERENCES

[1] Hatada, K. & Cuadra, C., Dynamic response of low-rise building with different base isolation devices. *Proceedings of the 15th European Conference on Earthquake Engineering*, Istanbul, Turkey, Paper No.397, 2014.

[2] The Japan Society of Seismic Isolation, Data on the number of seismic isolated buildings and vibration controlled buildings, 2019. (In Japanese.) https://www.jssi.or.jp/menshin/doc/ms_ss_data.pdf.

[3] Cuadra, C. & Meguro, J., Evaluation of the dynamic characteristics of a base-isolated low-rise RC building after the Great East Japan Earthquake. *Proceedings of the 15th World Conference on Earthquake Engineering*, 24–28 September, Lisbon, Portugal, Paper No. 1284, 2012.
[4] Miwada, G., Komaki, J., Sato, K., Sano, T., Katsumata, H., Takiyama, N. & Hayashi, Y., Experiments and simulation analysis of collision to retaining wall with real scale base-isolated building. *The Architectural Institute of Japan's Journal of Structural and Construction Engineering*, **76**(663), 2011. (In Japanese.)
[5] Miwada, G., Sano, T. & Katsumata, H., Experiments of collision to retaining wall with real-scale base-isolated building. Research Institute of Technology, Obayashi Corporation, Report No.74, 2010. (In Japanese.)
[6] Sasaki, T., Sato, E., Fukuyama, K. & Kajiwara, K., Enhancement of base-isolation based on e-defense full scale shake table experiments: Dynamic response of base-isolated building under impact due to pounding. *Proceedings of the 16th World Conference on Earthquake, 16WCEE 2017*, Santiago Chile, 9–13 January, Paper No. 4082, 2017.
[7] Nakamura, S., Suzuki, T., Tojo, T., Asahara, S., Kinoshita, T. & Nakamura, N., Analytical study of the evaluation procedure and effect when base-isolated building collides with retaining wall. *Proceedings of the 16th World Conference on Earthquake, 16WCEE 2017*, Santiago Chile, 9–13 January, Paper No. 1467, 2017.
[8] Cuadra, C. & Hasegawa, J., Safety of base-isolated building in case of impact against retaining wall. *International Journal of Safety and Security Engineering*, 9(3), pp. 269–279, 2019.
[9] Kaneko, M., Nakamura, Y., Kambara, H. & Tamura, K., Seismic safety evaluation method for building contents. *Proceedings of the 14th World Conference on Earthquake Engineering*, 12–17 October, Beijing, China, Paper No. 14_S20-002, 2008.

MECHANICAL ANALYSIS OF A PIEZOELECTRIC SENSOR USED FOR DAMAGE DETECTION IN A STEEL STRUCTURE JOINT

NOBUHIRO SHIMOI, CARLOS CUADRA & HIDETO KANNO
Faculty of System Science and Technology, Akita Prefectural University, Japan

ABSTRACT

Recently, sensors that use piezoelectric materials are studied to detect impacts, number of impact occurrences, vibrations, etc. in structures. One of them uses a piezoelectric film attached to a thin steel plate that is installed on a steel column-beam joint to investigate its mechanical behaviour. Since the piezoelectric material reacts when sudden changes occur, the piezoelectric film is covered by a thin glass plate, and it is expected that when glass plate cracks due to an applied load stress, the film will react by emitting a voltage signal. In this way it might be possible to detect certain level of damage in a structural steel joint. Using this sensor system, static tests of steel beam-column joints were carried out and the applicability of the sensor was verified. However, the sensor reaction begins at high levels of deformation and the cracking of the glass does not occurs in the central part of the sensor plate as it was expected. Then is this research a structural analysis of the sensor steel plate is performed to analyse the stress distribution and the possible causes of glass failure at the ends of the sensor plate. The load test was found to produce a bending moment on the sensor plate and this moment has a great influence on the location of the glass crack. Also, for reverse loading, the steel plate is subject to compressive stresses and buckling failure must be considered.

Keywords: piezoelectric sensor, seismic damages, compression failure, steel structure failure.

1 INTRODUCTION

During the 1995 Kobe earthquake, many steel structures were damaged due to large lateral displacements caused by input motions. Assessing the level of damage is an important task in estimating the security and safety of a building after an earthquake. If the number of buildings to be evaluated is very large, it is desirable to have an indicator of the level of stress or displacement experienced. This indicator can be obtained if appropriate sensors are installed in the building.

In steel frame buildings, generally the connections are bolted or welded joints. In the case of bolt fixing, when a dynamic external force such as impact, vibration, or thermal load (expansion) acts on the bolted joint, the bolt often loses its fixing force due to loosening of the nut and therefore, its fixing function is lost. In the case of welded joints, there are few accidents due to vibrations and loosening of the nuts, however, due to the thermal effects during welding, brittleness occurs around joints.

Inspection methods for welded joints include non-destructive inspection techniques and visual inspection, including x-ray analysis techniques, image analysis techniques using ultrasonic waves and magnetism, and hammering. Visual inspection techniques are also effective tools.

Using the finite element method, it is difficult to realistically express crack propagation and fracture analysis even in the case of the initial state of the weld. Therefore, it is also difficult to model the aging effects of a welded joint. In Japan, which has experienced the Great East Japan Earthquake, when a large earthquake occurs with a seismic intensity of six or more, according to the architectural design standards the entire building is plasticized without collapse of the building, absorbing the seismic energy and saving human lives. In other developed countries that have also experienced major earthquakes, there is no

WIT Transactions on The Built Environment, Vol 206, © 2021 WIT Press
www.witpress.com, ISSN 1743-3509 (on-line)
doi:10.2495/SAFE210101

monitoring report of long-term robustness using precise measurement focusing only on joints.

In this research a piezoelectric sensor is studied for the detection of damage. The sensor consists of a piezoelectric film mounted on a thin steel plate and covered with a small, thin glass plate. The idea is that when a large deformation occurs, the glass plate breaks, at which point the piezoelectric material emits a voltage signal that is indicative of a certain level of damage. To investigate the stress distribution that affects the sensor response, a structural analysis of the sensor's steel plate is performed. This analysis helps explain the reasons for the location of the glass failure at the ends of the sensor plate. From the analysis it was found that the load test produces a bending moment on the sensor plate and this moment has a great influence on the location of the glass crack. For reverse loading, the steel plate is subjected to a compressive stress and the stress distribution is also affected by buckling.

2 EXPERIMENTAL TESTS

A piezoelectric sensor developed to detect the level of damage in the joint of steel structures was experimentally studied through a series of static test. These tests consisted of full-scale test on beam-column joint steel specimen and automated tests using a robotic system.

2.1 Sensor characteristics

In this research, the developed sensor consists of a piezoelectric film that is placed on a steel plate and covered with a layer of glass, as shown in Fig. 1. A special glue is used to fix the glass. This glue is a silicone that requires the application of ultraviolet rays to harden.

The steel base or plate of the sensor allows the detection system to be installed in a structural joint as can be seen in Fig. 2.

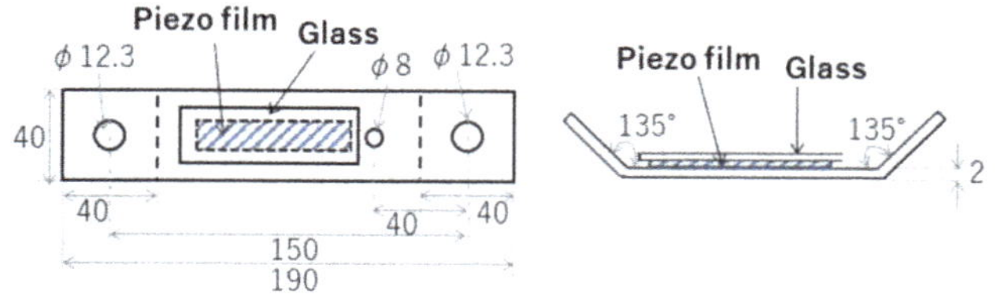

Figure 1: Piezoelectric sensor systems.

Figure 2: Sensor systems installed at beam-column joint.

The proposed piezoelectric sensors use piezoelectric films. This film could be, for example, DT2-028K film (16 mm × 73 mm) and LDT1-028K film (16 mm × 73 mm) shown in Fig. 3. These are commercial multipurpose piezoelectric sensors to detect physical phenomena such as vibrations or impacts. The piezoelectric film element is laminated to a polyester sheet and produces a usable electrical signal output when forces are applied to the sensing area. The dual wire lead attached to the sensor allows easy connection to a monitoring circuit or device to process the signal.

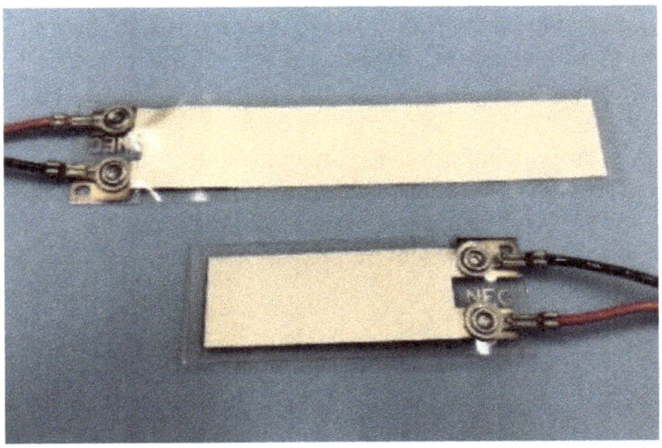

Figure 3: Piezoelectric films.

2.2 Static test on full scale specimen

The applicability of the proposed sensor was investigated by lateral loading experimental tests on full-scale specimens. The cross section of steel elements (beam and column) is a 100 mm × 100 mm box section with a 6 mm thin. This steel section is specified by the Japanese Industrial Standards as carbon steel square-tube for general structure (STKR400). The column and beam elements are connected perpendicularly using a fillet weld. The resulting test specimen has an inverted T shape as shown in Fig. 4.

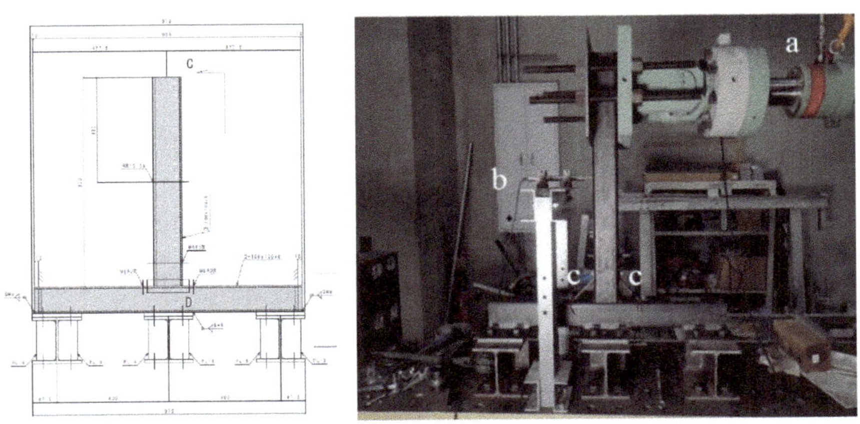

Figure 4: Test specimen and full-scale static test setup.

The horizontal part of the specimen is attached to a reaction system while the lateral force is applied to the free end. Piezoelectric sensors were installed at both corners of the specimen connecting the beam and column via the small diagonal steel plate that contains the piezoelectric sensor. Then, when lateral force is applied to the specimen it is assumed that this plate works in tension or compression depending on its position. Free end lateral displacement and lateral load were measured using displacements transducers and load cell, respectively.

Monotonic lateral load was applied reaching a maximum load of the order of 16 kN and the maximum displacement was 35 mm which corresponds to a drift angle of 1/20. Results for load history and sensor response are show in Fig. 5. Piezoelectric sensors show a first response at a drift angle of the order of 1/100 (Crack 1), and a second response is clearly observed at a drift angle of 1/50 (Crack 2). The drift angles at which piezoelectric sensors respond are relatively large. Therefore, it is necessary to improve the setup of sensors to obtain responses at early steps of loading. Although these preliminary results permit to verify the response of sensors for large deformation it can be concluded that it is possible to estimate the level of deformations using these simple piezoelectric sensors.

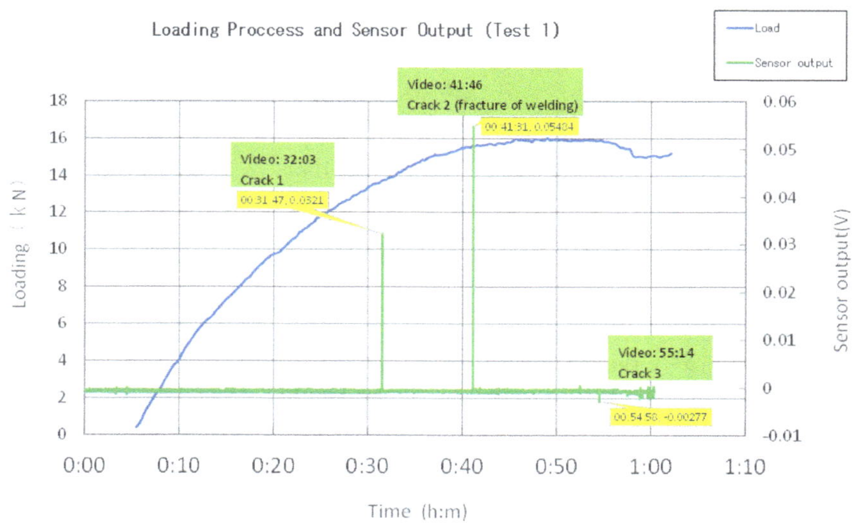

Figure 5: Test specimen and full-scale static test setup.

2.3 Static test on systematic test device (robot SALLY)

Full scale test to study the response of the piezoelectric sensor at the same time as the response of the structure is not an optimal procedure. Therefore, to study only the sensor response, an automated procedure was designed. This automated test allows studying the behaviour of the piezoelectric sensor system under different load conditions and allows a variation of the dimensions and shape of the sensor. The automated system shown in Fig. 6 is called as a Sally robot that consists of a horizontal beam and a vertical element connected by a pin joint with the horizontal element. The load is applied at the top of the vertical element by a screw system controlled from a computer and the load is measured by an electric load cell and the displacement stroke is measured by electrical transducers. For positive load as shown in Fig. 6, the sensor is supposed to work in tension and for negative load direction, the sensor should work in compression.

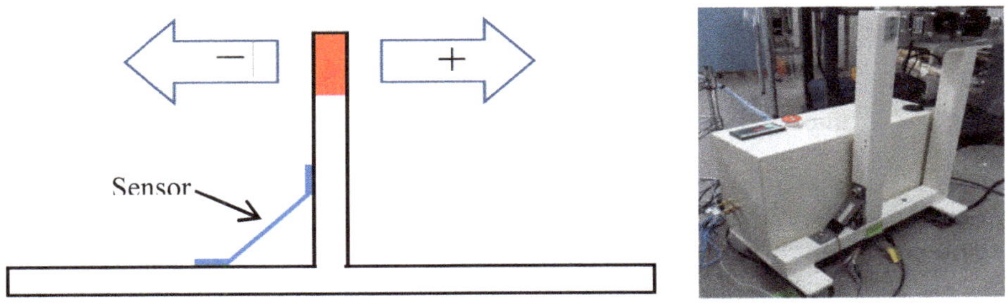

Figure 6: Automatised test system (robot Sally).

The results for loading in positive and negative direction are shown in Figs 7 and 8. The sensor response is plotted in red colour and the sensor reaction (or glass breakage) occurs at 8 mm and 10 mm for positive and negative directions respectively. This is a large level of displacement corresponding to a drift angle of 1/75 to 1/50, which are important values for structural deformations. Therefore, it is necessary to redesign the dimensions of the sensor system to obtain reactions to smaller deformations.

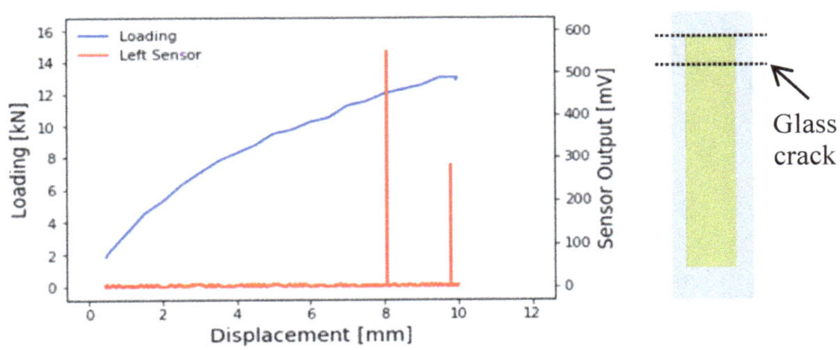

Figure 7: Results for loading in positive direction.

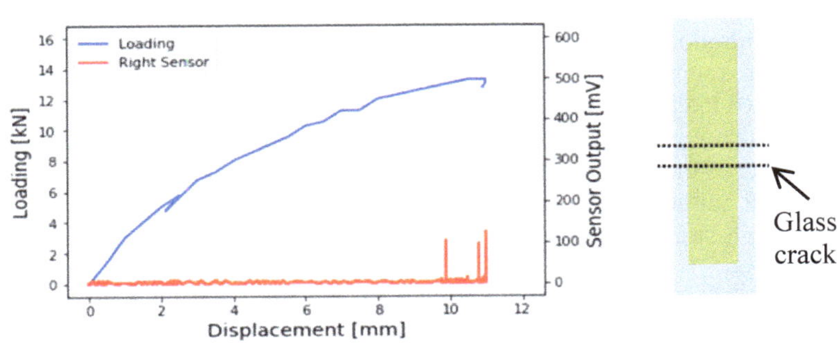

Figure 8: Results for loading in negative direction.

On the right side of Figs 7 and 8, the location of the glass cracking is indicated by dotted lines. It can be observed that in case of positive load (the sensor plate is supposed works in tension) the cracks are located at the upper part of the sensor plate and not at the central part as originally planned. In the case of negative load, the sensor plate works mainly in compression and the cracks are located in the central part of the plate. To ensure that cracks always occur in central part of the plate, it is necessary to decrease the cross section in the central part to induce a stress concentration in this part.

3 FEM ANALYSIS

In case of automated tests, a finite element analysis is performed to investigate the stress distribution on the sensor plate due to both load patterns. The model of the sensor plate is shown in Fig. 9. Plane shell elements are use and it is also considered the hole that is used to pass the electric cables. The objective is also to investigate the effect of this hole in the stress distribution. Both ends of the plate are considered pinned joints. The sensor plate is fixed to the horizontal and vertical element by means of bolts and therefore it is considered a good approximation to model these boundaries as pinned joints. Then, deformation of the upper right part of the plate is applied as external action that causes the stress in the central plate.

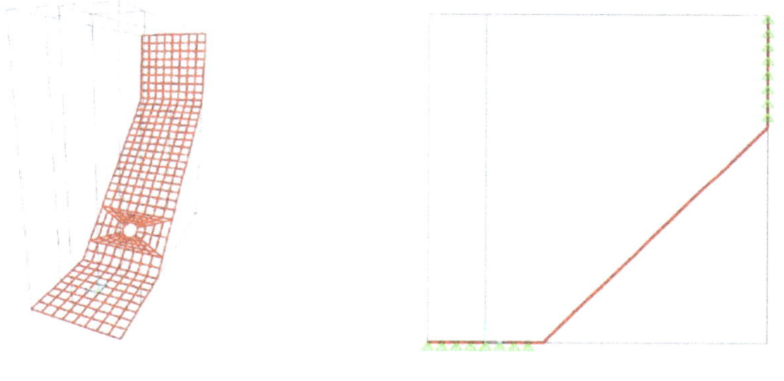

Figure 9: Finite element model of sensor plate.

The distribution of stress and deformation due to positive load is shown in Fig. 10. Since the displacement with linear distribution is applied in the upper right part of the plate, this part rotates as a rigid body. The deformation shows that flexion due to bending moment is concentrated at upper part of the sensor plate. Therefore, this moment causes the stresses on the plate. The concentration of this stress (which is tensile stress) causes the cracking of glass at the upper part of the sensor.

In case of negative load, compression stress is originated at upper part as is show in Fig. 11. However, in this case since a thin plate is subjected to compression stress, it is necessary to consider the effect of the buckling behaviour. The results considering the buckling effect are shown in Fig. 12 and it can be observed that the stresses are concentrated in the central part of the sensor. This stress is then believed to cause cracking of the glass in the central part of the plate.

From the analysis of both load cases, it was observed that the small hole in the plate does not significantly affect the stress distribution.

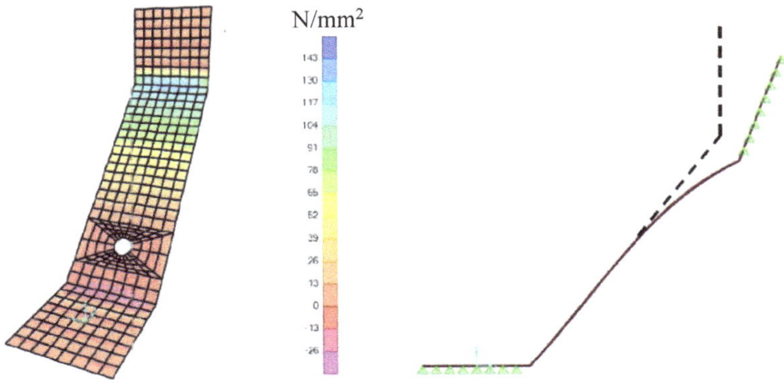

Figure 10: Results for loading in positive direction.

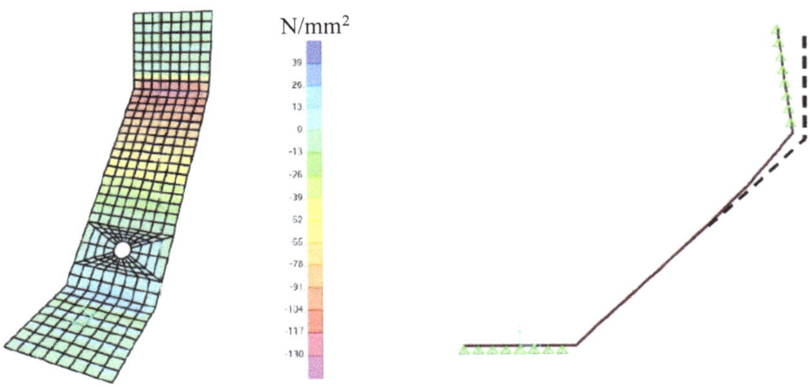

Figure 11: Results for loading in negative direction.

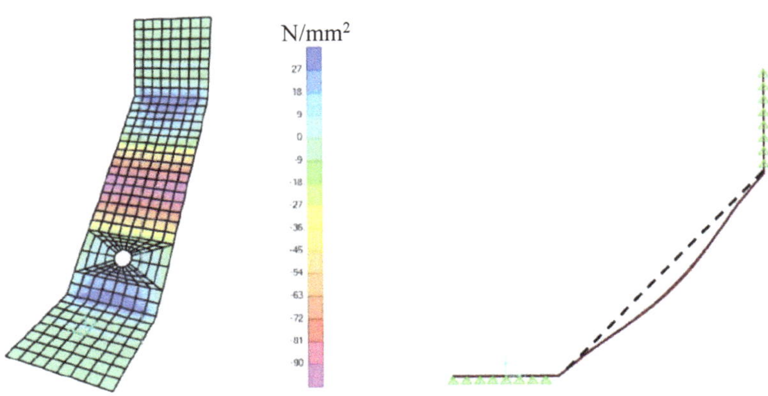

Figure 12: Results for loading in negative direction and considering buckling effect.

4 CONCLUSIONS

In this research, analysis of a new piezoelectric sensor system was performed to identify its behaviour pattern. The sensing system consists of a thin rectangular piezoelectric film mounted on a steel base plate and covered by a piece of glass. Cracking of the glass during loading allows a reaction of the piezoelectric material and this reaction indicates some level of damages of a structure. The applicability of this piezoelectric sensor system to detect the occurrence of failure was experimentally verified using a steel specimen subjected to lateral loading. Sensor reaction was also investigated using an automated robotic test setup.

Although preliminary results allow to verify the response of sensors for large deformations, it can be concluded that it is possible to estimate the level of deformations using this simple piezoelectric sensor.

The finite element analysis method was used to investigate the stress distribution affecting the sensor reaction and the location of the cracks in the part of the sensing glass. It was found that the steel base plate works in bending and the stresses originate by bending moment. In the case of compressive stresses, it was shown that it is important to consider the buckling effect.

For a regular and stable sensor response, it might be desirable to have the stress concentration in the middle of the sensor plate. This concentration could be achieved by changing the cross section of the plate near the centre.

ACKNOWLEDGEMENT

The authors acknowledge Japan Society for the Promotion of Science for the Grants-in-Aid for Simple Autonomous Health Monitoring System for Infrastructure Using Piezoelectric Limit Sensors, under which this research was conducted.

REFERENCES

[1] Cuadra, C., Shimoi, N., Sasaki, T. & Taguchi, T., Preliminary evaluation of piezoelectric sensors for the prediction of compression failure of brick masonry components. *WIT Transactions on The Built Environment*, vol. 153, WIT Press: Southampton and Boston, 2015. DOI 10.2495/STR150501.

[2] Shimoi, N., Nishida, T., Obata, A., Nakasho, K., Madokoro, H. & Cuadra, C., Comparison of displacement measurements in exposed type column base using piezoelectric dynamic sensors and static sensors. *American Journal of Remote Sensing* **4**(5), pp. 23–32, 2016.

[3] Cuadra, C., Shimoi, N. & Saijo, M., Dynamic characteristics of a bridge estimated with new bolt-type sensor, ambient vibration measurements and finite element analysis. *International Journal of Safety and Security Engineering*, **6**(1), pp. 40–52, 2016.

SECTION 5
HUMAN FACTORS
AND RISK PERCEPTION

EMOTIONAL REACTIONS TO RISK PERCEPTION IN THE ROYAL PALACE OF CASERTA, ITALY

FABIO GARZIA[1,2,3,4,5], FRANCESCO BORGHINI[1], ALBERTO BRUNI[4,5,6],
MARA LOMBARDI[1], ILARIA ESPOSITO[1] & SOODAMANI RAMALINGAM[7]
[1]Safety and Security Engineering Group – DICMA, SAPIENZA – University of Rome, Italy
[2]Wessex Institute of Technology, UK
[3]European Academy of Sciences and Arts, Austria
[4]Safety and Security Project, Pompeii Archaeological Park, Italy
[5]Smart@Pompeii Project, Pompeii Archaeological Park, Italy
[6]Pompeii Archaeological Park, Ministry of Culture, Italy
[7]School of Physics, Engineering and Computer Sciences, University of Hertfordshire, UK

ABSTRACT

The evaluation of the perception of risk of people represents a vital element for security management. Each individual trusts other people's opinions to do a selection and the Internet represents the place where these opinions are mostly explored, obtained, and evaluated. Opinion mining and sentiment analysis embodies valuable means. They were primarily used as market investigation means to collect opinions about products and they have later turned out to be appropriate in other areas such as safety and security. Security managers can utilize the opinions communicated by individuals to find out the unforeseen vulnerabilities of a monitored location or at least the security assessment of persons which occasionally cannot be the same of the real level of security of a given site. Opinion mining can be useful when requiring continuous feedback about risk perception and to determine when an appropriate and prompt action is required based on the perceived risk and security. Gathering the opinions to be utilized for this objective requires seeking in different open sources (OSINT (Open Source INTelligence)) and therefore processing huge amount of digital data where information and knowledge must be extracted from. The objective of this paper is to study the perception of risk in the Royal Palace of Caserta (UNESCO World Heritage Site since 1997), focusing on the emotional elements, utilizing the semantic examination of the textual contents present in Twitter.

Keywords: perceived risk assessment for security, open-source intelligent techniques for security, OSINT, opinion mining for security, sentiment analysis for security, cultural heritage security.

1 INTRODUCTION

The evaluation of the perception of risk of people represents a vital element for security management.

Each individual trusts other people's opinion to do a selection and the Internet represents the place where these opinions are mostly explored, obtained, and evaluated. Opinion mining and sentiment analysis embodies valuable means. They were primarily used as market investigation means to collect opinions about products and they have later turned out to be appropriate in other areas such as safety and security. Security managers can utilize the opinions communicated by individuals to find out the unforeseen vulnerabilities of a monitored location or at least the security assessment of persons which occasionally cannot be the same of the real level of security of a given site.

Opinion mining can be useful when requiring continuous feedback about risk perception and to determine when an appropriate and prompt action is required based on the perceived risk and security.

Gathering the opinions to be utilized for this objective requires seeking in different open sources (OSINT (Open Source INTelligence)) and therefore processing huge amount of digital data where information and knowledge must be extracted from [1]–[14]. In the present work, Twitter has been employed as resource and a suitable evaluation of keywords, enclosed

www.witpress.com, ISSN 1743-3509 (on-line)
doi:10.2495/SAFE210111

into different tweets, was achieved. Keywords were chosen bearing in mind that the evaluation of the perceived risk is deeply correlated to psychological features by means of emotional responses aroused from a given location, represented, in this case, by the Royal Palace of Caserta (RPC) in Italy that has been a UNESCO World Heritage Site since 1997.

The objective of this paper is to study the perception of risk in the Royal Palace of Caserta, focusing on the emotional elements, utilizing the semantic examination of the textual matters present in Twitter.

2 THE ROYAL PALACE OF CASERTA

The Royal Palace of Caserta, shown in Fig. 1 [15], is the largest royal residence in the world that belonged to the Bourbon dynasty that in the years 1734–1735 led by Charles invaded and conquered the kingdoms of Naples and Sicily which were until then under Austrian rule.

Figure 1: View of the facade of the Royal Palace of Caserta from outside.

Charles of Bourbon aimed to give the kingdom a new capital far from the sea and the threats that could come from it. The baroque royal residence par excellence was designed by the architect Luigi Vanvitelli on behalf of the sovereign in 1752. It was later completed by his son Carlo Vanvitelli and other architects trained in the same school.

The residence on the outside has a contained elegance, inside it is luxurious and with great attention to detail. Spread over four monumental courtyards, the building, which occupies 45,000 m^2, is faced by the spectacular park which is now a destination for thousands of tourists. With its five floors occupied by 1,200 rooms, 34 stairways, 1,742 windows it reaches a height of 36 m. The ground floor leads to the first floor with a beautiful and scenic central flight of stairs full of marble that reaches the royal apartments. All the rooms, besides the royal apartments, are of exceptional beauty, such as the Theatre, the Pictures Gallery, and the library. Inside the Palace everything is a triumph of marble, gilded decorations, bas-reliefs, and paintings. The Throne saloon is shown in Fig. 2 [15].

Figure 2: Views of the Throne saloon of the Royal Palace of Caserta.

The Park is a splendid and grandiose work of art and contributes to making the Royal Palace of Caserta one of the most beautiful in the world. It extends for 3 km in length over 120 hectares and includes two gardens: the Italian one and the English one. The first, articulated according to imposing geometric patterns, is a succession of pools, waterfalls, and fountains that it reaches the culmination in the Great Cascade. The second, less symmetrical and ordinary than the Italian one, leans towards a natural freedom of greenery with plants of great value, ponds, and streams.

The Royal Palace of Caserta is an ideal stage for any artistic and cultural event, a natural theatre for concerts, an exceptional space for art exhibitions. For this reason, it has been used as a film set for many famous films. In 1997 the Royal Palace of Caserta with the adjoining park was declared a World Heritage Site by UNESCO. The cultural site is visited on average by 800,000 tourists a year with a steadily increasing number of visitors before COVID-19 pandemic.

This exceptional location certainly provokes emotions in visitors affecting their awareness of the risk in this site. Inimitable and composite cultural heritage sites, as the considered one, need a noteworthy endeavour to guarantee security and safety, cultural heritage conservation and defence as well as openness for visitors, with special attention to visitors with disabilities, and for personnel generally present for site administration. These objectives can be attained by means of integrated systems [16], [17] and advanced technologies, as Internet of Things (IoT)/Internet of Everything (IoE) that allows to link people, things (mobile devices, smart sensors, apparatuses, actuators, wearable tools, etc.), data/information/knowledge and processes [17], [18]. The IoT/IoE system have to implement and sustain an integrated multidisciplinary model for security and safety management (IMMSSM) for this location [20], [21] that includes a proper risk assessment [22]. For this reason, a suitable project named "Safety and Security for the Royal Palace of Caserta" has been activated.

The objective of this paper is to examine the risk perception in the Royal Palace of Caserta, providing importance to the emotional aspects, applying the semantic analysis of the textual matters present in Twitter. It operates as a valuable instrument serving key purposes, namely (i) supplying continuous feedbacks about the perception of risk; (ii) support planning forthcoming actions designed to increase real or perceived security as counter measures; and (iii) to acquire the outcome on the perceived security and risk.

3 OPINION MINING

The semantic knowledge included in the data can be obtained by means of sentiment analysis, also known as opinion mining [1]–[14]. This kind of evaluation obtains the personal judgments of the users about an entity of interest via the assessment of the polarity of the feeling (that can be positive, negative, or neutral) utilized to illustrate the experience with the latter. In general, the sentiment analysis is achieved following the different phases shown in Fig. 3 [3].

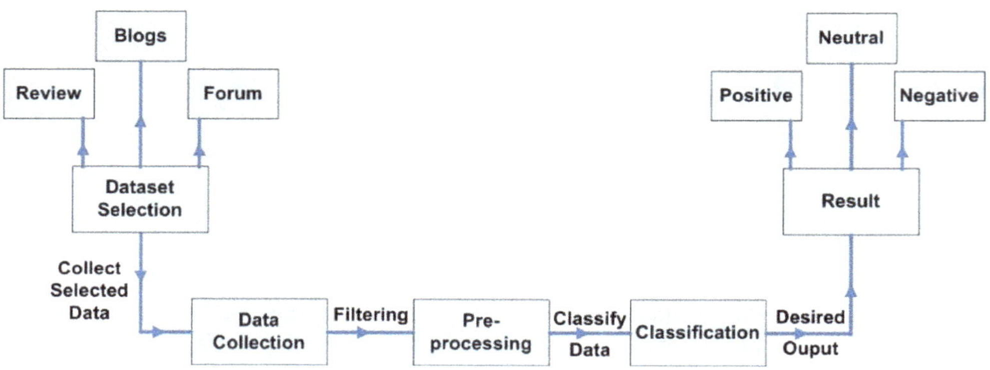

Figure 3: General opinion mining steps.

The study concentrated on the evaluation of the tweets (equally in English and Italian) of RPC with the purpose of obtaining the emotions and opinions of people to evaluate their risk perception of the place. Twitter has been selected as the supplier of data from which to extract writings since it is one of the most utilized social networks in the world and since it has the highest percentage of public profiles with respect to others (Facebook® profiles and comments are normally not public). Further, the text is the principal element (Instagram® includes mostly pictures) whose briefness, 280 characters/tweet, permits to evaluate each tweet in a more effortlessly and successfully way.

There is much software offered for social network evaluation, but NodeXL is usually the most popular since it does not need the expertise of any programming language. NodeXL is a free plug-in for Excel and the essential edition, employed in this work, lets import data and process it employing metrics. The restricted Twitter API of the basic version allows to extract up to 2,000 tweets or at least tweets no older than 7–8 days. For longer period or for a greater number of tweets it is necessary to use the pro version. In this situation, the metric "Words and word pairs" has been employed, that counts the words and the pairs of words which occurred in the tweets. Basically, this metric achieves the sentiment analysis just by examining the text, distinguishing the words that are included in the two "Lists of sentiment words" with the number of occurrences. The keywords employed as seek term are shown in the Table 1.

Table 1: Summary of the English and Italian keywords used as search terms.

Keywords	Language
Caserta Palace	English
The Royal Palace of Caserta	English
Reggia Caserta	Italian

Furthermore, the listing of positive and negative words, a third group called "Awareness words" was implemented. It embodies a group fitting to show up if and how the site is cited in the mass media. The words are hence names of newspapers, TV channels, broadcasting companies and events that took place in the Royal Palace of Caserta.

To ensure a meaningful outcome of the evaluation, the word listing, that characterizes the categorization in one of the three groups, is produced step by step, trusting on the default English settings of the software. It is recognized that dictionary for the English and Italian language suitable for this type of evaluation are already accessible but, as some words are unique for the considered situation, it was chosen to generate a novel vocabulary of sentiments.

At this point, each word (positive or negative) was linked with the major emotion aroused by it, to get the emotional reactions of visitors. To achieve this objective, appropriate questionnaires were formed and compiled by two casually collected statistical reference sample (one composed by English native speakers and one composed by Italian native speakers) consisting of 41 people, equally divided between male and female, aged between 17 and 70. Once the outcomes of the surveys were gathered and examined, it was feasible to associate the prevalent emotions with each word.

4 RESULTS

The treating of the data to be processed have therefore generated a vast quantity of information condensed in the Table 2 while Fig. 4 shows the distribution of the number of tweets across the considered timeline.

From Table 2 it is possible to note that the number of words belonging to the list of non-categorized words, that is, uncategorized words, is very great. The main reason is because in this list it has been chosen to consider all the keywords and their combination (Palace, Royal etc.) present in numerous users' tweets. Furthermore, in this list there are also words with an

Table 2: Total number of words used for the evaluation.

Word	Count
Words in sentiment list#1: Positive	1,877
Words in sentiment list#2: Negative	129
Words in sentiment list#3: Awareness	1,696
Non-categorized words	27,391
Total words	31,093

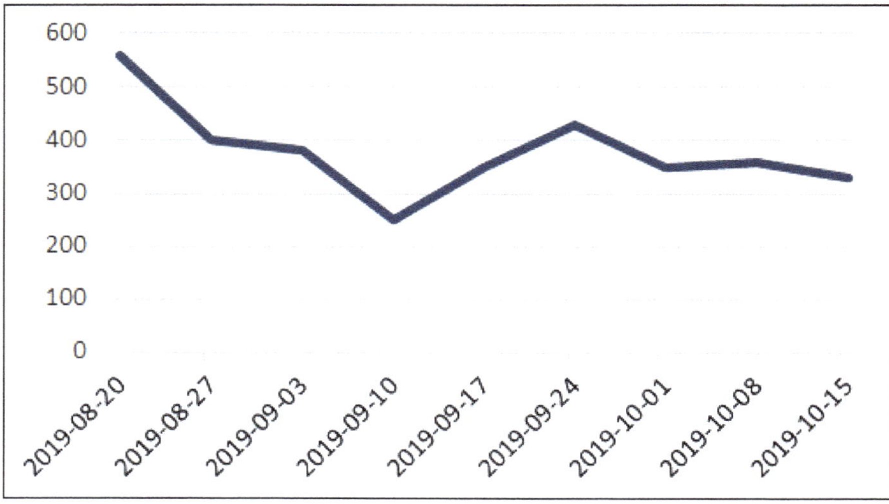

Figure 4: Considered timeline of tweets.

ambiguous meaning that do not express an evident polarity, be it positive or negative, to define belonging to the positive and negative categories.

From Fig. 4, it is possible to note some peaks that are due to events that were present in the period of data collecting in the Royal Palace of Caserta.

Considering 571 English words related to emotional reactions, however, between emotions with positive or negative valence, it emerges that "Attraction" is the dominant one, since it is about 45% of the total, as shown in Figs 5 and 6.

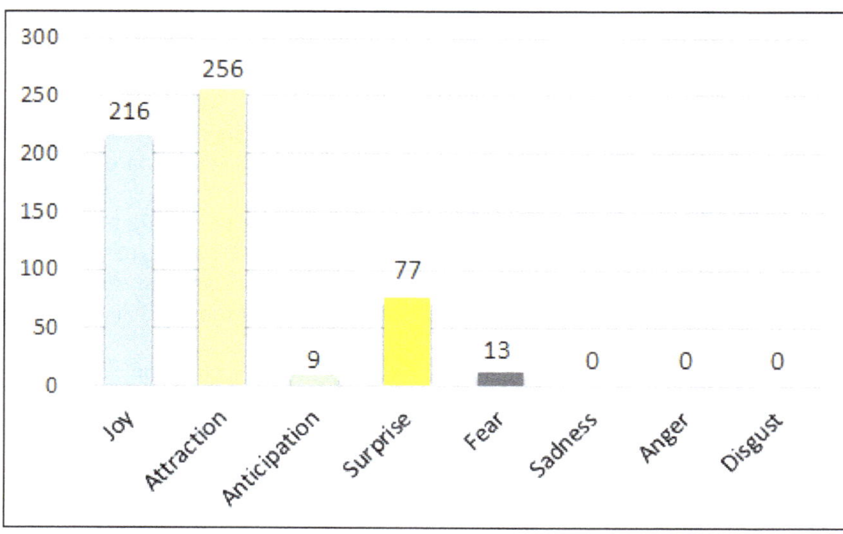

Figure 5: Incidence of each primary emotion for English words.

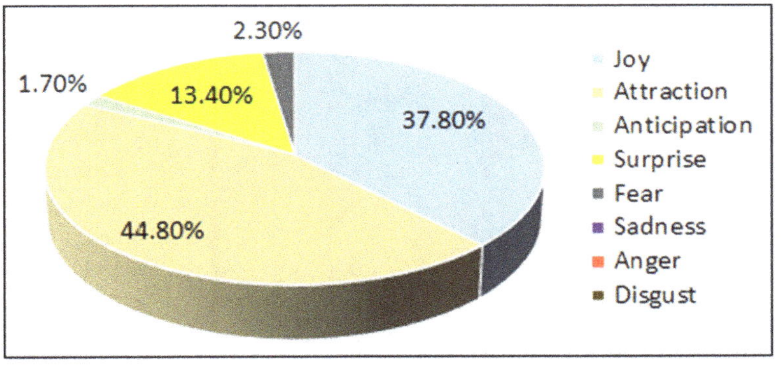

Figure 6: Percentage distribution of each primary emotion for English words.

Considering 1,435 Italian words related to emotional reactions, however, between emotions with positive or negative valence, it emerges that "Attraction" is the dominant one, as the previous case, since it is about 46% of the total, as shown in Figs 7 and 8.

Considering all the English and Italian words, between emotions with positive or negative valence, it emerges that the most dominant positive emotion is represented by "Attraction", as shown in Figs 9 and 10.

It is thus conceivable to affirm that the overwhelming emotion provoked by the experience of visiting the Royal Palace of Caserta is Attraction (45.6%), followed by Joy (40.9%), Surprise (4.7%), Anger (3.3%), Anticipation (2.3%), Fear (2.2%), Sadness (0.6%), Disgust (0.4%).

Assessing suitably the opinions regarding the experience of the visit to the Royal Palace of Caserta, it is found that the word most extensively utilized in the English language is represented by "Great", followed by "Art", "Site", "Discovering", "Treasure", "Amazing" "Paintings" and "World" which relates to emotions aroused by the beauty and the positive environment visited, as shown in Fig. 11(a).

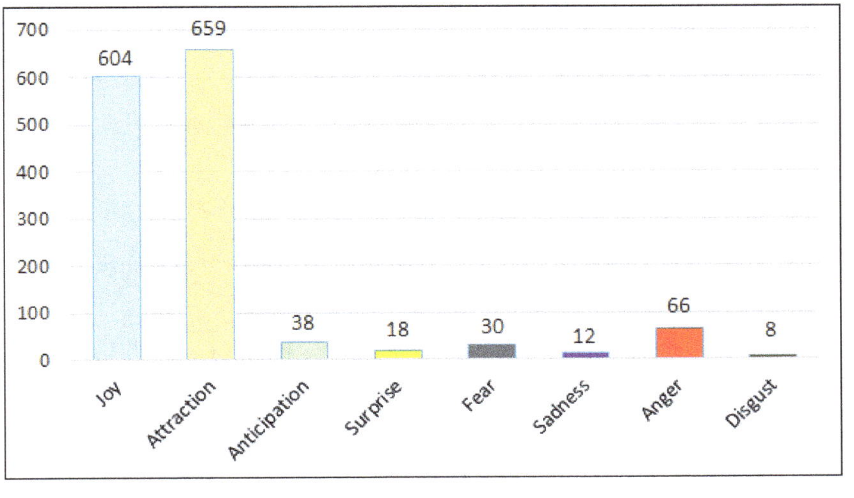

Figure 7: Incidence of each primary emotion for Italian words.

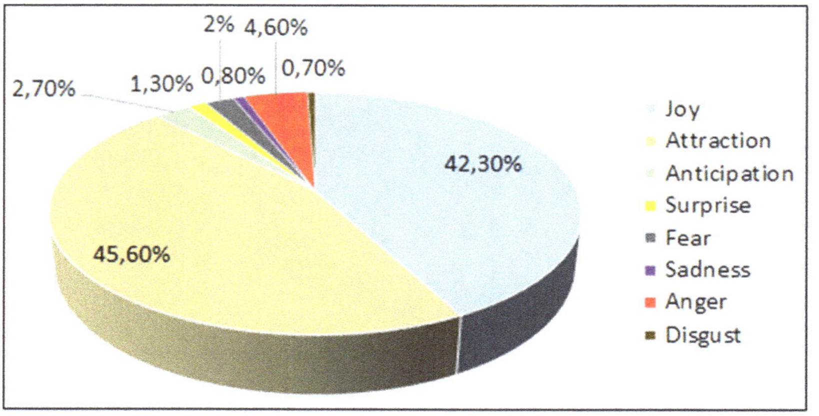

Figure 8: Percentage distribution of each primary emotion for Italian words.

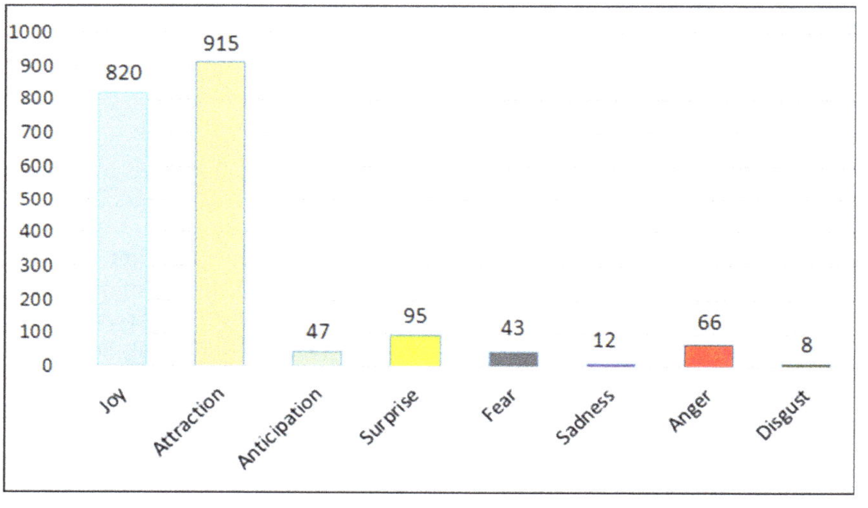

Figure 9: Incidence of each primary emotion for English plus Italian words.

The negative words used with low occurrence by English-language visitors instead are represented only by "Blitz" and "Attempt", as shown in Fig. 12(a).

In the Italian language instead, it is found that the word most widely utilized is represented by "Summer" (due to the period of data collecting), followed by "Exhibition", "Art", "Beautiful", "UNESCO", "Great", "Sunday", and "Wish", as shown in Fig. 11(b).

The negative word most used, with low occurrence, by Italian-language visitors instead is represented by "Bullying", followed by the other words characterized with a lower occurrence, as shown in Fig. 12(b).

It is therefore possible to state that the Royal Palace of Caserta, according to the direct and unaffected judgment of the visitors who visited this place, certainly aroused positive emotions.

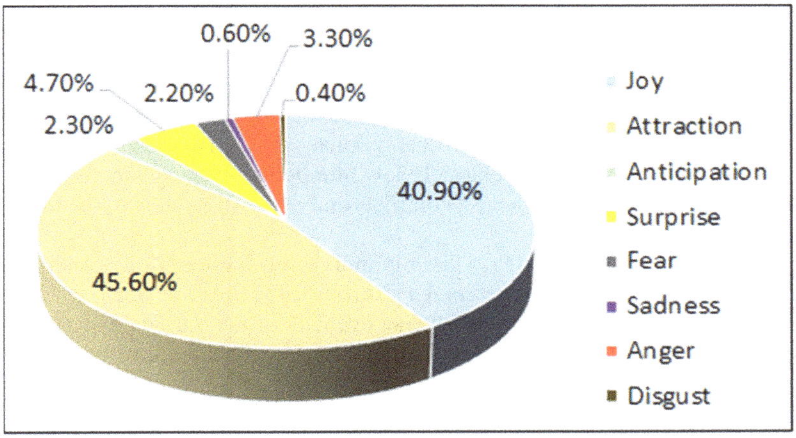

Figure 10: Percentage distribution of each primary emotion for English plus Italian words.

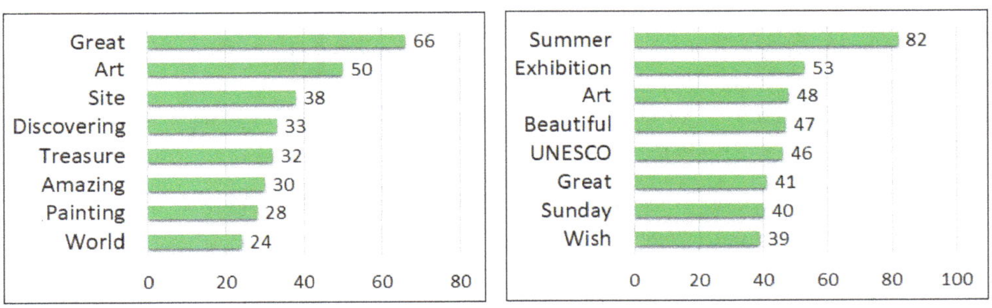

(a) (b)

Figure 11: Ranking of the most mentioned positive words in: (a) English; and (b) Italian.

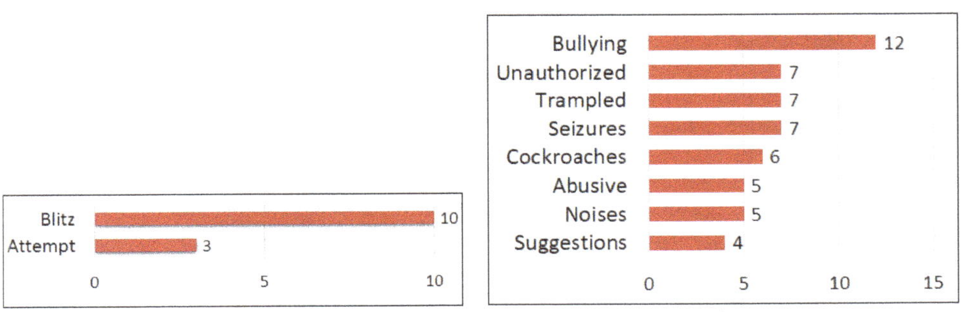

(a) (b)

Figure 12: Ranking of the most mentioned negative words in: (a) English; and (b) Italian.

The predominant emotion aroused was the "Attraction" that reached 45.6%, slightly detached from "Joy" with 40.9%. It is clear that, without reason for criticism, there was a notable homogeneity of judgment regarding the words; much more than half of the judgments and evaluations on the locality are associated to experiences with a positive valence.

Out of 2,006 words in total, only 129, therefore decidedly contained, have a negative value. This unfavourable judgment is aroused not so much by the place visited as by factors external to it or by completely personal influences that are not significant for the purposes of the work performed. Despite these relatively low numbers, it was considered appropriate to take them into account, in order to have a complete and exhaustive view of the final judgment of the considered site.

It pursues that the perception of risk is minimal, as all the emotions considered widely positive can confirm the absence of perceived risk, while the negative ones can be considered as a demonstration of the fear of some adverse events that the visitors experienced outside the considered place. In fact, it is well known that the intensity and the value of the perceived risk are deeply connected to the affective aspects. Beginning from sensory stimuli, via emotions, one is led to the perception of security as feeling-need. It is conceivable to achieve the complete lack of risk perception when the meaning of the lived experience is in total harmony with the existential expectation of the person. For this reason, if a person who visits a place can ascertain the presence of satisfactory security measures, he feels a state of absolute risk-freeness.

Furthermore, the assessment of the achieved results (also considering the words of sentiment lists of Table 2) lets state that there is an extensive sharing of positive emotions and sentiments, that is not an ordinary result, and to state the existence of two general average emotions, represented by "Attraction" and "Joy", confirming Royal Palace of Caserta to be a strong catalysing environment.

5 COMPARISON WITH RESULTS OBTAINED IN POMPEII ARCHEOLOGICAL PARK AND HERCULANEUM ARCHAEOLOGICAL PARK

It is interesting to compare the attained results with the results deriving from a similar study done in Pompeii Archaeological Park [3] and Herculaneum Archaeological Park [4], that are shown in Fig. 13.

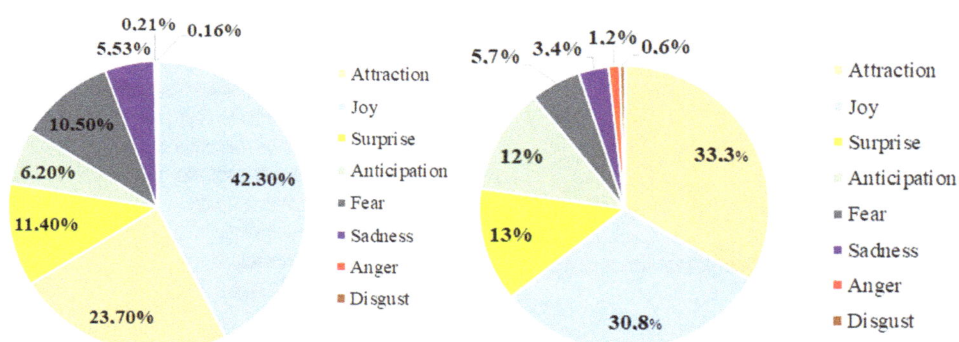

(a) (b)

Figure 13: Percentage distribution of each primary emotion for English plus Italian words in: (a) Pompeii Archaeological Park; and (b) Herculaneum Archaeological Park.

As is it possible to see from Fig. 13, comparing it with previous figures related to the results obtained in the present work, in Pompeii Archaeological Park (PAP) the dominant emotion is represented by "Joy" (42.30%), in Herculaneum Archeological Park (HAP) "Joy" and "Attraction" are comparable and equal to about 30% each, while in the Royal Palace of Caserta (RPC) "Attraction" is the dominant emotion (45.60%). Further, it is important to note that the sum of percentages of "Joy" and "Attraction" is equal to 66% in PAP, 64.1% in HAP and 86.5% in the Royal Palace of Caserta. This is probably due to the absence of negative emotions aroused by the looming presence of Vesuvius volcano and the related remind of its terrible eruption that buried both PAP and HAP in 79 AD.

6 CONCLUSIONS

A methodology for the evaluation of the perception of risk in the Royal Palace of Caserta in Italy has been illustrated, providing importance to the emotional factors, by means of the semantic analysis of the textual matters present in Twitter.

It embodies a valuable means as it allows to receive continuously feedbacks about the perception of risk and when upcoming actions targeted at increasing effective or perceived security is put up, so that it is feasible to get the influence on the perceived risk and security. The obtained results have been also compared with results already achieved in the Pompeii Archaeological Park and the Herculaneum Archaeological Park.

REFERENCES

[1] Borghini, F., Garzia, F., Borghini, G. & Borghini, A., *The Psychology of Security, Emergency and Risk,* WIT Press, 2016.

[2] Garzia, F., Cusani, R., Borghini, F., Saltini, B., Lombardi, M. & Ramalingam, S., Perceived risk assessment through open-source intelligent techniques for opinion mining and sentiment analysis: The case study of the Papal Basilica and Sacred Convent of Saint Francis in Assisi, Italy. *Proc. of IEEE Int. Carnahan Conf. on Security Technologies,* Montreal (Canada), pp. 11–15, 2018. DOI: 10.1109/CCST.2018.8585519.

[3] Garzia, F., Borghini, F., Bruni, A., Lombardi, M., Mighetto, P., Ramalingam, S. & Russo, S.B., Emotional reactions to the perception of risk in the Pompeii Archaeological Park. *International Journal of Safety and Security Engineering,* **10**(1), pp. 11–16, 2020. DOI: 10.18280/ijsse.100102.

[4] Garzia, F., Borghini, F., Bruni, A., Lombardi, M., Minò, L. & Ramalingam, S., Emotional reactions to the perception of risk in the Hercolaneum Archaeological Park. *Proc. of IEEE Int. Carnahan Conf. on Security Technologies,* Hatfield, UK, 2021.

[5] Liu, B., *Sentiment Analysis: Mining Opinions, Sentiments, and Emotions,* Cambridge University Press, 2015.

[6] Vijarania, M., *Sentiment Analysis: A Combined Approach*, Lambert Academic Publishing, 2020.

[7] Medhat, W., Hassan, A. & Korashy, H., Sentiment analysis algorithms and applications: A survey. *Ain Shams Engineering Journal*, pp. 1093–1113, 2014.

[8] Ramalingam, S., Metadata extraction and classification of YouTube videos using sentiment analysis. *Proc. of IEEE International Carnahan Conference on Security Technology,* Orlando, USA, 2016.

[9] Cambria, E., Schuller, B., Xia, Y. & Havasi, C., New avenues in opinion mining and sentiment analysis. *IEEE Intelligent Systems,* **28**(2), pp. 15–21, 2013.

[10] Borghini, F., Garzia, F., Lombardi, M., Mete, M., Perruzza, R. & Tartaglia, R., Human factor analysis inside a peculiar job environment at the Gran Sasso Mountain Underground Laboratory of Italian National Institute for nuclear physics. *Int. J. of Safety and Security Engineering,* **8**(3), pp. 390–405, 2018.

[11] Park, S.M., Baik, D.K. & Kim, Y.G., Sentiment user profile analysis based on forgetting curve in mobile environments. *Proc. IEEE 15th International Conference on Cognitive Informatics and Cognitive Computing,* pp. 207–211, 2016.

[12] Anastasia, S. & Budi, I., Twitter sentiment analysis of online transportation service providers. *Proc. of Int. Conf. on Advanced Computer Science and Information Systems,* pp. 359–365, 2016.

[13] Akter, S. & Aziz, M., Sentiment analysis on Facebook group using lexicon based approach. *Proc. of 3rd International Conference on Electrical Engineering and Information Communication Technology*, pp. 1–4, 2016.

[14] Shaikh, T. & Deshpande, D., A review on opinion mining and sentiment analysis. *International Journal of Computer Applications (0975-8887)*, 2016.

[15] https://it.wikipedia.org/wiki/Reggia_di_Caserta.

[16] Garzia, F., Sammarco, E. & Cusani, R., The integrated security system of the Vatican City State. *Int. J. of Safety and Security Engineering,* **1**(1), pp. 1–17, 2011.

[17] Contardi, G., Garzia, F. & Cusani, R., The integrated security system of the Senate of the Italian Republic. *Int. J. of Safety and Security Engineering,* **1**(3), pp. 219–246, 2011.

[18] Garzia, F., Implementing an Internet of Everything system in the archaeological area of Quintili's Villa in the Ancient Appia route park in Rome. *WIT Transactions on the Built Environment*, vol. 174, WIT Press: Southampton and Boston, pp. 261–272, 2017. DOI: 10.2495/SAFE170241.

[19] Garzia, F. & Sant'Andrea, L., The Internet of Everything based integrated security system of the World War I commemorative museum of Fogliano Redipuglia in Italy. *Proc. of IEEE Int. Carnahan Conf. on Security Technologies*, Orlando, USA, pp. 56–63, 2016.

[20] Garzia, F., An integrated multidisciplinary model for security management and related supporting integrated technological system. *Proc. of IEEE Int. Carnahan Conf. on Security Technologies,* Orlando, USA, pp. 107–114, 2016.

[21] Garzia, F. & Lombardi, M., Safety and security management through an integrated multidisciplinary model and related integrated technological framework. *WIT Transactions on The Built Environment,* vol. 174, WIT Press: Southampton and Boston, pp. 285–296, 2017. DOI: 10.2495/SAFE170261.

[22] Garzia, F., Novel risk assessment methodology for cultural heritage sites. *Proc. of STREMAH 2021 – 17th International Conference on Structural Studies, Repairs and Maintenance of Heritage Architecture*, pp. 149–160, 2021. DOI:10.2495/STR210131.

PREVENTION OF OCCUPATIONAL RISKS IN GEOTECHNICAL DRILLING WORKS THROUGH VIRTUAL REALITY TRAINING

AITOR FERNÁNDEZ[1], FELIPE MUÑOZ-LA RIVERA[1,2,3] & JAVIER MORA-SERRANO[1,2]
[1]School of Civil Engineering, Universitat Politècnica de Catalunya, Spain
[2]International Centre for Numerical Methods in Engineering (CIMNE), Spain
[3]School of Civil Engineering, Pontificia Universidad Católica de Valparaíso, Chile

ABSTRACT

The architecture, engineering, construction and operations industry has high accident rates as a result of characteristics of the construction site and work activities, the heavy machinery used and a lack of safety culture. In particular, geotechnical drilling worksites involve the use of high-powered machinery and workers are exposed to different risks when using them. However, risk prevention training courses are not well-specialised for this type of work. There is a lack of adequate learning content specific to drilling works, meaning the heavy machinery, how to use it and how to prevent accidents due to these drilling tasks are poorly understood. Therefore, this research explores the potential of virtual reality technology as a tool for analysing the risks associated with geotechnical drilling works and as complementary training content to traditional courses. A specific use case, a geotechnical drilling machine is modelled in 3D and integrated into a realistic virtual reality environment where the movements of the machine are recreated, in addition to interactions so that workers can simulate the geotechnical drilling process, identify risks and prevent accidents, and, moreover, be trained in best practices for machinery usage, according to previous real-world experiences from senior workers.
Keywords: virtual reality, BIM, geotechnical drilling works, occupational risks, storytelling, construction site, training.

1 INTRODUCTION

Geotechnical works are an important element at the beginning of construction projects. Unlike the later stages of construction, where a large number of professionals, roles, and work packages interact at the site simultaneously, during the geotechnical work stages, the use of heavy machinery predominates. These machines operate at high pressures, exerting cyclic forces on the ground for aspects such as drilling, anchor placement and soil compaction, among others [1].

During this type of work, workers are continuously exposed to risks, where the misuse of machinery or carelessness in the workplace can cause accidents of varying degrees [2]. In drilling works, the machinery can exert thrust forces of between 2 and 9 tons and rotational forces between 100 and 1,200 rpm. These characteristics cause risks of accidents associated with: (1) hitting the machine head (this element moves vertically at low speed but with high forces and could hit a worker's head); (2) entrapment of clothing, hair or a worker's limb by the rod; (3) crushing and/or blows from the rod, which can potentially break and fall from a height onto the worker; and (4) entrapment by machine jaws, which trap and hold the entire weight of the rod and can cause a worker's limb to be crushed [3].

High accident rates characterize the construction sector [4]. One of the most important aspects to help reduce these rates is training workers in occupational risk prevention. This process aims to create a culture of prevention through training that generates genuine awareness among workers [4]. However, training in accident prevention in construction has not been entirely effective, with deficiencies identified in both the content taught and the teaching–learning methods [5]. In the context of construction 4.0 and digitization trends in

WIT Transactions on The Built Environment, Vol 206, © 2021 WIT Press
www.witpress.com, ISSN 1743-3509 (on-line)
doi:10.2495/SAFE210121

the construction industry aligned to building information modelling (BIM) [6], virtual reality (VR) and game engines provide successful alternatives to improve regular teaching of accident prevention in construction [7], by creating customised contents for each specific construction site, machinery to be used, tasks and profiles of the workers. In the field of geotechnics, virtual reality has been used to train workers in the handling of earthmoving machinery, the functions performed with these machines, and elements of accident prevention associated with their use [8]–[10].

Thus, considering the risks associated with geotechnical drilling works, the training needs and the possibilities offered by virtual training for these purposes, this study aims to create a virtual training environment to prevent occupational hazards in geotechnical drilling works. It contains a minimum work team, comprising a drilling machine and two operators, and recreates some of the most common accident types in these works. The simulation seeks to complement the current theoretical and digital training by providing an experience closer to reality thereby making it easier and more intuitive to apply the knowledge acquired through training to a real working environment.

2 RESEARCH METHODOLOGY

The Design Science Research Methodology (DSRM) has been adapted to guide the entire process of project justification, development and testing [11], with the following four stages: (1) identifying observed problems and motivations; (2) defining the objectives of a potential solution; (3) design and development; and (4) demonstration. Fig. 1 shows a diagram of the research methodology; in the first stage, a background review was carried out to identify the main risks involved in geotechnical drilling works, the types of training used and their deficiencies, in addition to how VR can improve the training. This literature search was conducted using the Scopus and Web of Science databases, together with handbooks and technical guides. In the second stage, the objective of a potential solution was defined: how the use of VR can help security training to geotechnical drilling works. In the third stage, the conceptual proposal was created, and the scenario and tools for virtual interaction were developed. Finally, in the fourth stage, the VR experience was implemented and evaluated.

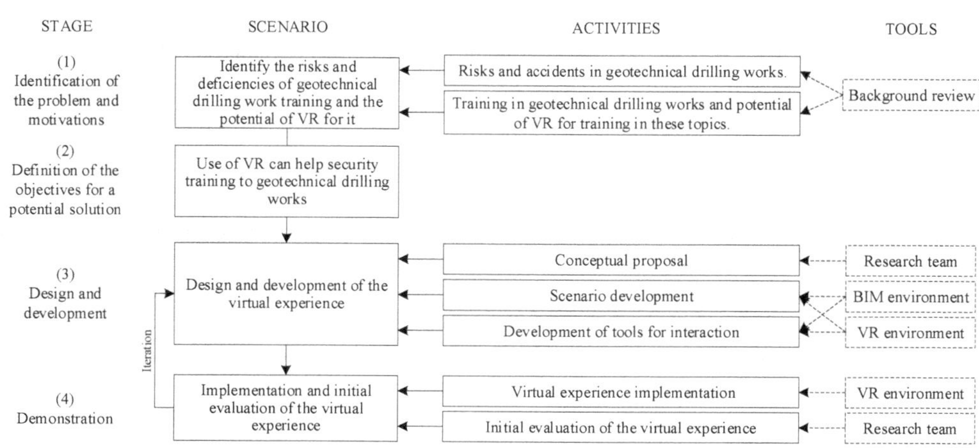

Figure 1: Research methodology.

3 DESIGN AND DEVELOPMENT OF THE VIRTUAL EXPERIENCE

The experience recreates a geotechnical drilling scenario. The user must perform the work of drilling and extracting soil cores, assisted by a virtual supervisor (driller) who will give the user the instructions of the process. The worker will then perform the process as they would in a real environment; however, the supervisor will warn of actions that could lead to an accident. In addition, if the worker suffers an accident in the virtual scenario, the supervisor will then provide recommendations to avoid the accident and perform the job.

3.1 Conceptual design, framework and tools for implementation

Fig. 2 shows the result of the conceptual design of the virtual experience, including the development of a narrative and related cause–effect relationship tree. In the initial main tab, the user will find the instructions and controls for how to use the simulation. At the beginning of the simulation, the driller then prompts the user to follow a series of instructions to finish drilling. The simulation is divided into two main processes: drilling and extraction.

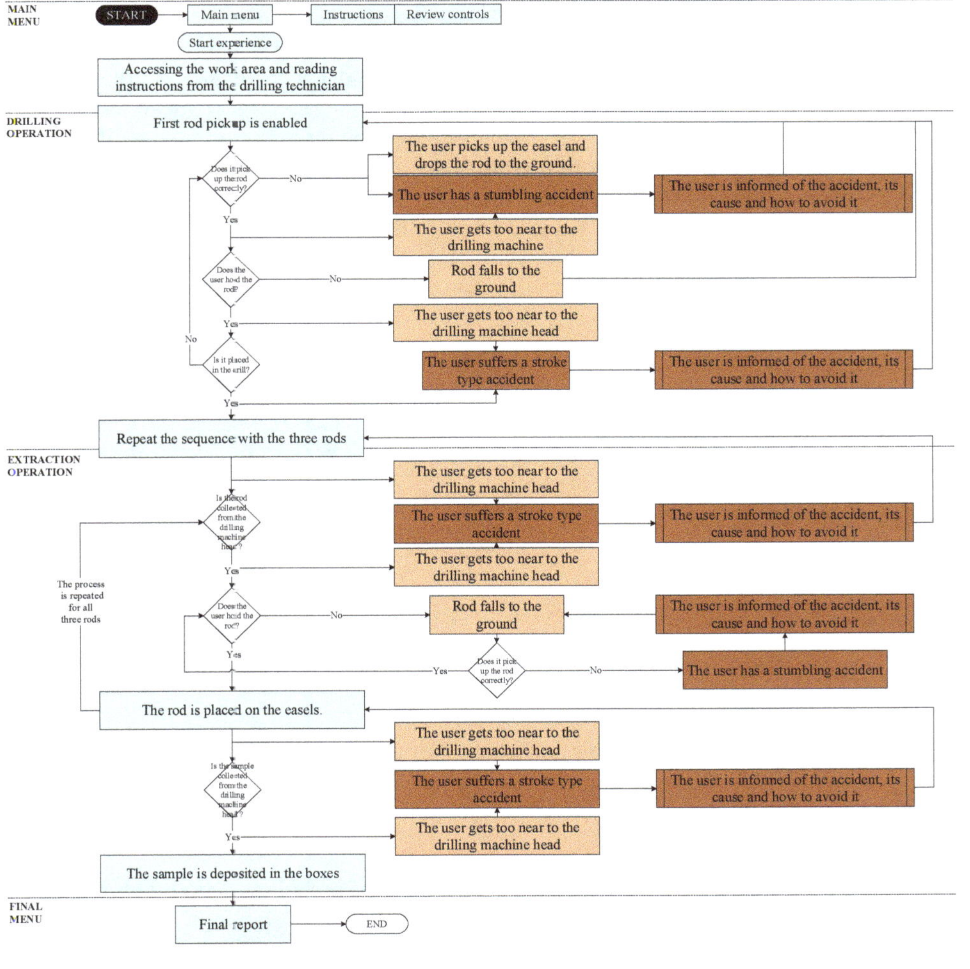

Figure 2: Conceptual design of the virtual experience.

In the simulated drilling process, the user must search for and locate three rods, in the order indicated by the driller, strategically distributed around the working perimeter to finish placing them in the drill head. This could be changed depending on the kind of work the trainer would like to teach. Each rod has a potential trip hazard associated with it. In addition, rods two and three add a further associated risk; in locating the second one, there is the possibility of being run over by the machine. For the third rod, it is necessary to position oneself correctly so as not to catch the trestle instead of the rod, which would cause the rest of the rods to fall to the ground. In each instance when the user approaches to place the rods, they risk being hit by the machine head if they get too close. Then, in the extraction process, all the rods should be removed and the drilled sample (in the form of a hard rock core) recovered to be placed in boxes. The user must pick up the rods from the drill and place them on the trestles. The process of approaching the drill involves a risk of hitting the drill head. Finally, the rock core must be retrieved and deposited in the sample boxes; this action involves the potential risk of tripping over the boxes themselves.

Fig. 3 shows the steps to implementation of the VR experience. In step one, we obtained technical documentation of geotechnical drilling machinery uses and associated risks. Based on this information, the conceptual design shown in Fig. 2 is created. In step two, a drilling machine is modelled, considering realism in terms of its appearance and accurate joints for realistic movement, selecting the level of detail as a balance between completeness and lightness of the application. In step three, a real construction scenario is generated in the VR development environment, incorporating that machinery, and programming the different user interactions to operate it, sequences of possible user actions and interactions with the virtual environment in general. Finally, in step four, the virtual experience is compiled for the specific platform (PC, smartphone, VR glasses). For this particular development process, the tools used included Autodesk Revit, Sketchup 3D, Unity 3D and Visual Studio (C#).

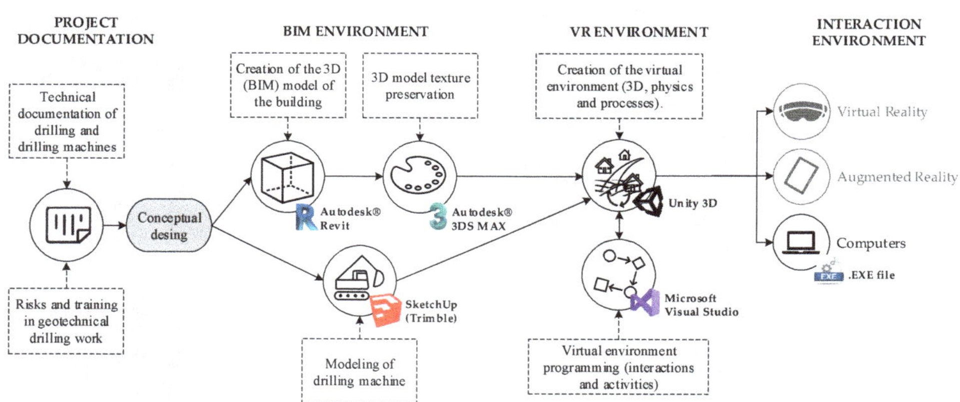

Figure 3: Workflow for creation of the virtual experience.

3.2 Drilling machine model

Since the main element of the training experience is the user's interaction with the drilling machine, a specific and typical drilling machine is modelled, as shown in Fig. 4. According to the objects defined for the virtual experience developed, several elements have not been considered. In this sense, with the objective of training in risk prevention associated with the use of machinery, six main parts of the model are highlighted and differentiated:

- Rotation head, responsible for transmitting the rotation and thrust forces to the drill string, and therefore, to the drilled ground.
- Penetrometer, a mobile sub-component that allows execution of the SPT (standard penetration test), sometimes also used for drilling.
- Rod assembly, composed of the drilling battery and the rest of the rod, which allows the transfer of forces from the rotating head to the ground and extraction of the drilled sample.
- Control panel, where each of the levers and regulators have been modelled, allowing their movement as necessary; these levers are responsible for driving almost all the functionality and movement of the machine.
- Clamps and retainers, where each of the four pistons and their respective clamping pads were modelled independently, which allow the rod assembly to be held during the drilling and extraction process, as well as screwing and unscrewing the rod when necessary.
- Running gear, made of rubber and modelled independently of the chassis, and reducer, present in the drilling machine to allow relative movement during displacement of the machine.

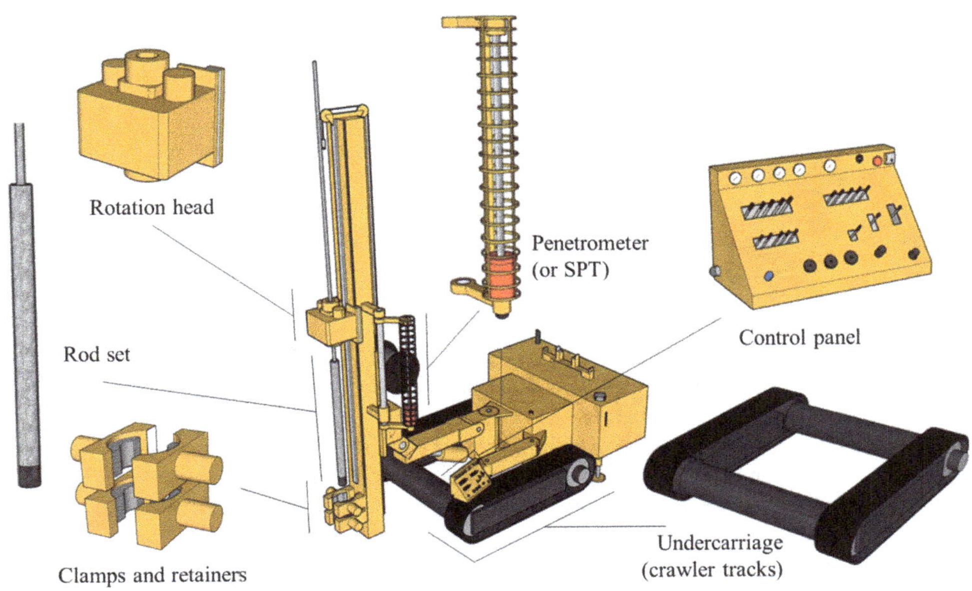

Figure 4: Drilling machine model in Sketchup®.

The second key virtual model is the construction scenario where the drilling machine is placed. In this case, different zones are marked depending on the risks (not visible to the user, but to code the different worker's interaction), as shown by the letters in Fig. 5(a), together with the environment of the construction site in Fig. 5(b)

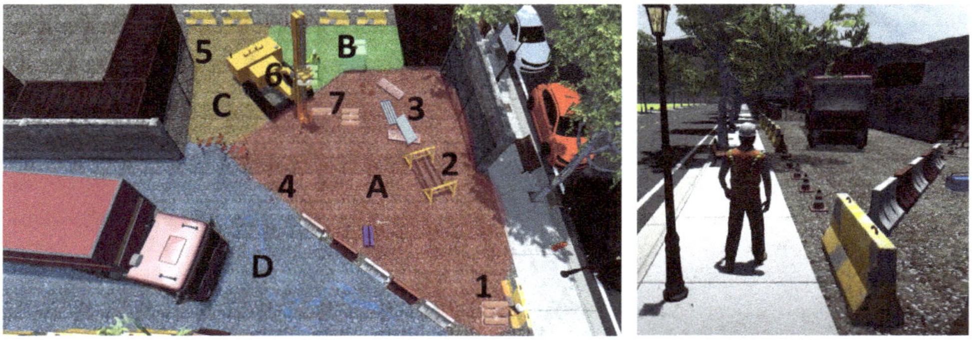

(a) (b)

Figure 5: Virtual scenario. (a) Work zones; and (b) General environment.

The four zones are as follows: (A) Drilling zone: located at the front of the drilling machine, this area is the main work zone of the assistant and should be the main attention zone of the driller; (B) Control area: this area contains the control panel that controls almost all the functions of the drilling machine; the driller is in charge of managing these controls and must remain in this area to ensure control of the machine at all times during drilling. The rest of the perimeter of the drilling machine is only frequented by the assistant occasionally, including zone (C), i.e. the low visibility zone. This zone does not require continuous attention by the operators for this reason, and even though the machine is considered static when in working order, this zone is also included in the safety perimeter of the drill; (D) Work zone outside the drilling area. Zones A, B and C form the minimum safety perimeter where access is forbidden to any person except the drill site operators. Based on the zones shown in Fig. 5, Table 1 provides a description of each of these zones and the potential accidents associated with each of them, as considered in the simulation.

It is important to note that this classification can be easily modified and extended by the programmer and prevention technician, depending on the building site, kind of machinery and workers profile.

4 DEMONSTRATION OF THE VIRTUAL EXPERIENCE

Fig. 6 shows the simulation sequence to illustrate the final result of the implementation and potential of the tool to train workers.

After the typical settings and instructions from the main menu, the worker enters the work zone and receives instructions from the supervising driller (Fig. 6(a)). Three incorrectly performed user actions are then shown: in Fig. 6(b), a trestle has been picked up instead of a rod, causing the rod to fall to the ground and in Fig. 6(c) and 6(d), a tripping accident has occurred by approaching too close to the boxes and a rod respectively. In both accidents, a message is displayed at the bottom of the screen explaining what happened, the cause of the accident and how to avoid it. In Fig. 6(e), the user has correctly picked up the rod from the floor and is about to place it in the drill. Fig. 6(f) and 6(g) show the sequence of a hit-and-run accident, where the user is initially positioned in an area of poor visibility and is warned by a message in the lower area of the screen. In the latter image, the effect is shown of a user continuing to approach the drill until impacting with it, suffering an accident that is then notified on-screen. Fig. 6(h) shows the correctly executed action of depositing a rod on the trestles to keep the work area tidy. Fig. 6(i) and 6(j) show, once again, the sequence of a hit-

Table 1: Zones and potential accidents in virtual environments.

N	Name zone	Description	Associated accidents
A	Drilling area	Work area containing tools, soil core boxes and rods.	Stumbling type, with rods and boxes. Tapping type, with the drilling head.
B	Control area	Usual area of the drilling technician.	No associated risks.
C	Low visibility zone	Opposite to the helper and drilling technician, with respect to the drilling machine.	Contains a rod potentially liable to cause an accident due to low visibility.
D	Outside work zone	Zone of other construction site interactions.	No associated risks.
1	Helper	Character with which the user interacts.	Not a risk in itself.
2	Trestles and rods	Rods must be collected to keep the work area tidy.	A tripping accident can occur, caused by the rod falling after picking up the trestle.
3	Soil core boxes	Used to store the final soil core obtained by the character.	Incautiously approaching will cause a tripping accident.
4	Frontal risk zone	Warns the user to increase their caution.	Not a risk in itself.
5	Rear risk zone	Warns the user of the danger zone.	Approaching the machine too closely will result in a run-over accident.
6	Drilling machine	Drilling and extracting soil core samples.	Accident type involving striking by the drilling head.
			Accident type hitting with the drilling head.
			Rear run over type accident.
7	Rod	Rod must be picked up and put into the drilling machine.	Excessive approach without picking up the rod will result in a tripping accident.
	Rod	Rod must be picked up and put into the drilling machine.	Approaching too close without picking up the rod will result in a tripping accident, as well as being located in a low visibility area associated with a potential run-over accident.

and-run accident, in this case involving the drilling head. Fig. 6(k) and 6(l) show the user retrieving the last rock core from the borehole and placing it in the boxes; in addition, the final message from the driller notifying the user of completion of the borehole is also shown.

Once the user has completed the entire simulation, an ideal route can be viewed in which no accidents occur, and the one taken by the user. Notably, there are also alternative scenarios in which various accidents can occur, up to a total of three different types, i.e. tripping, being

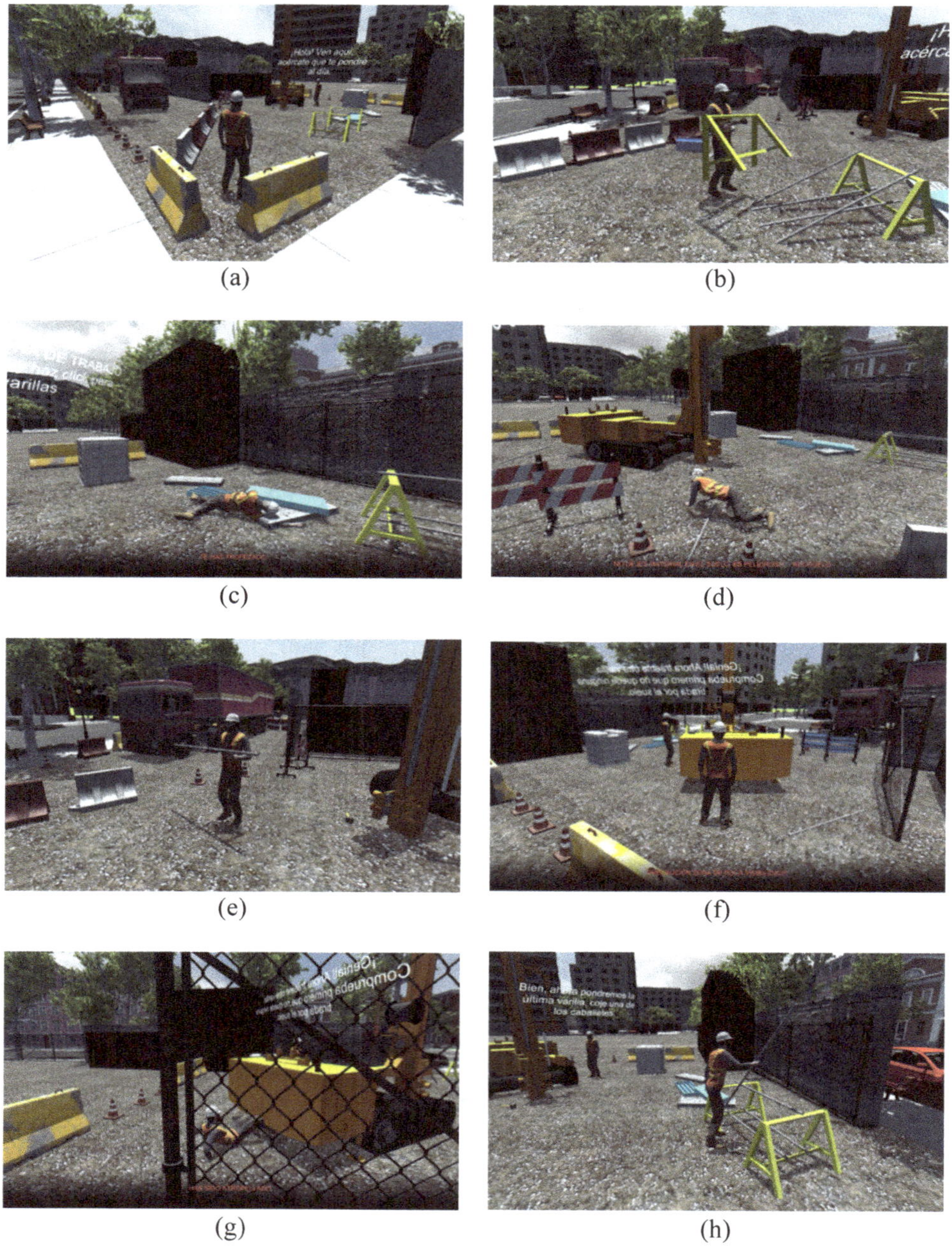

Figure 6: Scenes in virtual experience route. (a) Beginning of the simulation and drill operator's instructions; (b) Lifting the trestle; (c) Falling with the boxes; (d) Falling with a rod; (e) Lifting a rod; (f) Rear area of the drill; (g) Accident with the drill; (h) Animation leaving a rod in the trestles; (i) Rod extraction process; (j) Accident with a drill head; (k) Obtaining the rock core; and (l) Completion message.

(i)

(j)

(k)

(l)

Figure 6: Continued.

run over and being hit. In the case of such accidents, the user is informed and warned about how to avoid them in future. The simulation can also be started or terminated at any time.

5 CONCLUSIONS

A virtual training experience for geotechnical drilling work was designed and implemented. Three relevant aspects are highlighted: the realism and functionality of the geotechnical drilling machine; the generation of a realistic work environment that allows realistic immersion; and the implementation of user interaction with designed actions and potential accidents. As a main conclusion, customised training criteria for the use of the machinery, cleanliness of the working area, accident prevention, and mastery of the work area are all considered. The potential for adapting virtual reality tools for training in diverse contexts and types of work in the construction sector has been demonstrated, thanks to the re-use of BIM models with multiplatform videogames platforms. To sum up, this research shows the conceptual and technological development of customised virtual training for geotechnical drilling works.

ACKNOWLEDGEMENTS

This work has been supported by the CONICYT through its economic support of Felipe Muñoz, beneficiary of a pre-doctoral grant (CONICYT – PCHA/International Doctorate/2019-72200306) and by the Ministry of Science, Innovation and Universities of Spain (MICIU) through the BIMIoTICa project (RTC-2017-6454-7). The authors also acknowledge financial support from the Spanish Ministry of Economy and Competitiveness, through the "Severo Ochoa Programme for Centres of Excellence in R&D (CEX2018-000797-S)".

REFERENCES

[1] Qian, Q. & Lin, P., Safety risk management of underground engineering in China: Progress, challenges and strategies. *J. Rock Mech. Geotech. Eng.*, **8**(4), pp. 423–442, 2016. DOI: 10.1016/j.jrmge.2016.04.001.

[2] Luo, Z., Zeng, L., Pan, H., Hu, Q., Liang, B. & Han, J., Research on construction safety risk assessment of new subway station close-attached undercrossing the existing operating station. *Math. Probl. Eng.*, **2019**(March), 2019. DOI: 10.1155/2019/3215219.

[3] Kale, Ö.A. & Eskisar, T., Work-related injuries and fatalities in the geotechnical site works. *Ind. Health*, **56**(5), pp. 394–406, 2018. DOI: 10.2486/indhealth.2017-0166.

[4] Teizer, J., Right-time vs real-time pro-active construction safety and health system architecture. *Constr. Innov.*, **16**(3), pp. 253–280, 2016. DOI: 10.1108/CI-10-2015-0049.

[5] Kassem, M., Benomran, L. & Teizer, J., Virtual environments for safety learning in construction and engineering: Seeking evidence and identifying gaps for future research. *Vis. Eng.*, pp. 1–15, 2017. DOI: 10.1186/s40327-017-0054-1.

[6] Muñoz-La Rivera, F., Mora-Serrano, J., Valero, I. & Oñate, E., Methodological–technological framework for construction 4.0. *Arch. Comput. Methods Eng.*, **28**(2), pp. 689–711, 2021. DOI: 10.1007/s11831-020-09455-9.

[7] Mora-Serrano, J., Muñoz-La Rivera, F. & Valero, I., Factors for the automation of the creation of virtual reality experiences to raise awareness of occupational hazards on construction sites, pp. 1–21, 2021.

[8] Kayhani, N., Taghaddos, H., Noghabaee, M. & Hermann, U., Utilization of virtual reality visualizations on heavy mobile crane planning for modular construction. *ISARC 2018 – 35th Int. Symp. Autom. Robot. Constr. Int. AEC/FM Hackathon Futur. Build. Things*, 2018. DOI: 10.22260/isarc2018/0170.

[9] Vahdatikhaki, F., El Ammari, K., Langroodi, A.K., Miller, S., Hammad, A. & Doree, A., Beyond data visualization: A context-realistic construction equipment training simulators. *Autom. Constr.*, **106**(May), 102853, 2019. DOI: 10.1016/j.autcon.2019.102853.

[10] Su, X., Dunston, P.S., Proctor, R.W. & Wang, X., Influence of training schedule on development of perceptual-motor control skills for construction equipment operators in a virtual training system. *Autom. Constr.*, **35**, pp. 439–447, 2013. DOI: 10.1016/j.autcon.2013.05.029.

[11] Peffers, K., Tuunanen, T. & Gengler, C., The design science research process: A model for producing and presenting information systems research. *J. Manag. Inf. Syst.*, pp. 83–106, 2006.

TRUST CASE AND THE LINK TO SAFETY CASE

TOR STÅLHANE[1] & THOR MYKLEBUST[2]
[1]NTNU, Norway
[2]SINTEF, Norway

ABSTRACT
The TrustMe project develops a safety case for autonomous buses. A safety case is mostly based on information from the developers and refers to one or more relevant safety standards. The bases for a safety case are the defined safety standards and proof of compliance, based on the paper trails left by each required activity. A trust case is different and trust and safety assessments are not necessarily correlated. In order to make self-driving buses a success they need to be considered trustworthy. Thus, we need a "Trust case". To ensure that the vehicles are safe and to inform the public, we have developed both a developer safety case and safety case for the public. To take care of the remaining factors we have developed a "Trust case". The trust case has been developed as part of literature studies, surveys and interviews. We have made a survey of 311 passengers and interviewed 18 autonomous bus passengers. Based on literature studies, surveys and interviews, we have proposed a set of issues that should be included into a "Trust case". By providing the public with a "Trust case" together with a "safety case for the public" we will help manufacturers of autonomous vehicles and operators to gain public trust.
Keywords: trust, technology acceptance models, autonomous vehicles, safety.

1 INTRODUCTION

Both safety cases and trust cases are assurance cases. Piovesan and Griffor [1] define an Assurance case as "A structured argument, supported by evidence, intended to justify that a system is acceptably assured relative to a concern". The trust case is one type of justification – the justification of trust in travelling with autonomous buses. ISO/IEC 15026-1:2013 – Systems and software assurance – defines an assurance case as a reasoned, auditable artefact that supports the contention that its top-level claim or set of claims are satisfied. This includes systematic argumentation and its underlying evidence and explicit assumptions that support the claim(s). An assurance case contains the following:

- one or more claims about properties;
- arguments that logically link the evidence and any assumptions to the claim(s);
- a body of evidence and possibly assumptions supporting these arguments for the claim(s);
- justification of the choice of the top-level claim and the method of reasoning.

Trust is used in several ways, depending on the application area. According to Frederiksen [2] trust and risk should be considered different ways to manage uncertainty. According to Perrow, trust may be used in two ways [3]:

- Reliability trust: trust using experiences made in former interactions to assess the degree of uncertainty that is associated with a specific transaction partner, e.g., the subjective probability that a transaction with this partner, e.g., a trip with a bus, will be successful.
- Decision trust: the extent to which an entity is willing to enter into a transaction (interaction) with another.

Trust is in our case mainly reliability trust or lack thereof. The degree of trust will decide whether you will use a particular service or not. In some fora, trust is used in a rather informal

WIT Transactions on The Built Environment, Vol 206, © 2021 WIT Press
www.witpress.com, ISSN 1743-3509 (on-line)
doi:10.2495/SAFE210131

way so that for instance trust, reliability and reliance are all used to identify the same thing. Trust can also be seen as a person-to-person relationship. In this case, trust is a relationship to social actors such as designers, creators and operators of technology.

If trust is a person-to-person relationship, we need to identify the persons that the users of self-driving vehicles need to trust. We also need to extend trust to include organizations such as the service provider or the software development company. For autonomous buses, this is often the vehicle manufacturer. As a starting point, we could choose the service provider, e.g., the bus operator, since he is the legally responsible person. As an alternative, we might consider the personnel at the company that built the software.

In everyday speech, safety and trust are different. As is explained in the Risk Communication Guidelines for Public Officials [4] "Safety is connected to a statistical safety analysis and is hard to grasp. For example, a scientist uses a one-in-a-million comparison to convey a specific risk measurement. Health experts understand this to mean that, given one million persons, there is one person who is at risk. To the non-technical person, however, the one person may be someone they know. The public will often personalize risk with the same conviction that most scientists depersonalize it". Trust is different from reliability – it cannot be estimated but can be based on previous experiences – own and others. In this paper we try to show a way to assess trust. Based on the trust-model use by Man et al. in their TAM model, we will build a trust case analogous to a safety case [5]. In addition to the trust case, the TrustMe project is also developing a safety case – aiming at the bus manufacturer – and a safety case for the public (see Myklebust et al. [6]).

2 THE TAM MODEL OF VENKATESH AND DAVIS

The ESREL 2021 paper on trust and self-driving buses [7] used the extended TAM-model by Venkatesh and Davis [8] (see Fig. 1). The main problem, as the TrustMe team sees it, is that perceived safety is not involved – at least not directly.

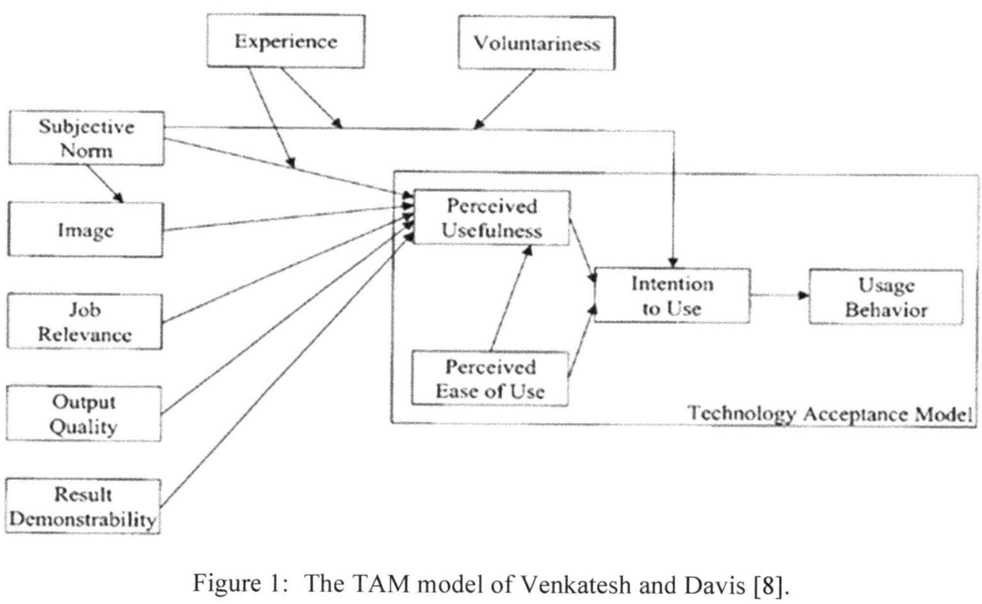

Figure 1: The TAM model of Venkatesh and Davis [8].

The term output quality needs to be clarified. Venkatesh and Davis use the following definition: "Output Quality is related to the tasks a system is capable of performing and the degree to which those tasks match their job goals (job relevance). People will take into consideration how well the system performs its tasks, which we refer to as perceptions of output quality. Empirically, the relationship between perceived output quality and perceived usefulness has been shown before. We expect output quality to be empirically distinct from, and to explain significant unique variance in, perceived usefulness over and above job relevance because a different underlying judgmental process is involved".

In addition to the issues included in output quality by Venkatesh and Davis, the TrustMe project wants to also include relevance for vacations and leisure time and thus replace "Job relevance" with "Transport needs relevance".

3 THE TAM-MODEL OF MAN ET AL.

One weakness with the TAM model of Venkatesh and Davis is that it does not consider trust except as a part of output quality. The TAM-model of Man et al. contains trust as a separate component which makes it easier to discuss this part of technology acceptance.

The dotted connection lines in Fig. 2 are connections that turned out to not be significant based on available data. Note that perceived privacy risk has no significant influence, neither on trust nor on perceived usefulness.

- Compatibility can be defined as the degree to which a technology complies with the needs and lifestyles of users.
- System quality refers to overall consumer perceptions of the excellence and effectiveness of a particular system. Note that this not the same as the quality-in-use as defined by ISO/IEC 25010:2011 – Systems and software engineering – since the standard includes "Freedom from risk" as part of quality in use. According to SQuaRE (Systems and software Quality Requirements and Evaluation), "freedom from risk" has the following components:
    - Economic risk mitigation – a product or system's ability to mitigate the potential risk to financial status, efficient operation, commercial property, reputation or other resources in the intended contexts of use.
    - Health and safety risk mitigation – a product or system's ability to mitigates the potential risk to people in the intended contexts of use.
    - Environmental risk mitigation – a product or system's ability to mitigate the potential risk to property or the environment in the intended contexts of use.

Note the difference between compatibility and system quality. Compatibility is related to customer needs while system quality is related to how well the system meets these needs. For the sake of comparison between the models of Man et al. and the model of Venkatesh and Davis, we will assume that:

- Compatibility is equivalent to subjective norm, image and job relevance. As mentioned in Section 1, TrustMe adds vacations and leisure time so that job relevance is replaced by transport-need relevance.
- System quality is equivalent to output quality (result) and demonstrability (seeing is believing).

In order to make it easier to compare the two previous models and based on the two assumptions above, we can rearrange the model of Venkatesh and Davis shown in Fig. 1 to the one shown in Fig. 3.

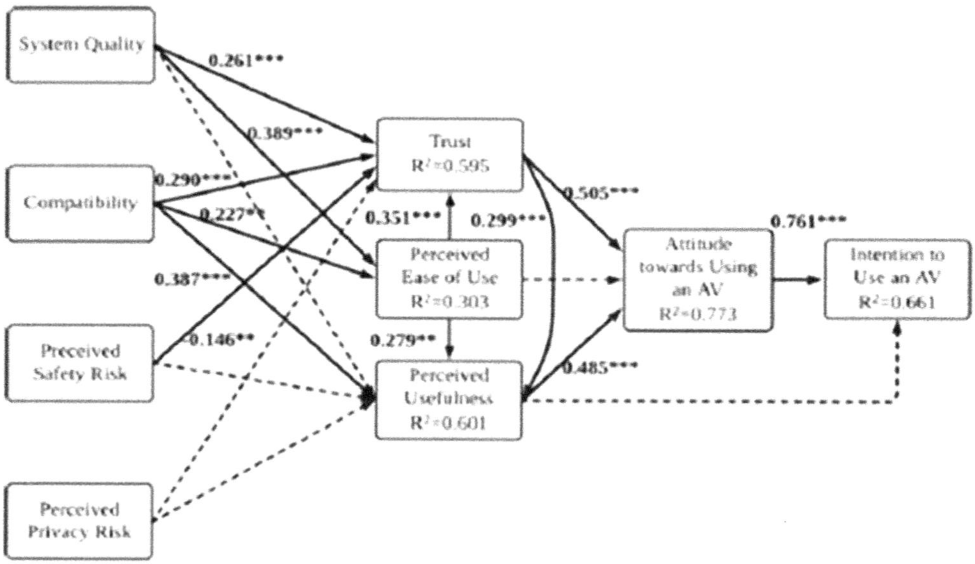

Figure 2: The TAM-model of Man et al [10].

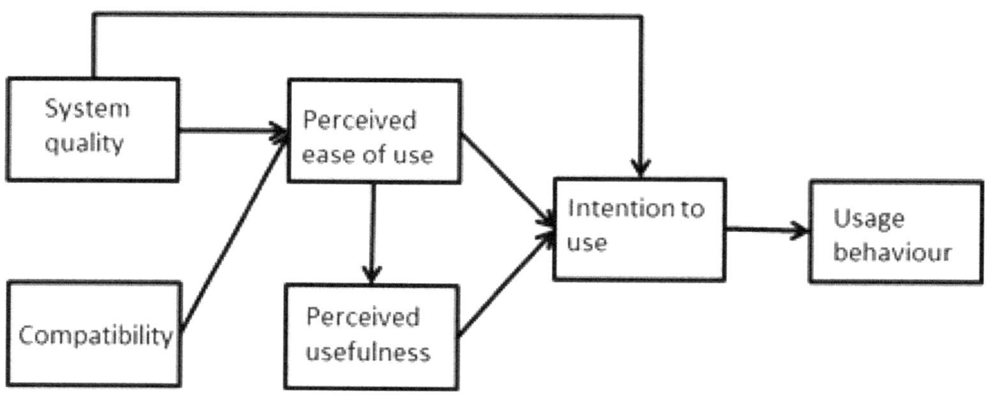

Figure 3: The TAM model of Venkatesh and Davis rearranged [8].

Health and safety risk mitigation is close to what we have called trust. There has, however, been little attention paid to environmental consequences. On the other hand, the potential for autonomous buses to create environmental damage is considered to be low.

4 TRUST ACCORDING TO ISO TR 24028

Earlier, safety and adhering to the relevant standards' requirements was considered to be sufficient for self-driving cars. With the arrival of artificial intelligence (AI) and machine learning (ML) in self-driving vehicles, trust has become an issue also here. According to the ISO TR 24028 "Information technology – Artificial intelligence – Overview of trustworthiness in artificial intelligence" standard, trust is a combination of:

- Physical trust – often synonymous with the combination of reliability and safety. This will be equivalent to "the risk to safety is low" in Bezai's mode (see Section. 5).
- Cyber trust – concerns often shift to IT infrastructure security requirements. According to the experiments of Zhang et al. [9] and Man et al. [10] neither security nor privacy is considered relevant for peoples' trust. However, Zhang et al. add that "…future studies should explore the role of other factors such as reliability, perceived cyber-security risk, liability concerns, and driving pleasure on acceptance".
- Social trust – based on a person's way of life, belief, character. In our case this is related to the service provider.

Social trust is vaguely related to subjective norms and image in the original TAM model by Venkatesh and Davis (see Fig. 1). However, this perspective is left out in the TAM of Man et al. and is replaced by system quality, which refers to overall consumer perceptions of the excellence and effectiveness of a particular system.

5 UNIFYING THE MODELS OF VENKATESH AND MAN

By adding trust and perceived safety risk to Venkatesh's model, we get the unified model shown in Fig. 4. The two models referred in Figs 1 and 2 allow us to include perceived safety and trust as model factors. The dotted lines show how trust and perceived risk are added to Venkatesh's model. From Fig. 4 we see that trust, in our opinion, depends on four factors:

- system quality;
- relevance – compatibility;
- perceived ease of use;
- perceived safety risk.

Even though some psychologists and sociologists separate trust, reliability and reliance, we will follow those who claim that (1) trust is confidence in or reliance on some person, organization or quality and (2) risk assessment and rational calculations reduce uncertainty and thus increase trust. However, this does not reduce the risks. For an engineer's point of view, risk mitigation will also play an important role but this will not affect the general public.

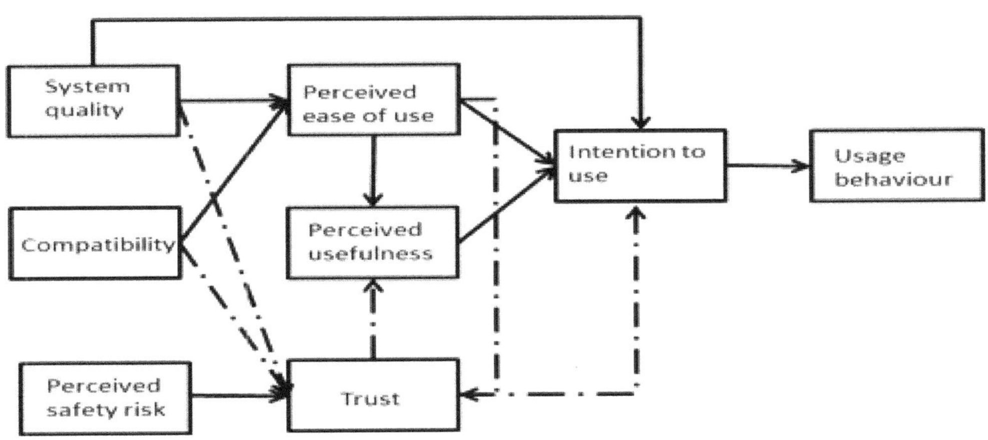

Figure 4: An attempt to unify the models of Venkatesh and Davis [8] and Man et al. [10].

The engineering view is simple and uses trust synonymous with reliability – an engineer trusts a component or system with high reliability. Thus, to build a trust case – analogous to a safety case – we need to convince the users that the four factors summed up above as influencing trust have been taken care of.

Bezai et al. [11] have developed a model that can be used to describe user's acceptance and behaviour. As we see, their model, shown in Fig. 5 has three main components – perception, vehicle performance and usage and cost. Our suggested model for a trust case has several ideas in common with this model, mostly related to vehicle performance and usage.

- Vehicle performance and usage: System quality, that the system satisfies our needs and that the system is easy to use
- Perception – the risk to safety is low and safety feeling condition

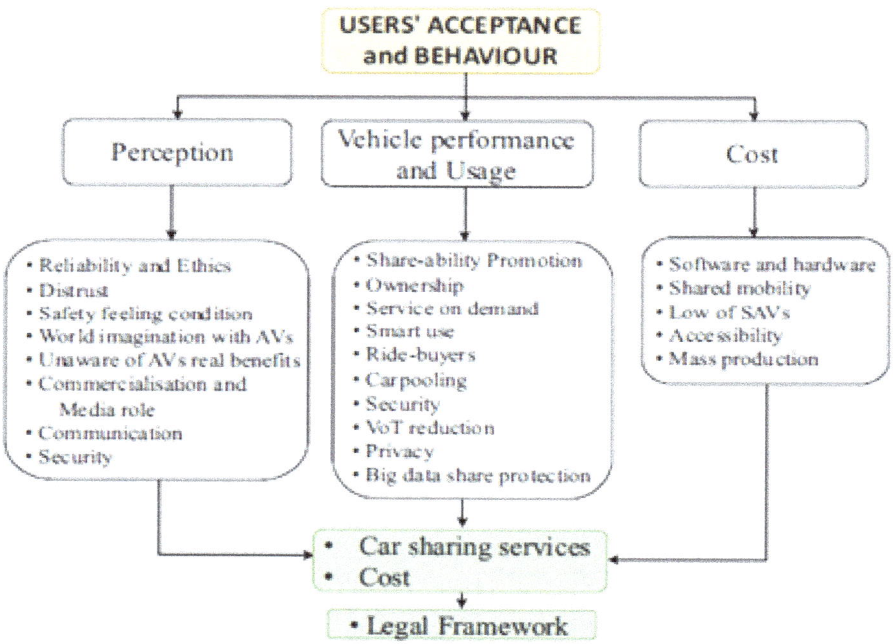

Figure 5: Bezai et al.'s model for user acceptance and behaviour [11].

The model of Bezai is useful, since it points out issues that we have not yet considered, e.g., VoT (Value of Time), ethics, commercialisation and media's role. Ethics is not relevant here for two reasons: (1) the autonomous buses considered here are level 3 (conditional driving automation) but will later move to level 4 (High driving automation). Most discussions on ethics are related to when the system should keep control and when the control should be handed over to the driver – a concept that is not relevant for self-driving buses. Ethics might, however, be relevant later when more public focus might be on the producers of autonomous buses (see also [12]). The model shown below also includes costs, which is not all that relevant for autonomous buses. In addition, the cost component in Bezai's model is mostly related to vehicle production costs.

Fig. 6 shows the four main components of trust according to the model in Fig. 3. The model may be refined by adding concepts related to each main component in the table below each main component. None of the TAM models mention environment. We need it here to cater to such things as weather, traffic density and road quality. Environment in Fig. 4 is equivalent to the ODD (operational design domain). When we use the term "The system is easy to use" we are referring to the system controlled by the computer, not the computer system itself.

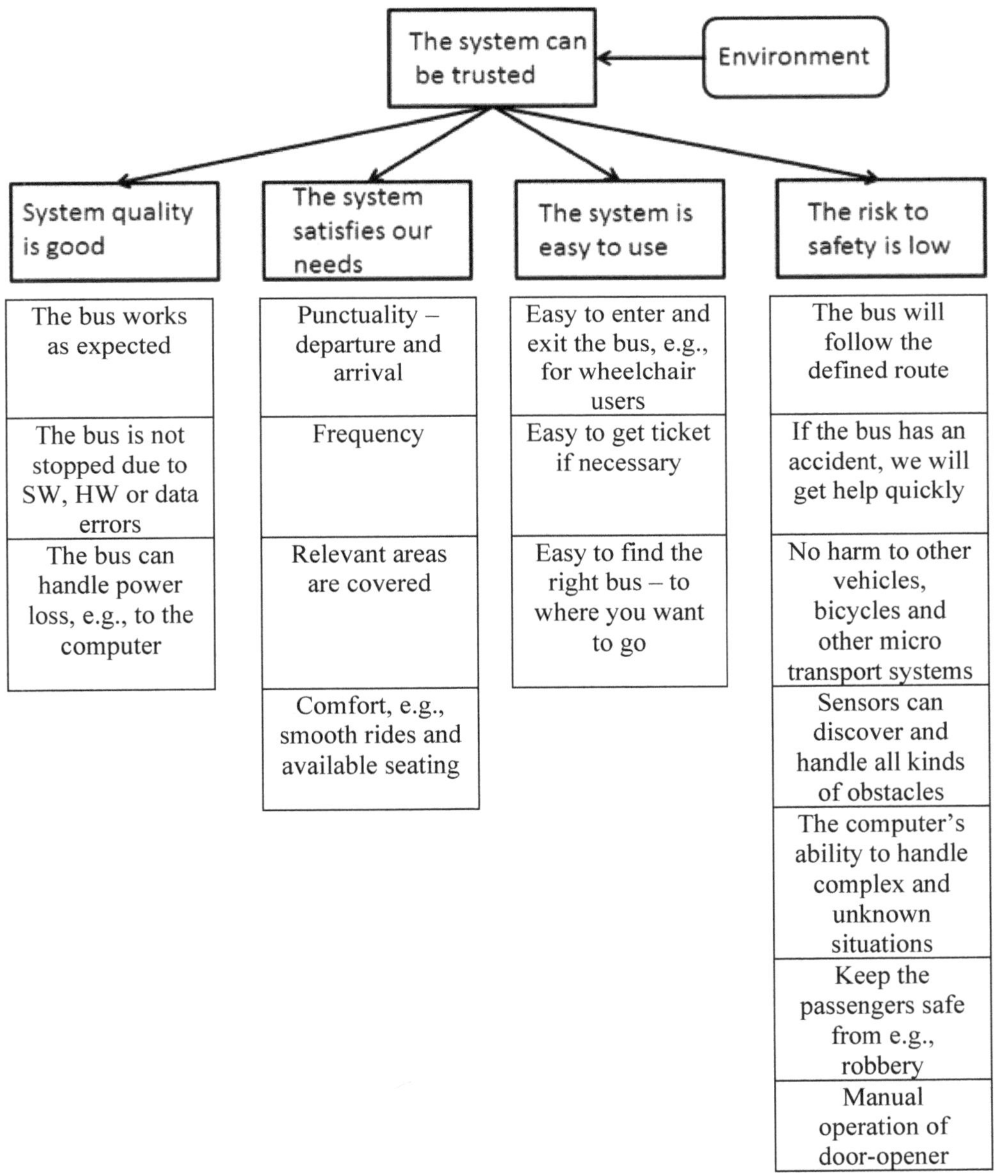

Figure 6: The trust case and the relevant topic.

It is tempting to equate "The risk to safety is low" with the safety case or the safety case for the public, but it is important to bear in mind that risk here is really reliability trust, i.e., the subjective probability that a transaction with this partner, e.g., a bus, will be successful while reliability in a safety case is related to proof of compliance with a standard. However, the model shown in Fig. 5 might be included in the safety case for the public as a spate component.

Note that security is not included in Fig. 6. This is mainly based on the results from Zhang et al. [9] and Man et al. [10] who included security and privacy in their experiment but its influence on neither trust nor perceived usefulness was statistically significant. In addition, security was mentioned only once during the two focus group meetings. In Fig. 5, we have included the issues identified during two focus group meetings in 2020 [13]. The meetings were arranged by AtB – a local bus service provider in Trondheim. As you can see, the majority of the issues are related to the risk to safety. If we want to combine the trust case and the safety case the trust case component "the risk to safety is low" will be the component that bridge trust case and safety case. The environment component will be part of both cases. Trust Case will be further developed, for instance in a 1–2 pages information sheet similar to the safety case for the public [6].

The experiments described in Zhang et al. [9] and Man et al. [10] both shows that security and privacy are not considered important for trust or perceived usefulness. The results are surprising, considering the strong focus on privacy and security we have lately seen in the media. Thus, we will perform a new survey on autonomous buses and be more explicit when it comes to security challenges.

6 CONCLUSIONS

One of the goals of the TrustMe project is to develop a complete Trust case for the public for autonomous buses. The model presented in Fig. 5 is the first step for this work. Both the technological and psychological factors are important. However, for the general public, the psychological factors are the most important ones, see, for instance, the quote from Risk Communication Guidelines for Public Officials [4]: "A scientist uses a one-in-a-million comparison to convey a specific risk measurement. Health experts understand this to mean that, given one million persons, there is one person who is at risk. To the non-technical person, however, the one person may be someone they know. The public will often personalize risk with the same conviction that most scientists depersonalize it".

Thus we need to focus more on this side of the acceptance of autonomous buses. The psychological factors are included in a trust case, as shown in Fig. 6.

ACKNOWLEDGEMENT

The "TrustMe" project is supported by The Norwegian Research Council, project number 309207 – IPOFFENTLIG19.

REFERENCES

[1] Piovesan, A. & Griffor, E., Reasoning about safety and security. *Handbook of System Safety and Security*, ed. E. Griffor, Elsevier, Chapter 7, 2017.

[2] Frederiksen, M., Trust in the face of uncertainty: A qualitative study of inter-subjective trust and risk. *International Review of Sociology*, **24**(1), 130–144, 2014.

[3] Perrow, C., *Normal Accidents: Living with High Risk Technologies*, Basic Books, p. 326, 1984.

[4] U.S. Department of Health and Human Services, Communicating in a crises: Risk Communication Guidelines for Public Officials. Public Health Service: Rockville.

[5] Myklebust, T. & Stålhane, T., *The Agile Safety Case*, Springer, 2018.
[6] Myklebust, T., Stålhane, T. & Jensen, G., *Safety Case for the Public*, ESREL, 2021.
[7] Stålhane, T., Myklebust, T. & Haug, I.S., Trust and acceptance of self-driving buses. Submitted to ESREL, 2021.
[8] Venkatesh, V. & Davis, F.D., A theoretical extension of the technology acceptance model: Four longitudinal field studies. *Management Science*, **46**(2), pp. 186–204, 2000.
[9] Zhang, T., Tao, D., Qu, X., Zhang, X., Lin, R. & Zhang, W., The roles of initial trust and perceived risk in public's acceptance of automated vehicles. *Transportation Research Part C*, 2019.
[10] Man, S., Xiong, W., Chang, F. & Chan, A.H.S., Critical factors influencing acceptance of automated vehicles by Hong Kong, *IEEE Access*, 2020.
[11] Bezai, N.E., Medjdouba, B., Al-Habaibeha, A., Chalal, M.L. & Fadli, F., Future cities and autonomous vehicles: Analysis of the barriers to full adoption. *Energy and Built Environment*, **2**, 2021.
[12] Narayanan, A., When is it right and good for an intelligent autonomous vehicle to take over control (and hand it back)? Department of Computer Science School of Engineering, Computer and Mathematical Sciences, Auckland University of Technology, New Zealand.
[13] Haug, I.S., The customers' experiences with self-driving buses, November 2020. (In Norwegian.)

SECTION 6
MODELLING AND EXPERIMENTS

3-D PRINTED PALM ADAPTORS FOR HAND–ARM VIBRATION MEASURES: TRANSMISSIBILITY PROPERTIES AND MISALIGNMENT REDUCTION WHEN USED WITH GLOVES

ANDREA ANTONUCCI[1], GIANLUCA COLTRINARI[1], PIETRO NATALETTI[2], RAOUL DI GIOVANNI[2] & DARIO LIPPIELLO[1]
[1]University of Rome 3, Italy
[2]Italian Workers Compensation Authority (INAIL), Italy

ABSTRACT
The estimation of the transmissibility of antivibration gloves requires the use of handheld adaptors for holding the accelerometer and assure that it is correctly positioned at the palm level of the hand. The marketplace proposes only few adaptors, specially made for hand–arm vibration measures, so researchers often had to make their proper homemade adaptors to performing their experiments on the gloves' transmissibility. A way to easily build hard-plastic adaptors is given by the possibility to use 3D printers, but in the literature, there is a lack of information about their properties. This paper investigates the properties of 3-D printed adaptors, confronting different materials among the most used (PLA, ABS, and Nylon), and evaluates possible changes of their performance at different infill densities. Results show that 3-D printed adaptors can be a valid alternative to those present on the market. At middle–low frequencies (10 Hz–400 Hz) their transmissibility values are inside the range ± 0.05. No significant differences were found between the transmissibility of those printed at 50% or 100% of infill, especially when the acceleration is W_h weighted. It has been also experimented a simple solution that aims to reduce possible misalignments of the adaptor when inserted inside the glove. Two external landmarks were designed in order to have a visual check of the position of the adaptor. The first, coming out from the glove through a little slit, was made for checking about the rotation on the Y–Z plane, while the second, parallel to the wrist, was made for checking rotation on the X–Z plane. The statistical analysis, among 30 measures performed by three operators holding the instrumented handle of a mono-axial shaker, supports the idea that the first landmark can help to achieve a better alignment of the adaptor, limiting its rotation along the Y–Z plane. Lesser evidence resulted on the reduction of the misalignment on the X–Z plane, due to the use of the second landmark. A possible reason is the small thickness of the ordinary leather glove used in the tests, which produces too little misalignments angles for being checked by the adaptor's landmark. Some further tests can be carried out using anti-vibration gloves, generally made of thicker and softer materials.
Keywords: antivibration gloves, hand–arm vibration, vibration transmissibility, handheld adaptors, 3D printers, PLA.

1 INTRODUCTION

Prolonged exposure to hand–arm vibration can be at the origin of several illnesses, generating a set of symptoms generally called hand–arm-vibration syndromes (HAVS). Many tools usually used in workplaces can generate high levels of vibration such as chainsaws, pneumatic rock drills, pneumatic breakers, impact wrenches, orbital sanders, etc. [1]–[3], especially when workers need to use these tools for several hours a day. It thus results difficult to limit the exposure to vibration, calculated as recommended by the ISO Standard 5349-1 [4], below the limits required in many countries [5], [6].

The use of personal protection equipment, such as antivibration gloves (AV), can help to protect workers. The ISO Standard 10819 [7] specifies the properties that gloves must-have in terms of transmissibility, evaluated as the ratio between the vibration measured at the palm level inside the glove and that measured in the same conditions in a bare-adaptor test. To be

WIT Transactions on The Built Environment, Vol 206, © 2021 WIT Press
www.witpress.com, ISSN 1743-3509 (on-line)
doi:10.2495/SAFE210141

labeled as an AV glove, the transmissibility must be lesser than or equal to 0.9 in the middle-frequency range (25–200 Hz) and lesser than or equal to 0.6 in the high-frequency range (200–1,250 Hz). Anyway, the efficacy of AV gloves is controversial, since it depends on many factors, often difficult to take all into account, both when the tests are performed as specified by the ISO 10819:2013 Standard, then much more, when the measurements are taken using real tools [8], [9].

Among all the different factors that can influence the evaluation of the transmissibility of gloves, two of them can introduce systematic relevant errors that can invalidate the results of the measures. These are the geometrical and mechanical characteristics of the adaptor used to locate the accelerometer at the palm level [10], [11], and the possible misalignments that the adaptor can have with respect to the direction of the vibration [12], [13].

The ISO 10819:2013 specifies the characteristics (shape, dimensions, weight, materials, and transmissibility) that handheld adaptors must have for being used in the measures of gloves' transmissibility. Among the allowed materials it is admitted also the use of hard plastic.

The marketplace proposes few handheld adaptors, specially made for hand–arm vibration measures, as specified by the ISO 5349:2001. For this reason, these adaptors often have dimensions dissimilar from those specified by the ISO 10819:2003 standard. Moreover, in most cases, they are designed to hold only a specific accelerometer. The errors introduced by these adaptors, even when used for hand–arm vibration measurements, are often greater than those made for gloves' transmissibility tests [10], [14], [15]. It follows that researchers often had to resort to making homemade adaptors to perform their experiments on gloves' transmissibility, often using metal materials [12], [14].

A way to easily build hard-plastic adaptors is the use of 3-D printers. Even though this solution has been already experimented in some tests on hand–arm vibration [11], [16], in the literature there is a lack of information regarding the properties that 3-D printed adaptors have in terms of transmissibility, both when different filaments are used and when changes the infill percentage, that is one of the main parameters required to be set when 3-D printers are used.

In the light of the previous considerations, this paper aims to investigate more deeply the properties of 3-D printed adaptors, confronting different materials among the most used, and evaluating possible changes of their performance at different infill densities.

Regarding the adaptors' misalignments and their consequences on the transmissibility's evaluation, several studies were made along the years, both estimating the magnitude of the errors [1], [13], and trying to find out solutions for their limitation, such as the use of triaxial accelerometers [17]–[20]. In this paper, we also experimented a simple solution that aims to reduce possible misalignment of the adaptor when inserted inside the glove for antivibration transmissibility tests.

2 MATERIALS AND METHODS

Starting from the dimensions provided by the ISO 10819:2013 [7], an adaptor with little modifications (1.45 mm higher) was designed in order to be able to contain a 10 mm cube accelerometer. Fig. 1 reports both the specified dimensions of the ISO Standard (in black) and the modifications made (in red). All the quotes indicated as "ISO" are the same prescribed by the 10819:2013 Standard. A 3-D model was then drawn and printed with an FDM (Fused Deposition Modeling) printer, using a 1.75 mm filament and 0.4 mm nozzle. Three different materials (PLA, ABS, and Nylon), each one printed with 2 different infill densities (50% and 100%), were used for making the adaptors. These materials, commonly used with 3-D printers, have different mechanical and chemical properties [21]. PLA

(Polylactic Acid) is a vegetable-based fully biodegradable plastic material, probably the most commonly used for FDM printers. It has higher stiffness compared to the others (Young's Modulus (E) = 800 MPa), but less strength (Ultimate Tensile Strength (σUTS) = 25 MPa). The ABS (Acrylonitrile Butadiene Styrene), is a thermoplastic polymer, more flexible than the PLA (E = 1.300 MPa), but it generally provides higher strength (σUTS = 45 MPa); while nylon provides the least stiffness (E = 1.500 MPa) but the highest strength (σUTS = 50 MPa). The weights of the printed adaptors at 100% of infill including the accelerometer, are as follow: PLA 14.3 g; ABS 12.4 g; Nylon 13.3 g.

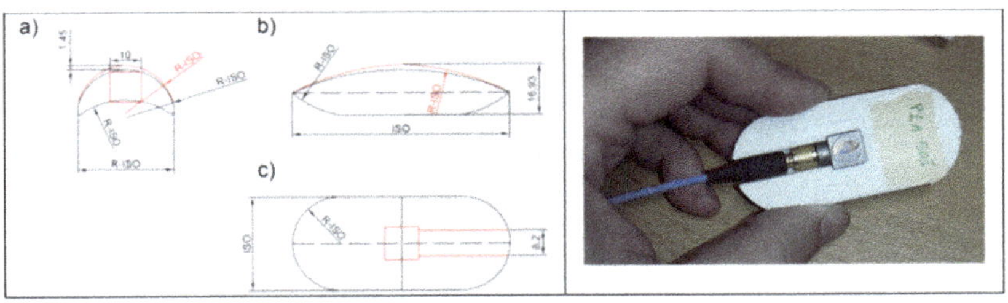

Figure 1: Dimensions of the printed adaptor.

We tested also other two adaptors. The first one was made of polypropylene (σUTS = 75 MPa; E = 1470 MPa; 10.5 g) fabricated using the injection molding technique, with sizes conforming to the prescribed dimensions of the ISO 10819:2013 Standard. The second was a commercial T-shape fingers-held metal adaptor, made of anodized aluminum, with dimensions of 90 × 22.5 mm, and a weight of 24 g. This adaptor is one of those generally provided on the market for hand–arm vibration measurements according to the Standard ISO 5349-2 [22].

A single-axis shaker (RMS-SW-1508/1) was used for the tests of the transmissibility. It was set to generate a broadband random vibration spectrum from 10 to 400 Hz, according to the acceleration spectrum shown in Fig. 2. Although the fundamental natural frequency of the instrumented handle of the shaker was indicated to be above the frequencies range prescribed by the ISO 5349-1 [4], during the set-up a resonance phenomenon was found on the instrumented handle around 800 Hz. For this reason, we limited the spectrum of the tests to 400 Hz, considering the results valid only for low–middle frequency vibrations.

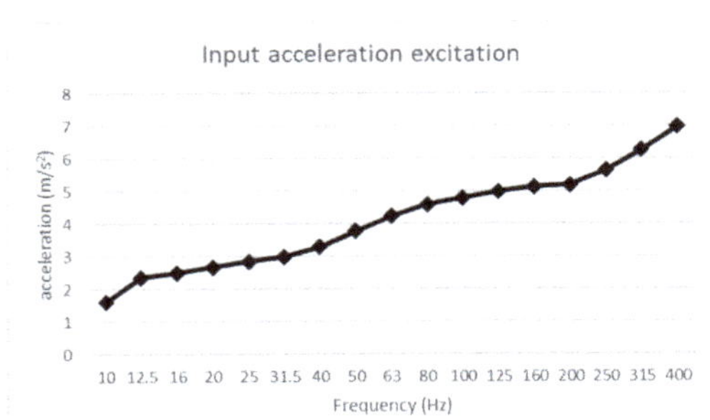

Figure 2: Broadband random vibration spectrum of the shaker.

Although the frequency range used in this experiment does not cover the one required by the ISO Standard, it results quite sufficient for the purposes of this study, since the vast majority of powered hand tools fall within this frequency range [23]. Moreover, the frequency weighting at the frequencies above 400 Hz is less than 4% of the peak weighting value.

Each adaptor was fixed to the instrumented handle using a hose metal clamp, putting a little rubber layer between both the contact surfaces, as shown in Fig. 3. The tightening force was fixed at 80 N through the use of the screw of the clamp. The value of the force was checked by two force sensors located inside the instrumented handle.

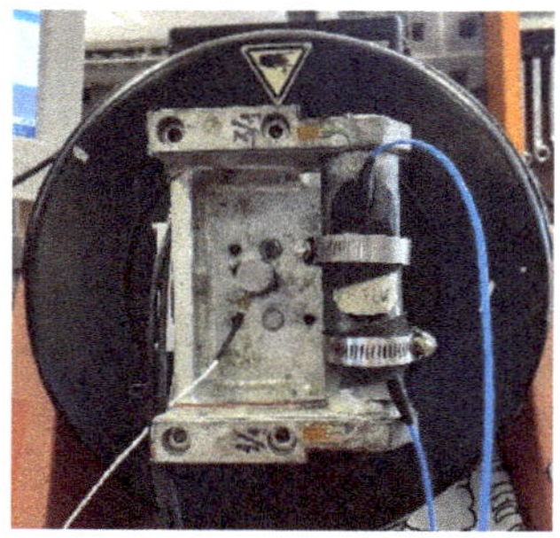

Figure 3: Instrumented handle.

For the measures of the vibration of all the printed adaptors and the metal one, a three axial 10 mm cubic accelerometer (SEN 040 F) was used, weighing 5 g; while for the test on the adaptor in polypropylene, it was used a three-axial accelerometer (SEN 026) 7 × 12 × 12 mm, weighing 3.1 g. In order to have a better estimation of the acceleration produced by the shaker, other than that located on the plate of the shaker, another 3 axes accelerometer (SEN 40 F; 10.2 × 19.6 × 10.2 mm; 5 g) was fixed directly on the external surface of the instrumented handle, close to the adaptor.

A total of five tests were performed for each of the adaptors, each one lasting 30 seconds. Although we used three-axial accelerometers, after having verified that the components of the acceleration along the X and Y axes were not significant, we considered only the Z-axis acceleration component for the estimation of the transmissibility of the adaptors. For each test, the baseline transmissibility at each frequency was computed from:

$$T_{R(f)\_A} = \frac{a_{z(f)\_Bare\_Adaptor}}{a_{z(f)\_Handle}}, \tag{1}$$

where $T_{R(f)\text{-}A}$ is the transmissibility of the adaptor at the single frequency (f), and for the same frequency, $a_{z(f)\text{-}Bare\text{-}Adaptor}$ is the acceleration along the Z axis measured by the accelerometer positioned inside the adaptor, and $a_{z(f)\text{-}Handle}$ is the acceleration along the Z axis of the accelerometer positioned on the instrumented handle.

The total average transmissibility among all the frequencies was then calculated using the following equations, in both of the cases of linear and W_h weighted acceleration:

$$\bar{T}_{RL\_A} = \frac{\sqrt{\sum_f \bar{a}^2_{z(f)\_Bare\_Adaptor}}}{\sqrt{\sum_f \bar{a}^2_{z(f)\_Handle}}}, \tag{2}$$

$$\bar{T}_{RWh\_A} = \frac{\sqrt{\sum_f \bar{a}^2_{z(f)\_Bare\_Adaptor} \cdot W^2_{h(f)}}}{\sqrt{\sum_f \bar{a}^2_{z(f)\_Handle} \cdot W^2_{h(f)}}}, \tag{3}$$

where in this case $\bar{a}^2{}_{z(f)}$ is the average acceleration of all the five tests for each material in the same conditions (bare adaptor and handle), and $W_{h(f)}$ is the frequency weighting factor defined in the ISO 5349-1 (2001) [4].

Another series of tests were also made to reduce the possible misalignment of the adaptors when used inside gloves. To facilitate the correct position of the accelerometer, when used with gloves, the ISO 10819:2013 suggests to place a slit in the seam of the glove between the thumb and index finger for visually check the position of the adaptor inside the glove. Starting from what suggested, and considering what already done in another experiment that needed external references to achieve a good position of the adaptor [11], we drew a protrusion along the longitudinal axis of the PLA 70% infill adaptor, for making easier to check for possible misalignments respect to the X–Z plane (Fig. 4).

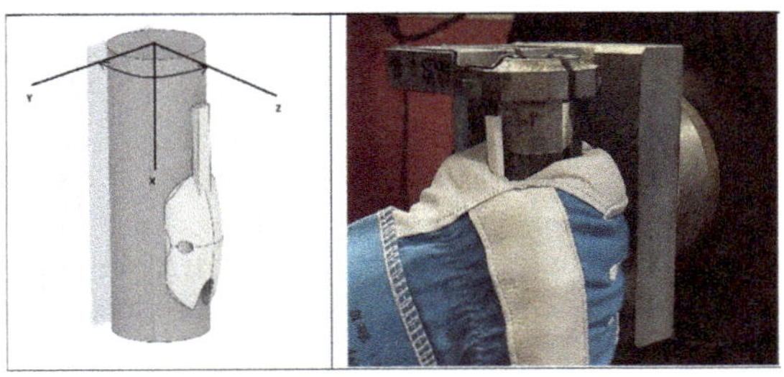

Figure 4: Adaptor's longitudinal landmark

Some other tests were also made using a second protrusion, draw along the transversal axis of the adaptor. In this case, the external referment was made for checking about possible misalignments with respect to the Y–Z plane Fig. 5. These misalignments could mainly happen due to the compression of the soft materials of which gloves are made [19].

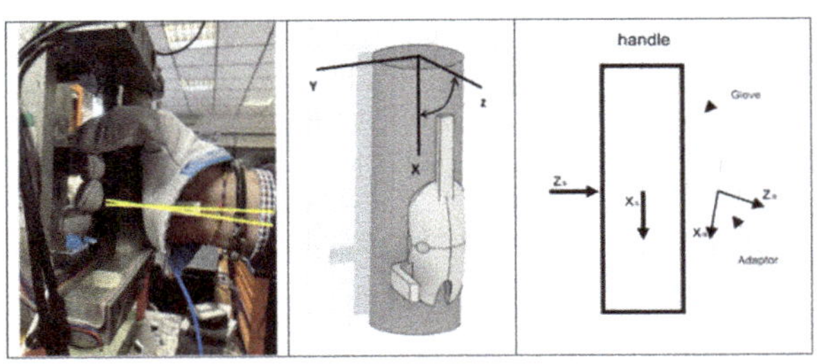

Figure 5: Adaptor's longitudinal and transversal landmarks.

Three healthy male subjects having hand size between 7 and 10, as specified by the EN 420-2003 [24], were recruited for the tests. The shaker was set to produce a vibration of 24 m/s^2 along the Zs axis at 200 Hz. The subjects were asked to maintain the same body and hand–arm postures as those required in the standardized glove test [7]. Each subject was asked to maintain a grip force of 80 ± 5 N during the tests, without applying any feed force.

The decision to not apply the feed force was taken for minimizing the non-axial vibration produced when the instrumental handle is coupled with the human hand together with a force plate [19]. A visual display of the force signals helped the subjects to maintain the same grip during the tests. An ordinary working leather glove was used for the tests, since its transmissibility is generally close to the unit. Moreover, this glove is not expected to attenuate low–middle frequency vibration [25].

In the first series of tests, the subjects were asked to put the adaptor inside the glove, at the palm level, as recommended by the ISO Standard [7]. No cut on the glove was made in this case for helping to visualize the position of the adaptor. A researcher was in charge to assist the subjects during this phase, providing help in positioning the adaptor inside the glove, so to be aligned as much as possible to the axis of the vibration excitation (Zs). All the subjects and the tests were randomized. Ten tests were made for each of the subjects, each time repeating the entire procedure.

The same protocol was then repeated other two times in order to test the efficacy of both the external referments connected to the adaptors. A little slit was made along the seam of the glove for allowing the longitudinal protrusion to come off. For the other external referment, it was not necessary to cut the glove, since it comes out along the wrist. Also in this case, a three-axial accelerometer was used (SEN 040 F) to measure the vibration at the palm level inside the glove.

Some preliminary tests were made for checking the transmissibility of the bare hand adaptor to be sure to minimize the non-axial vibration produced when the instrumented handle is coupled with the human hand. The misalignment of the adaptor, when positioned inside the glove, was evaluated confronting the values of the vibration along Xa and Ya axes, both in the case of the use and in the one of no use of the external adaptor protrusions.

3 RESULTS

3.1 Adaptor's transmissibility

Table 1 reports the average linear and W_h weighted transmissibility among the considered frequencies for all the tested adaptors, calculated using eqns (2) and (3). Every tested adaptor shows a good response both in linear and in weighted acceleration. The bigger error in linear acceleration was found on the metal adaptor (+5.2%), while the one made in polypropylene performed the best result (+0.3%). For the printed adaptors the errors were between +1% for the Nylon-50% and +2.2% for the Nylon-100%.

Table 1: Average linear and W_h weighted transmissibility of the tested adaptors.

	PLA 100%	PLA 50%	ABS 100%	ABS 50%	NYL 100%	NYL 50%	POLY	MET
$T_{RL\text{-}A}$	1.014	1.019	1.018	1.021	1.022	1.010	1.003	1.052
$T_{RWh\text{-}A}$	0.997	1.005	1.004	1.006	1.002	0.999	0.963	1.033

Regarding the weighted acceleration, the better transmissibility was performed by the printed adaptors. In this case, the range of error was between –0.1% for the Nylon-50% and +0.6% for the ABS-50%. The polypropylene adaptor showed the worst behavior producing an underestimation of the total transmissibility of –3.7%, while the metal adaptor performed an overestimation of +3.3%.

Among the printed adaptors, statistical analysis (ANOVA) reveals no differences at different infill densities for the PLA ($p = 0.06$) and ABS ($p = 0.65$), while differences were found among Nylon at 50% and 100% of infill density ($p < 0.015$). Anyway, as reported in Table 1, the differences are more formal than substantial, especially when the acceleration is W_h weighted.

Regarding the materials, ANOVA analysis confirmed the differences between PLA and Nylon ($p = 0.0029$) as well as between ABS and Nylon ($p = 0.0024$) but only at 50% of infill, while no difference was found between PLA-50% and ABS-50% ($p = 0.75$). No differences were found at 100% infill ($p = 0.15$) among all the printed materials.

Fig. 6 reports the average transmissibility values ($TR_{(f)\text{-}A}$) of all the tested adaptors at the single frequencies between 10 and 400 Hz. The printed adaptors showed a good response in terms of transmissibility, resulting between 0.95 and 1.05, as prescribed by the Standard ISO 10819:2013, in quite all the analyzed frequency range, as well as the adaptor made in polypropylene. The only cases in which the average transmissibility resulted over 5% were for ABS-50% at 400Hz ($TR_{(f)\text{-}A}$= 1.06) and for Nylon-100% at 400 Hz ($TR_{(f)\text{-}A}$=1.07). The metal adaptor on the other hand, revealed an increase of transmissibility over 5% for most of the frequencies over 80 Hz, with two peaks at 100 and 400 Hz (TR(f)-A=1.08).

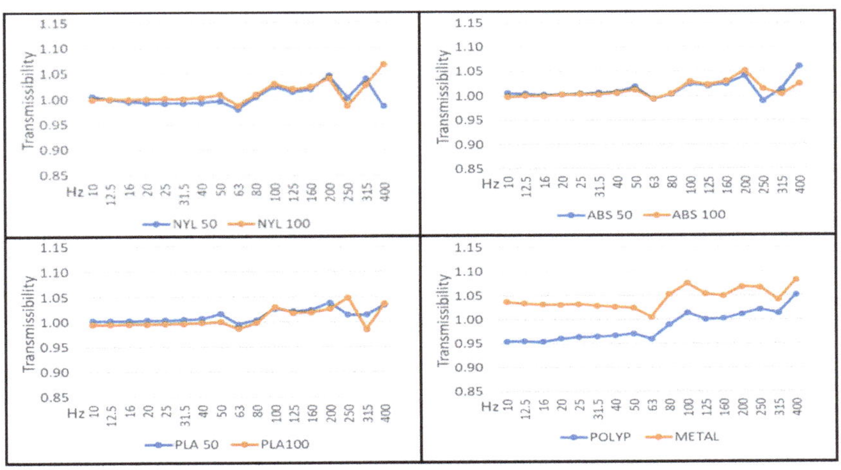

Figure 6: Transmissibility of the adaptors for all the tested frequencies.

All the printed adaptors have quite linear transmissibility, very close to the unit, from 10 to 80 Hz, then the values tend to be nonlinear, with a general tendency to increase as frequencies increase.

T-test was used for the frequency-stratified analysis. Also in this case, no statistical differences were found among the single frequencies between PLA-50% and PLA-100%, except at 315 Hz ($p = 0.001$). For ABS-50% and ABS-100% the only differences were found at 250 Hz and 400 Hz ($p < 0.004$), while in quite all the frequencies, except at 12.5 Hz, the transmissibility of the NYL-50% defers from NYL-100% ($p < 0.002$), showing, in this case, a little better transmissibility of the NYL-50% at lower frequencies.

The polypropylene adaptor tends to reduce the transmissibility in the range 10–63 Hz, as well as the metal adaptor amplifies the transmissibility in the range 10–50 Hz. The previous differences were confirmed by the statistical analysis ($p < 0.03$). In the range of frequencies above 100 Hz, the polypropylene adaptor seems to perform better transmissibility compared to those printed, even though the frequency-stratified statistical analysis does not confirm the differences in all the frequencies. The metal adaptor showed higher transmissibility compared to all the others, in quite all the frequency range ($p < 0.002$).

3.2 Adaptor's misalignments

Fig. 7 reports the component of the acceleration along the Y-axis (a_y) when the adaptor is positioned inside an ordinary working leather glove, for both the cases in which the external longitudinal protrusion was or was not used. As discussed above, the landmark connected to the adaptor should help the operator to limit the rotation of the accelerometer on the Y–Z plane. Even if some components along the Y-axis are always present when the instrumented handle is coupled with the hand [19], when the adaptor is correctly positioned inside the glove, the triaxial accelerometer should reveal mainly only the acceleration provided by the shaker along the Z-axis. On the contrary, if the adaptor is rotated along the longitudinal axis of the instrumented handle, the component of the acceleration increases proportionally along the Y direction [13]. Theoretically, the ratio of the acceleration in the Y direction (a_y) to that in the Z direction (a_z) is a tangent function of the yaw angle (α) [16] as reported in eqn (4).

$$\tan\alpha = \frac{\bar{a}_y}{\bar{a}_z}. \quad (4)$$

As it can be seen from the graphs reported in Fig. 7, for each of the operators, the component of the acceleration along the Y-axis was found to be greater when the longitudinal landmark was absent, in quite all the tests. Moreover, the acceleration along the Y-axis showed a greater variability among the tests when compared to that measured when the landmark is used.

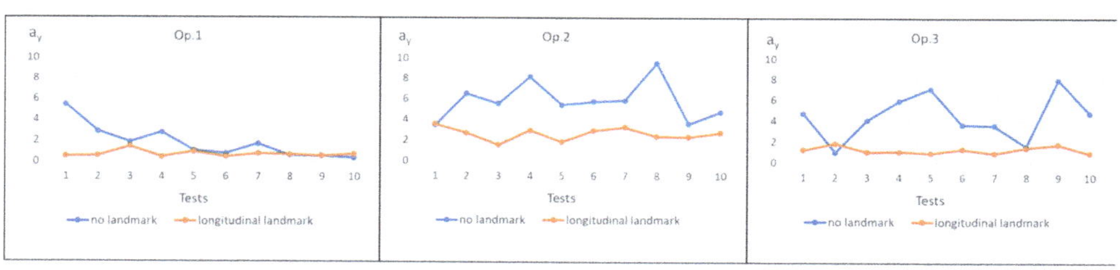

Figure 7: Component of the acceleration along the Y-axis (a_y) when the adaptor is positioned inside the glove.

The result supports the idea that the landmark can effectively help the operators for achieving a better alignment of the adaptor inside the glove, limiting the rotation along the Y–Z plane.

The ANOVA analysis confirmed the differences among the two series of tests for each of the subjects: operator 1 ($p = 0.043$); operator 2 ($p = 6.3E\text{-}5$); operator 3 ($p = 0.0002$).

Fig. 8 reports the component of the acceleration along the X-axis (a_x) when the adaptor was positioned inside the glove for both the situations: the one in which the external transversal protrusion was used and the one in which it was or not used. Note that the longitudinal protrusion was present also in these tests since it also helped in avoiding possible rotation of the adaptor on the X–Y plane (see Fig. 5).

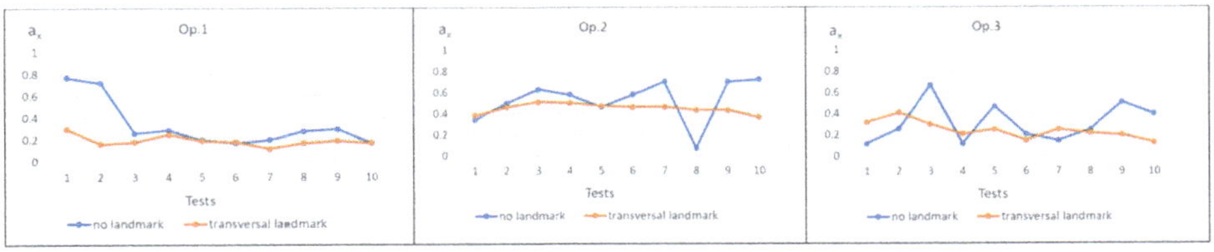

Figure 8: Component of the acceleration along the X-axis (a_x) when the adaptor is positioned inside the glove.

As shown in the graphs, the variability of the a_x component, among all the tests and for each of the operators, is lower when the transversal landmark is used, compared to the case in which it is not present. That means that the transversal protrusion of the adaptor could help the operator to put the accelerometer in the same position when more tests are performed. Anyway, in this case, the statistical analysis did not confirm the differences among the two series of tests for each of the subjects ($p > 0.22$).

A possible explanation is linked to the thickness of the glove used in the tests. In this case in fact, since the instrumented handle was perfectly straight along the X-axis, supposing that also the adaptor had only a very little rotation on the X–Y plane (this was checked looking at the rotation of the longitudinal landmark), possible misalignments can come only from a rotation of the adaptor on the Y–Z plane. This rotation can mainly result from the different compression of the material used in the glove for covering the palm of the hand, since the thickness of the glove at the palm level is homogeneous. Anyway, in our case, due to the small thickness of the working leather glove, also the angle between the hand and the forearm, checked locking at the transversal landmark, resulted too little to be easily estimated.

When anti-vibration (AV) gloves are used, since they are thicker and softer, the rotation of the adaptor along the X–Z plane could generate greater misalignment between the hand and the forearm, which could be easier checked through the use of the transversal landmark.

4 CONCLUSIONS

In the present study, adaptors for measuring hand–arm vibration transmissibility were analyzed when made of different plastic materials (PLA, ABS, and Nylon) at different infill densities (50% and 100%). Results show that 3-D printed adaptors, made using fused deposition modeling technique, can be a valid alternative to those on the market. At the middle–low frequencies range (10 Hz–400 Hz) the transmissibility values of the printed adaptors are in line with those of the adaptors found on the market or built using injection molding technique.

No significant differences were found between the transmissibility of those printed at 50% and 100% of infill, especially when the acceleration is W_h weighted. All the printed adaptors have quite linear transmissibility, very close to the unit, from 10 to 80 Hz, then the values tend to be nonlinear, with a general tendency to increase as frequencies increase. The average measured values of the transmissibility were generally inside the range ± 0.05 in all the tested frequencies.

Among the analyzed materials, statistical analysis did not show significant differences at 100% infill, while only the adaptor printed in Nylon showed a little better behavior at frequencies less than 160 Hz when printed at 50% infill. Anyway, these differences are very little for having practical repercussions.

Some solutions for better align the adaptor, when used inside gloves, were also evaluated. Starting from the dimensions and the shape reported in the ISO 10819:2013 Standard, a first modification was done, drawing a protrusion along the longitudinal axis of the adaptor. This landmark, coming out from the glove through a little slit made in the seam of the glove between the thumb and index finger, is meant to help for checking possible misalignments of the adaptor, mainly when it rotates on the Y–Z plane. Results obtained confronting the component of the acceleration along the y axis (a_y), in both the case of use and not use of the landmark, confirmed its utility. The landmark can also be useful for reducing possible misalignments on the Y–Z plane.

A second landmark drawn down along the transversal axis of the adaptor was also tested in order to limit possible misalignments due to a rotation on the X–Z plane. These misalignments are generally related to different localized compressions of the glove material at the level of the palm. In this case, even though the results showed a more homogeneous distribution of the acceleration component along the x-axis (a_x), used in this case as an indicator of possible adapter's misalignment, statistical analysis did not confirm the differences between the two series of tests. This can be due to the relatively small thickness of the glove used in the tests that produces little misalignments, too difficult to be checked by the landmark. Some further tests can be carried out using anti-vibration gloves, generally made of thicker and softer materials, to better evaluate the utility of the transversal protrusion applied to the adaptor.

REFERENCES

[1] Griffin, M.J., Measurement, evaluation, and assessment of occupational exposures to hand-transmitted vibration. *Occupational and Environmental Medicine,* **54**(2), pp. 73–89, 1997. https://doi.org/10.1136/oem.54.2.73.

[2] Alfaro Degan, G., Antonucci, A., Coltrinari, G. & Lippiello, D., Hand-arm vibration induced by the use of radial arm polisher machines. *International Multidisciplinary Scientific Geo-Conference Surveying Geology and Mining Ecology Management, SGEM,* **August**(1–2), pp. 127–134, 2020.

[3] Rempel, D., Antonucci, A., Barr, A., Cooper, M.R., Martin, B. & Neitzel, R.L., Pneumatic rock drills vs. electric hammer drill: Productivity, vibration, dust, and noise when drilling into concrete. *Applied Ergonomics,* **74**, pp. 31–36. 2019. https://doi.org/10.1016/j.apergo.2018.08.005.

[4] ISO 5349-1, Mechanical vibration: Measurements and evaluation of human exposure to hand-transmitted vibration – Part 1: General guidelines. International Standard Organization: Geneva, Switzerland, 2001.

[5] Directive 2002/44/EC of the European Parliament and the Council of 25 June 2002 on the minimum health and safety requirements regarding the exposure of workers to the risks arising from physical agents (vibration).

[6] ANSI S2.70, Guide for the measurement and evaluation of human exposure to vibration transmitted to hand (Revision of ANSI S.3.34-1986), American National Standards Institute (ANSI): New York, 2006.

[7] ISO 10819, Mechanical vibration and shock: Hand arm vibration – Method for the measurement and evaluation of the vibration transmissibility of gloves at the palm of the hand. International Organization for Standardization: Geneva, Switzerland, 2013.

[8] Dong, R.G., Welcome, D.E., Peterson, D.R., Xu, X.S., McDowell, T.W., Warren, C., Asaki, T. & Kudernatsch, S., Tool-specific performance of vibration-reducing gloves for attenuating palm-transmitted vibrations in three orthogonal directions. *International Journal of Industrial Ergonomics,* **44**, pp. 827–839, 2014. http://dx.doi.org/10.1016/j.ergon.2014.09.007.

[9] Budd, D. & House, R., Examining the usefulness of ISO 10819 antivibration glove certification. *Annals of Work Exposure and Health,* **61**(2), pp. 137–140, 2017. https://doi.org/10.1093/annweh/wxw018.

[10] Song, C.M., Jang, H.-K. & Chai, J., A comparison of the dynamic properties of transducer adapters for measurement of hand-transmitted vibration. *Journal of Mechanical Engineering Science – Part C,* **223**, pp. 345–351, 2009. https://doi.org/10.1243/09544062JMES1085.

[11] Xu, X.S., Dong, R.G., Welcome, D.E., Warren, C. & McDowell, T.W., An examination of an adapter method for measuring the vibration transmitted to the human arms. *Measurement,* **73**, pp. 318–334, 2015. http://dx.doi.org/10.1016/j.measurement.2015.05.039.

[12] McDowell, T.W., Dong, R.G., Welcome, D.E., Xu, X.S. & Warren, C., Vibration-reducing gloves: Transmissibility at the palm of the hand in three orthogonal directions. *Ergonomics,* **56**(12), pp. 1823–1840, 2013. http://dx.doi.org/10.1080/00140139.2013.838642.

[13] Hewitt, S., Assessing the performance of anti-vibration gloves: A possible alternative to ISO 10819-1996. *Annals of Occupational Hygiene,* **42**(4), pp. 245–252, 1998. https://doi.org/10.1016/S0003-4878(98)00030-1.

[14] Xu, X.S., Dong, R.G., Welcome, D.E., Warren, C. & McDowell, T.W., An examination of the handheld adapter approach for measuring hand-transmitted vibration exposure. *Measurement,* **47**, pp. 64–77, 2014. http://dx.doi.org/10.1016/j.measurement.2013.08.037.

[15] Moschioni, G., Saggin, B. & Tarabini, M., Uncertainty in hand arm vibration measurements due to the fixation method. *11th International Conference on Hand-Harm Vibration*, Bologna, Italy, 3–7 June, pp. 457–464, 2007.

[16] Xu, X.S., Welcome, D.E., Warren, C.M., McDowell, T.W. & Dong, R.G., Development of a finger adapter method for testing and evaluating vibration-reducing gloves and materials. *Measurement,* **137**, pp. 362–374, 2019. https://doi.org/10.1016/j.measurement.2019.01.034.

[17] Paddan, G.S., Effect of grip force and arm posture on the transmission of vibration through gloves. *Proceedings of the UK Informal Group Meeting on Human Response to Vibrations, MIRA*, September, Nuneaton, Warwickshire, UK, 1996.

[18] Smutz, W.P., Dong, R.G., Han, B., Shopper, A.W., Welcome, D.E. & Kashon, M.L., A method for reducing adaptor misalignment when testing gloves using ISO 10819. *Annals of Occupational Hygiene.* **46**(3), pp. 309–315, 2002. https://doi.org/10.1093/annhyg/mef037.

[19] Dong, R.G., Rakheja, S., Smutz, W.P., Shopper, A.W., Welcome, D. & Wu, J.Z., Effectiveness of a new method (TEAT) to access vibration transmissibility of gloves. *International Journal of Industrial Ergonomics,* **30**, pp. 33–48, 2002. https://doi.org/10.1016/S0169-8141(02)00076-8.

[20] Dong, R.G., Wu, J.Z., Xu, X.S., Welcome, D.E. & Krajnak, K., A review of hand-vibration studies conducted by US NIOSH since 2000. *Vibration*, **4**, pp. 482–528, 2021. https://doi.org/10.3390/vibration4020030.

[21] Lay, M., Thajudin, N.L.N., Hamid, Z.A.A., Rusli, A., Abdullah, M.K. & Shuib, R.K., Comparison of physical and mechanical properties of PLA, ABS, and Nylon6, fabricated using fused deposition modeling and injection molding. *Composites Part B,* **176**, pp. 1–8, 2019. https://doi.org/10.1016/j.compositesb.2019.107341.
[22] ISO 5349-2, Mechanical vibration: Measurements and evaluation of human exposure to hand-transmitted vibration – Part 2: Practical guidance for measurement at the workplace. International Standard Organization: Geneva, Switzerland, 2001.
[23] Pitts, P.M., Mason, H., Poole, K.A. & Young, C.E., Relative performance of frequency weighting W_h and candidates for alternative frequency weightings when used to predict the occurrence of hand-arm vibration induced injuries. *Industrial Health,* **50**(5), pp. 388–396, 2012. https://doi.org/10.2486/indhealth.MS1381.
[24] EN 420:2003+A1, Protective gloves: General requirements and test methods. The European Standards Body, 2009.
[25] McDowell, T.W., Welcome, D.E., Xu, X.S., Warren, C. & Dong, R.G., The transmissibility of vibration-reducing gloves at the palm of the hand in three orthogonal directions. *Proceedings of the 4th American Conference on Human Vibration*, Hartford, Connecticut, USA, June 13–17, 2012.

ONLINE HANDWRITING SIGNATURE: AN EXPERIMENTAL STUDY ON PRESSURE SIGNAL NORMALIZATION IN DIFFERENT DEVICES FOR BIOMETRIC ACQUISITION

FABIO GARZIA[1], MARIA CHIARA GIANNONE[1] & MARCOS FAUNDEZ-ZANUY[2]
[1]Safety and Security Engineering Group – DICMA, SAPIENZA – University of Rome, Italy
[2]Tecnocampus, Universitat Pompeu Fabra, Spain

ABSTRACT

Online handwriting signature is an important element in biometrics and therefore for security applications. In this paper the characterization of pressure for different acquisition devices for digital signature is presented, considering that technological evolution always produces new devices and that is why it is necessary to facilitate interoperability between them. The devices used are three different Wacom tablets: Wacom Intuos pro, Wacom Cintiq 16 and STU-530. Each of them has different pens and more than one nib and that is why for each of them over one case was analyzed in reference to the different pens used. It was experimented how, thanks to the respective software of the tablets, it is possible to acquire the pressure that is generated by the pen on the tablet and to connect it to the weight force exerted by them thanks to a balance placed under the tablet. The process used allowed to create for each tablet the pressure/weight plot that permitted to derive the different sensitivities of the tablets to pressure and also the different saturation values. In addition, it was also analyzed the variation of the thickness of the stroke exerted by the pen going to write on the tablet with different values of force applied. Thanks to the use of carbon copy paper, which allowed to reproduce the line generated by the pen on a sheet of paper, it was then possible to study under the microscope the trend of the thickness of the line drawn as the force varies. The obtained results allowed to obtain a direct comparison between the signals acquired by different devices thanks to the normalization processes that permit to increase the recognition performance.

Keywords: online handwriting signature, pressure, software, thickness, normalization processes.

1 INTRODUCTION

Online handwritten analysis presents many applications in e-security, signature biometrics being the most popular but not the only one [1]. From a human behavior and health condition perspective, online handwriting biometrics are more appealing and informative than other popular biometrics traits such as fingerprints or iris [2], [3]. It can be noted that nowadays, most of the published research in biometric signal processing is based on image and speech. This might be linked to the fact that these signals are simpler and cheaper to acquire compared with others such as online handwriting where specific digitizing tablets are needed [1].

Fortunately, online handwriting signals are more present in our society than a few years ago due to the increasing popularity of tactile devices and their corresponding cost reduction [4]. As a result, we forecast a considerable growth in applications in this field. Indeed, in a near future, these pen-based handheld devices (tablets, phablets, cellphones, etc.) will help to protect people's sensitive data, for example, incorporating automatic signature verification and writer identification [1].

Digitizing tablets for handwritten tasks are general devices that can be easily found in almost any computer shop. They are mainly sold for graphic design. However, some small devices exist for biometric online signature acquisition, for instance in a post office or for VISA/Mastercard payment. This later case are very small tablets with lower quality than the most popular ones for graphic design. On the other hand, scientific community has been using

WIT Transactions on The Built Environment, Vol 206, © 2021 WIT Press
www.witpress.com, ISSN 1743-3509 (on-line)
doi:10.2495/SAFE210151

all these tablets intensively for e-health and e-security analysis based on handwritten tasks. Calibrated tablets do not exist as they have not been developed for instrumentation. In one case, because only the graphic result and comfort for the user is important. In the other case, because it is just a substitute of the classic ink and paper method, and the goal is to save paper. The signature is not going to further treatment nor processing so accuracy of pressure is not really an issue. In fact, classic ink plus paper method does not consider pressure at all. Thus, rather than a gold standard or reference behaviour, a precision calibrated balance to obtain accurate measurements is used in the present work, as described in the following.

Regarding the safety of online handwritten, among the different signal functions, pressure is one of the most important because it has a discriminative power in e-security and e-health. In fact, in the medical field, the study of handwriting has proven to be useful in diagnosing and tracking some diseases of the nervous system [5].

In an online context, pressure can be measured directly from sensors embedded in the stylus, providing a continuous signal sampled at a given frequency and amplitude resolution. When the stylus is used with a tablet, x and y coordinates versus time are also available as well as tilt and azimuth angles. When the stylus is instrumented, other signals like acceleration can also be provided [6], even if previous works were made with limited resources also from precision of the used balance point of view (0.1 grams instead of 0.01 grams of the present work) [6]. Because of this, in this paper, the pressure signal normalization of three graphics tablets was analyzed. The chosen graphics tablets are from the Wacom company, which was founded in 1983. It is a graphic tablet and graphic design company based in Ōtone, Japan. Wacom's innovation lies in this active digitizer technology that supports up to 2056 levels of pressure, providing the Tablet PC with advanced digital ink capabilities and the ability to write naturally on the screen by resting your wrist on it. The three tablets chosen are the STU-530, Cintiq 16 and the Intuos Pro. Each has different features and different pens to use.

The study is focused on experimental tests that allow us to collect a set of pairs of values using the high precision balance FR-500, the Microscope OPTIKA XDS-3MET and the different acquisition software for each tablet. The values collected are the pressure level provided by the tablet and the weight measured in the balance. In addition, several tests have shown that as the pressure used to generate the line on the graphic tablet increases, the thickness of the line also changes, which allows us to understand how the mark is influenced by the force used to write on the tablet.

The study represents a significant improvement of previous works [1], [5], [6] as it is based on three different pens with seven different nibs instead of a single tablet and only two different pens (ink and plastic). Further, previous paper [6] had not deal a lot with ink trace measurement while the present work goes under very complete stroke analysis by means of an electronic microscope and these results are shown in the following. Moreover, a special wood device to measure the widths of both sides of the nib has been developed, while the previous paper [6] was limited to only one side (end).

The significance of this study is in e-health as some pathologies reveal a loss of pressure [7] and in e-security too, as some published papers [6] demonstrate the positive contribution of pressure information to biometric rates (identification and verification). In this paper the possibility to deal with signals acquired by different pens and tablets is opened. This is particularly important because current state of the art is mainly limited to situations where all the experiments are performed with same tablet and pen model, and this is could not represent a real situation [8].

The paper is organized as follows: Section 2 is devoted to the characterization of the three handwriting devices, Section 3 presents the methodology and data Sample analysis, Section 4 the experimental results, Section 5 the conclusions.

2 CHARACTERIZATION OF HANDWRITING DEVICES

For this study, three different Wacom digitizing tablets have been evaluated. Each model has several writing styluses (pens) to be tested, as shown in Table 1.

Table 1: Wacom tablets and pens analysed (lpi = lines per inch).

Characteristics\tablet	**STU-530**	**Intuos Pro**	**Cintiq 16**
Acquisition area	108 x 65 mm	430 x 287 mm	422 x 285 mm
Pressure range	[0, 1023]	[0, 7557]	[0, 32767]
Coordinates precision	±0.02 inches/ ± 0.5 mm (center)	–	–
Resolution of coordinates (non-interpolated)	2540 lpi 0.01 mm/pt	5080 lpi	5080 lpi
Report rate (non-interpolated)	200 points per second	200 points per second	–
Pens	Plastic1	Plastic2 Inkpen	Plastic3 Pro pen 3D

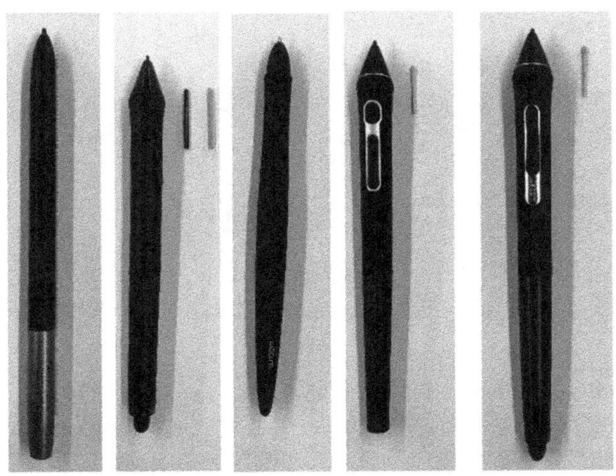

Figure 1: Pens and nibs used in the study.

3 METHODOLOGY AND DATA SAMPLE ANALYSIS

In this study, two different measurements were made for the respective graphic tablets. The first was made with the use of a FR-500 balance, which made it possible to measure the weight force exerted by the pen writing on the tablet and at the same time to capture, with the use of a software, the pressure exerted. The second was based on measuring the thickness of the line drawn by the pen thanks to the use of OPTIKA XDS-3MET Microscope and copying paper.

3.1 Experimental setup for pressure measurement

The first measurement is made by placing the tablet on the scale, to capture the weight of each measurement, as shown in Fig. 2, and in the meantime, we use the software to see the pressure generated on the computer. Thanks to this, it is possible to create a plot weight/Tablet Value that allows to see the trend of the curve and the saturation point of the pressure [9]–[11]. The test was done for each graphic tablet with the different pens described above.

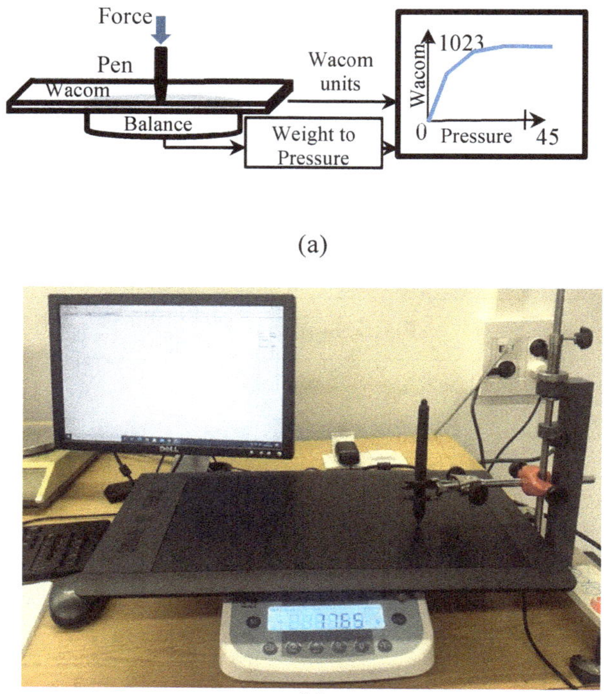

(a)

(b)

Figure 2: Configuration used for the analysis. (a) Schema for the experimental measurements to characterize the pen from [6]; and (b) Experimental setup (precision balance FR-500, stylus, and force actuator).

3.2 Line thickness measurement setup

The second study concerns the characterisation of pen nibs to determine the width of the strokes produced by different pens and tablets. A set of straights lines has been performed by hand trying to keep constant pressure during the whole realization. Then has been measured the width of each line with the help of the microscope OPTIKA XDS-3MET (https://www.optikamicroscopes.com/optikamicroscopes/product/im-3-series/) to evaluate and photograph the strokes.

To measure with the help of the microscope the width of the lines, were produced lines on white sheets of paper show in Fig. 3. In the case of the plastic pens, we have used a carbon copy paper on the top of the white paper. Also, the widths for the ink pen with and without carbon copy paper were measured, although it is not necessary to measure the width, as it is able to produce lines on paper directly. This permits to validate the process used with the plastic pen and carbon copy. So, in Fig. 4 we can see the image captured by the microscope to measure the thickness.

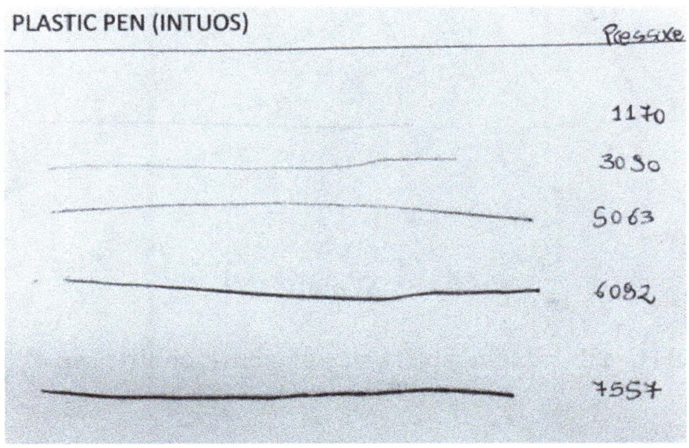

Figure 3: Example of the sheets (which are 13 × 9 cm).

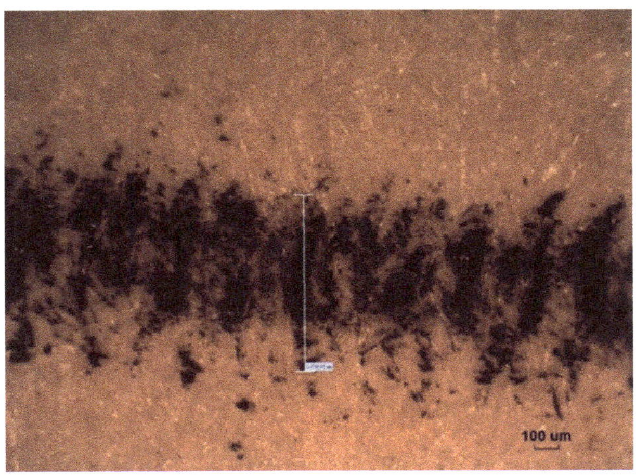

Figure 4: Straight line from the electronic microscope for the Intuos plastic pen.

4 EXPERIMENTAL RESULTS

The results obtained from these two studies have been plotted. For the first one, the tests carried out led to the construction of these graphs, while as far as the Intuos Pro plot is concerned, a study already carried out previously was taken [6].

So, comparing the weight obtained from the balance and the value obtained in the Wacom tablet, we find the experimental characteristic of the pressure sensor shown in Figs 5–7.

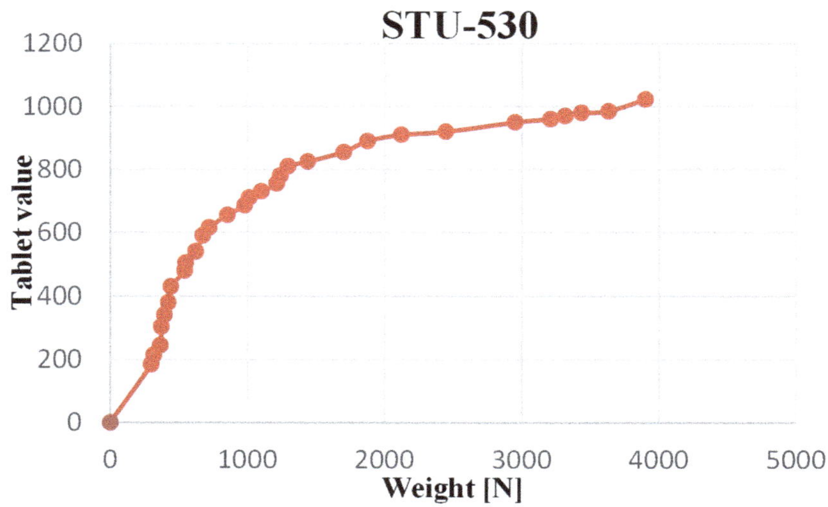

Figure 5: STU-530 Wacom tablet value obtained for different weight values.

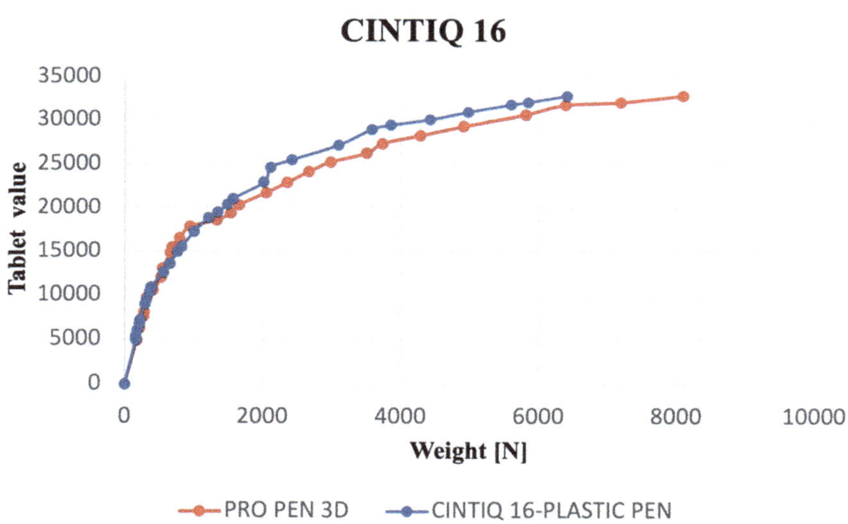

Figure 6: Cintiq 16 Wacom tablet value obtained for different weight values.

We observe in Fig. 7 that the plastic pen reaches saturation before the ink pen. Alternatively, the plastic pen has smaller dynamic range than the ink pen. This is not surprising since the ink pen has higher price than the plastic one, it must be purchased separately, and can be considered of better quality. However, its curve also exhibits a nonlinear behavior, which affects the obtained pressure values [6].

As for the microscopic study of the thickness of the line made with the different pens, it shows that in these graphs as the pressure increases, the thickness also increases.

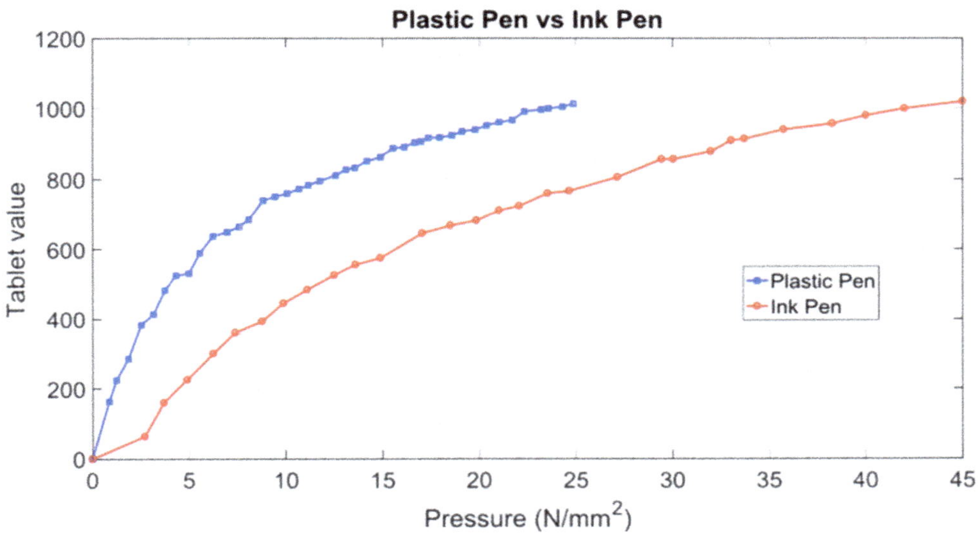

Figure 7: Intuos Wacom tablet value obtained for different pressure values [6].

Figs 8–10 show the measurements for different pens and exerted pressures. Two important observations can be remarked:

- It shows that the measurements for the case of the ink pen are very similar with and without carbon copy paper. This means that the use of carbon copy paper is not distorting the width measurements and validates the experimental procedure for the plastic pens.
- The ink pen shows better permanence on the line width for different pressure levels, which remains more constant than for the plastic pens.

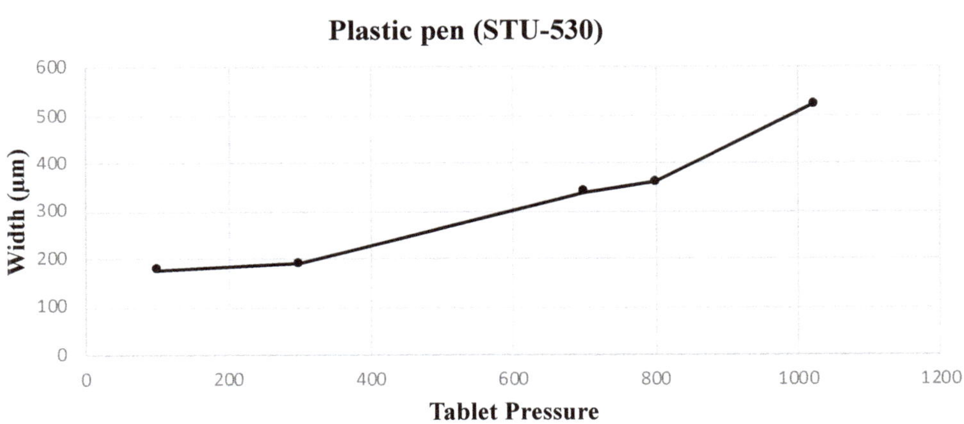

Figure 8: Width measurements for straight lines performed at different exerted pressures for the Wacom STU-530.

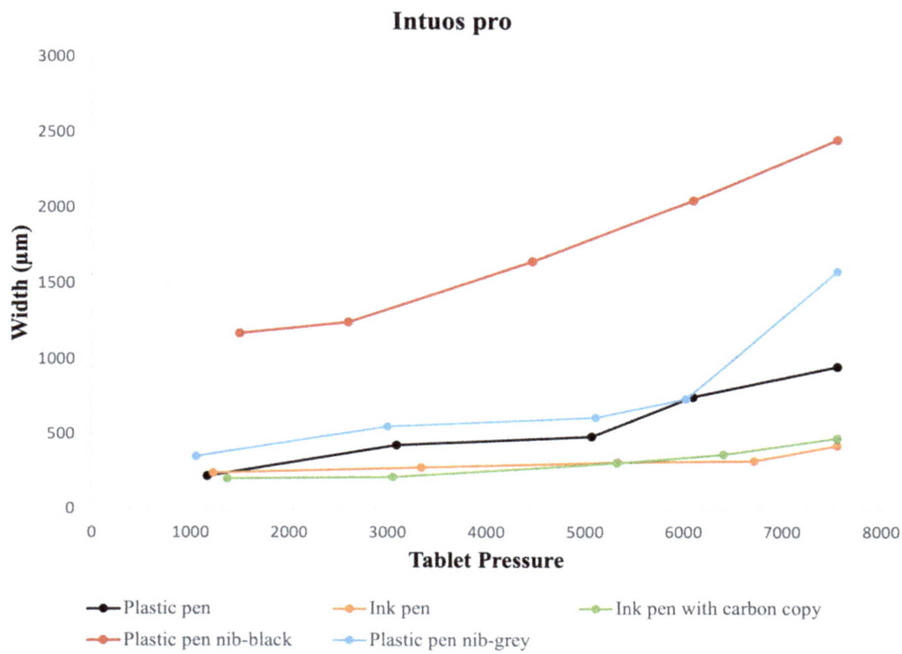

Figure 9: Width measurements for straight lines performed at different exerted pressures for Wacom Intuos Pro.

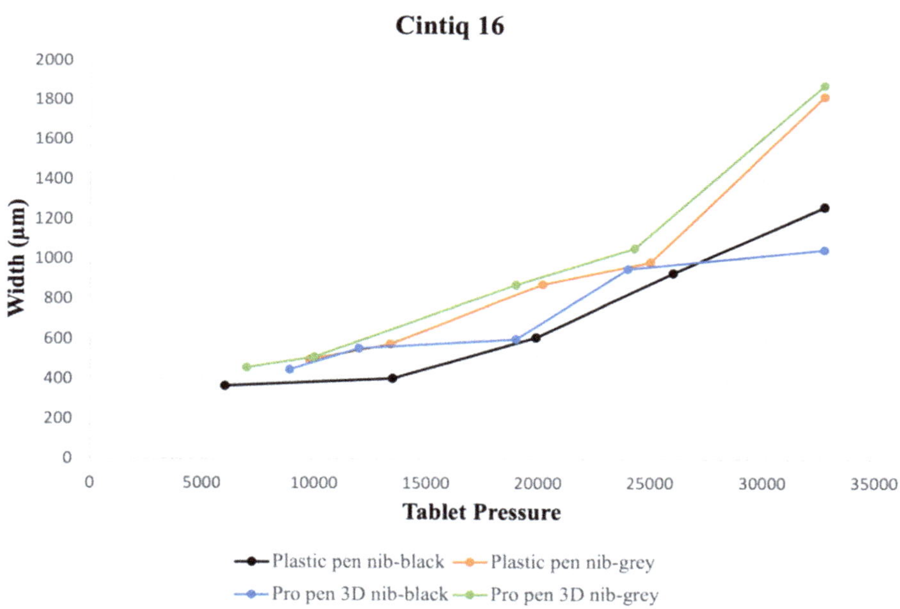

Figure 10: Width measurements for straight lines performed at different exerted pressures for Cintiq 16.

5 CONCLUSIONS

In this paper, we have presented an experimental setup for pressure and line thickness measurement for the three different Wacom graphics tablets. The main conclusions are:

- The curve that emerges from the studies of the pressure exerted by the stylus generates an exponential curve with important saturation values for high pressures.
- The Intuos Wacom plastic pen reaches saturation for lower pressure than the Wacom ink pen, which presents a higher dynamic range (almost two times higher) and thus, it is has a better pressure sensor [6].
- The Cintiq Wacom plastic pen reaches saturation for lower pressure that the Wacom pro pen 3D.
- The study of the thickness of the lines made with the different stylus shows that the increase of pressure also affects the thickness, which grows. The stylus which shows the best performance is the ink pen which remains almost constant.

ACKNOWLEDGEMENT

This work has been funded by Spanish grant Ministerio de ciencia e innovación PID2020-113242RB-I00.

REFERENCES

[1] Faundez-Zanuy, M., Fierrez, J., Ferrer, M.A. & Plamondon, R., Handwriting biometrics: Applications and future trends in e-security and e-health. *Cogn. Comput.*, **12**, pp. 940–953, 2020. DOI: 10.1007/s12559-020-09755-z.

[2] De Stefano, C., Fontanella, F., Impedovo, D., Pirlo, G. & Scotto di Freca, A., Handwriting analysis to support neurodegenerative diseases diagnosis: A review. *Pattern Recogn. Lett.*, **121**, pp. 37–45, 2019.

[3] Martinez-Diaz, M., Fierrez, J. & Hangai, S., Signature matching. *Encyclopedia of Biometrics*, 2nd ed., eds S.Z. Li & A.K. Jain, Springer, pp. 1382–1387, 2015.

[4] Tolosana, R., Vera-Rodriguez, R., Fierrez, J., Morales, A. & Ortega-Garcia, J., Benchmarking desktop and mobile handwriting across COTS devices: The e-BioSign biometric database. *PLoS One*, **12**(5), e0176792, 2017.

[5] Faundez-Zanuy, M. et al., A preliminary study of online drawings and dementia diagnose. *Neural Nets and Surroundings: Smart Innovation, Systems and Technologies*, eds B. Apolloni, S. Bassis, A. Esposito & F. Morabito, vol. 19, Springer, Berlin, Heidelberg. DOI: 10.1007/978-3-642-35467-0_36.

[6] Faundez-Zanuy, M., Brotons-Rufes, O., Paul-Recarens, C. & Plamondon, R., On handwriting pressure normalization for interoperability of different acquisition stylus. *IEEE Access*, **9**, pp. 18443–18453, 2021. DOI: 10.1109/ACCESS.2021.3053499.

[7] Faundez-Zanuy, M. et al., A preliminary study of online drawings and dementia diagnose. *Neural Nets and Surroundings: Smart Innovation, Systems and Technologies*, eds B. Apolloni, S. Bassis, A. Esposito & F. Morabito, vol. 19, Springer, Berlin, Heidelberg. DOI: 10.1007/978-3-642-35467-0_36.

[8] Faundez-Zanuy, M., On-line signature recognition based on VQ-DTW. *Pattern Recognit.*, **40**(3), pp. 981–992, 2007. DOI: 10.1016/j.patcog.2006.06.007.

[9] Schomakerand, L.R.B. & Plamondon, R., The relation between pen force and pen-point kinematics in handwriting. *Biol. Cybern.*, **63**(4), pp. 277–289, 1990. DOI: 10.1007/BF00203451.

[10] Alonso-Fernandez, F., Fierrez-Aguilar, J. & Ortega-Garcia, J., Sensor interoperability and fusion in signature verification: A case study using tablet PC. *Advances in Biometric Person Authentication* (Lecture Notes in Computer Science), vol. 3781, eds S.Z. Li, Z. Sun, T. Tan, S. Pankanti, G. Chollet & D. Zhang, Springer: Berlin, 2005, pp. 180–187. DOI: 10.1007/11569947_23.
[11] Antal, M. & Bandi, A., Finger or stylus: Their impact on the performance of on-line signature verification systems. *Proc. 5th Int. Conf. Recent Achievements Mechatron. Autom. Comput. Sci. Robot.*, pp. 11–22, 2017.

SECTION 7
RISK ANALYSIS
AND ASSESSMENT

INDIVIDUAL PREFERENCES IN SECURITY RISK DECISION MAKING: AN EXPLORATORY STUDY UNDER SECURITY PROFESSIONALS

JOHAN J. DE WIT[1], WOLTER PIETERS[2] & PIETER H. A. J. M. VAN GELDER[1]
[1]Delft Technical University, The Netherlands
[2]Radboud University, The Netherlands

ABSTRACT

Risk assessments in the (cyber) security domain are often, if not always, based on subjective expert judgement. For the first time, to the best of our knowledge, the individual preferences of professionals from the security domain are studied. In on online survey they are asked to mention, rate and rank their preferences when assessing a security risk. The survey setup allows to differentiate between easily accessible or "on top of mind" attributes and guided or stimulated attributes. The security professionals are also challenged to both non-compensatory and compensatory decision making on the relevance of the attributes. The results of this explorative study indicate a clear difference and shift in the individual perceived relevance of attributes in these different settings. Another remarkable finding of this study is the predominant focus on impact attributes by the respondents and the less significant position of likelihood or probability. The majority of professionals seem to ignore likelihood in their security risk assessment. This might be due to so called probability neglect as introduced by other scholars. the security in organisations and society is depending on the assessment and judgement of these professionals, understanding their preferences and the influence of cognitive biases is paramount. This study contributes to this body of knowledge and might raise attention to this important topic in both the academic and professional security domain.

Keywords: security risk assessment, decision making, risk management, decision biases, preferences, probability neglect.

1 INTRODUCTION

The security risk field is dealing with malicious, and therefore manmade, risks. Risks in general contain a level of uncertainty by nature as they involve a future state of affairs. The aspect of malicious intent of security risks add an extra dimension to this uncertainty. Malicious actions, like for example an intrusion, usually are meant to be unpredictable, concealed and evade existing risk controls.

The dynamic context of security risks, with ever changing modus operandi, in combination with the large variety of situations, both in location and time, add to the uncertainty. Because of this information about past security risks and events, if available, is often not sufficient to estimate or predict future security risk. The assessment of the uncertainty of security risks, therefore, is for a large part based on expert judgement rather than based on evidence or objective data.

The individuals assessing security risks, in this study referred to as security professionals, often, if not always, apply a risk management process of some sort to structure their assessment. The various risk management processes contain obvious process stages like: establishing the context, risk identification, risk analysis, risk evaluation and risk treatment.

So far little scientific studies are conducted exploring individual preferences and priorities guiding security professionals in their daily praxis of security risk decision making during these risk management processes. These professionals play a decisive or advisory role in security risk treatment, hence, they are determining or at least influencing the security in organisations and society. Understanding their individual preferences and priorities is of vital

WIT Transactions on The Built Environment, Vol 206, © 2021 WIT Press
www.witpress.com, ISSN 1743-3509 (on-line)
doi:10.2495/SAFE210161

importance to understand their security risk judgement. The purpose of this exploratory study is to examine the criteria, further referred to as attributes, and their priority, security professionals consider when assessing a security risk.

An online survey is conducted under security professionals of both the physical and cybersecurity domain. The survey set up is explained in more detail in the Method and Materials section.

This study is, for the first time, exploring security risk assessments by security professionals. What are the individual preferences and priorities of security professionals? Do they change after a "second thought"? Is individual expertise influencing these preferences? The purpose of this exploratory study is to enhance our understanding of individual decision making influencing the security in our society.

Section 2 presents a brief overview of the theoretical background of security risks, risk assessments and decision making. In Section 3 the research method is explained followed by the Results and Analysis section. This paper ends with conclusions and discussion in Section 5.

2 SECURITY RISK ASSESSMENTS AND DECISION MAKING

In this section first the characteristics of security, security risks and risk management processes are described followed by a theoretical background of decision making, cognitive biases, especially the availability heuristic.

2.1 Security, security risks and risk management

Keeping objects and organisations secure is the prime task of security professionals [1]. They have a decisive or advisory role in dealing with security risks. According to the ISO 31000, Risk management – Principles and guidelines, risk is "the effect of uncertainty on objectives". According to Hansson [2] "knowledge about a risk is knowledge about the unknown". This knowledge is in many cases incomplete and, therefore, will have to be supplemented or even replaced by expert judgment [3]. The latter is certainly the case in security related events.

Expert judgement is considered a degree of belief, based on tacit knowledge and expertise [4]. These subjective interpretations and assessments are not only based on "hard-to-measure" expertise but are also prone to numerous cognitive biases and heuristics. This has led many scholars to question the viability of such uncertainty assessments. Still in many domains, like security, there are no alternatives or objective procedures available. Therefore, intuitive judgements of uncertainty play an essential role in decisions [5].

2.2 Decision making

"A decision is a commitment to a course of action that is intended to produce a satisfying state of affairs" [6]. A decision involves a range of options for possible action or inaction. Decision options are further referred to as alternatives in this study. The decision agent is supposed to be equipped with a set of preferences based on objectives or goals.

In order to reach a final judgement and be able to select a possible decision alternative, the agent needs to analyse and differentiate the available alternatives [7]. Each alternative is, therefore, defined by a set of attributes associated with consequences when materializing. An attribute is defined as a certain aspect of an alternative. It is used to measure performance in relation to an objective.

Besides this more functional explanation of decision making, focussed on maximizing subjective expected utility, other functionalist metaphors, like accountability, influence human decision making. "Accountability refers to the implicit or explicit expectation that one may be called on to justify one's beliefs, feelings and actions to others" [8]. Due to the responsibility for managing something as important as security risks, the security professionals in this study can be expected to, consciously or unconsciously, consider accountability in their decisions.

The individual response to attributes consists of two main components: an affective response and a cognitive response. These relate to the so-called dual-process models. The most renown of these models is the system 1 and 2 model by Kahneman [9]. The affective response is related to system 1 which is considered to be more intuitive, automatic, fast, experience based and requires little cognitive effort. System 2, on the other hand, is considered deliberate, slow, concentrated, compensatory, and demands considerable cognitive effort. In the huge body of work on decision making that has evolved since the 1970s multiple heuristics and biases are identified and analysed. These heuristics and biases influence or even direct individual decision making.

This study focusses on availability (heuristic) which is considered one of the prominent general-purpose heuristics. A large body of research demonstrated that judgements in general are based on the information that is most accessible to the decision agent at the time of the judgement. Both ease of recall and content of recall (the number of associations) influences the estimation of likelihood and thus perceived risk.

Van der Pligt and Vliek [10] added a valuable observation to the availability heuristic. Combining the ease of recall and content of recall to decision attributes not only influences the estimation of possible frequencies, but also influences the judgement of prevalence or commonness of a situation. A prevalent situation or attribute is widely accepted or favoured and this leads these scholars to the observation that prevalence adds to the weight of an attribute. In other words: availability of information of an attribute leads to a higher priority of this attribute.

3 METHOD AND MATERIALS

This survey is committed to explore the attributes of security risks security professionals consider and prioritize when assessing security risks. The attributes of security risks which are considered by security professionals during their security risk assessment are, therefore, collected and analysed. The explorative results are retrieved via an online survey conducted between June 13, 2019 and August 28, 2019. Participation in the survey was promoted in both the IT and physical security professional community. It was promoted via LinkedIn and Twitter, both in general and in special interest groups like Security management, ASIS Europe and ASIS International, Dutch cybersecurity platform. Second, a direct email campaign was launched targeting the existing professional network of the researchers. Third, the survey was published via the website of The Hague Security Delta, a Dutch security cluster of businesses, governments and knowledge institutions. Finally, the survey was promoted on several conferences and meetings via flyers. The sample of respondents (N = 248) is regarded to be a convenience sample.

To challenge the respondents the survey starts with an open ended question. This question asks them to come up with the attributes (in the survey referred to as criteria) they consider when assessing security risks. These answers express what is "on top of mind" and quickly available for the respondents in a blind recall without prompting from external stimuli. This open ended question allows the respondents to answer based on their complete knowledge, perception and experience without restrictions. The question offers a maximum of 10 answer options (first field forced response). This question evokes the respondents to show their attitude based on the attributes they take into account when judging security risks and measures. The answers to these questions serve as an index of quickly or most available attributes. This availability of attributes is related to the well-known availability heuristic. The answers to these open ended questions reflect the priority of, in this case, attributes related to security risk assessment. They can be considered as most prominent by the security professionals at the point of time of answering the survey.

To be able to determine the priority of attributes in multiple attribute decision making in a fuzzy environment two subsequent processes are involved: rating and ranking of attributes [11]. In the second part of the survey the respondents are asked to assign a priority to a predefined list of 28 security risk attributes (see Table 1). Each of the presented attributes can be rated using a five point Likert scale: extremely important, very important, moderately important, slightly important, not at all important. As the rating is done per individual attribute the rating is non-compensatory. The predefined list of attributes is derived from risk assessment tools like the ISO 27005, Information security risk management and the ASIS International Risk assessment and the SCM model. Four attributes influencing risk perception are also added to the list dread (fear), knowledge of the risk, whether or not the exposure to a risk can be influenced and finally if the risk can be managed or controlled.

Table 1: Predefined security risk attributes.

Predefined security risk criteria:	
Context impact criteria: 1. Perceived impact (general) 2. Impact on health and safety of employees 3. Impact on health and safety of customers, or visitors 4. Impact on surroundings/community 5. Impact on business process (including IT downtime) 6. Impact on supply chain 7. Financial impact 8. Legal impact/liability 9. Environmental impact 10. Damage to the reputation of the organization 11. Impact on public opinion 12. Physical damage to assets 13. Data loss 14. Data disclosure (including privacy sensitive data) 15. Loss of data integrity 16. Disclosure or loss of intellectual property	**Individual/personal impact criteria:** 17. Personal responsibility/ accountability 18. Damage to personal reputation 19. Management attention 20. Personal liability 21. Personal financial loss 22. Personal conscience 23. Regret of no or inadequate action **Likelihood/probability:** 24. Probability/likelihood of risk (general) **Other/risk perception:** 25. Fear of a security risk 26. Involuntariness of risk taking 27. Uncontrollability of risk 28. Lack of knowledge about a risk

In the third part of the survey the respondents are forced to set priorities over the 28 predefined attributes. They are asked to rank their top 10 (1 is the most important attribute, etc.). To avoid order biases, or response order bias the list of predefined attributes is automatically randomized for each participant. At this point in the survey the respondents are asked to rethink their position on risk attributes for the third time and this time they even need to apply compensatory mental models. This is considered to be system 2 thinking. Comparing the answers to the open ended questions of the first part and the ranking answers to this third part is considered to show the difference in the judgement of security risk attributes between system 1 and system 2.

The survey ends with nine questions on individual characteristics: functional description, number of years in current position, number of years security expertise, number of years professional expertise, age, education level, specific security trainings, job sector, size of organization in number of employees.

Open ended questions usually lower the completion rate of a survey due to the required cognitive effort of the respondents. Taking survey fatigue into account the order of the survey questions is organized to start with the most demanding open-ended questions and lower the cognitive effort with each question. After agreeing the consent statement (N=248) 60% of the respondents stopped the survey at the start of the open ended questions. Of the remaining 99 participants 81 completed the entire survey.

4 RESULTS AND ANALYSIS

In this section the results of the survey are discussed in three parts: the result and analysis of the open ended questions, the results and analysis of the rating questions, the results and analysis of the ranking questions. Finally these results are combined and compared.

4.1 Part one: Open ended questions

The open ended question is answered in plain text by 99 respondents. Four of these did not answer seriously, their answers are excluded from the analysis. To answer the first question: "When you assess a security risk, which criteria do you consider or take into account during your assessment?", in total 463 free text fields are filled, containing 516 identifiable and interpretable answers. These answers are considered to be "on top of mind", easily available and primarily originating from system 1 thinking.

For a first interpretation of the answers the method of manual inductive or grounded coding is applied, the coding process thus allowed the main attributes and their structure to emerge. The coding frame that emerged from the manual inductive coding process revealed common risk components beyond the two expected general risk components following the narrow definition of risk: probability/likelihood and consequence/impact. The respondents seem to have included components of the risk management process leading to the final assessment of security risks. In the narrow scope as intended by the researchers risk assessment is forming a judgment of a security risk based on the two general attributes likelihood and consequence.

The observed categories are in line with the risk management process as detailed in security risk standards like the American National Standard: Risk Assessment, issued by ANSI, ASIS and RIMS [13], see Fig. 1.

As impact and consequence are not specifically defined in the survey the vast majority of the respondents used "impact", only four used the word "consequence". In the analysis of the answers in this study the categories impact an consequence are combined. This study focusses on the narrow definition of risk assessment (see Fig. 1).

The categories emerging from the inductive coding process fit the predefined risk attributes as presented in Section 3. The list of predefined risk attributes, however, contains

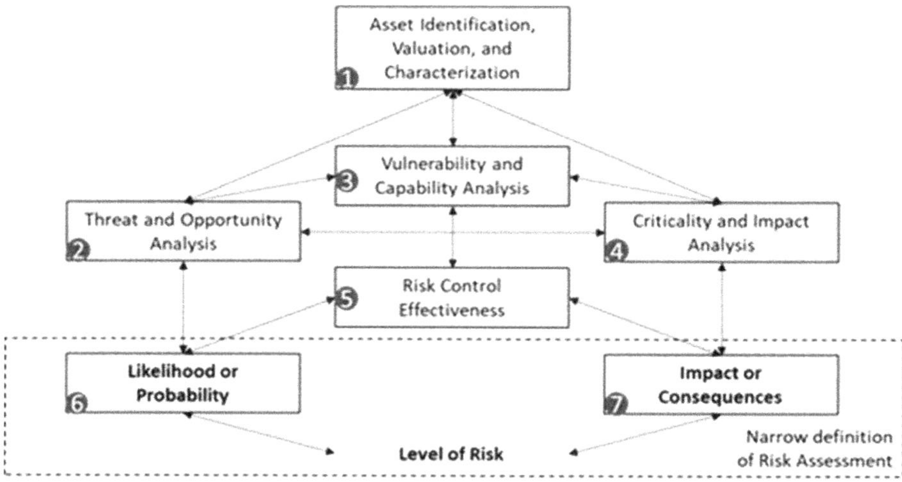

Figure 1: "Determining the level of risk", risk management process according to American National Standards Institute.

attributes that seem to be "not on top of mind" and they are not mentioned by the respondents. These attributes mainly concern individual/personal impact attributes and risk perception attributes. There are also three impact categories that some of the respondents pointed out that were not included in the predefined criteria list: Impact on trust, impact on/for customers, and political/national impact. The results of the open ended answers limited to the intended narrow definition of risk assessment are presented in Table 2.

A reliability analysis was carried out on the answers to the open ended questions. Cronbach's alpha showed the answers to reach low internal reliability, $\alpha = 0.469$.

Further correlation analysis showed various significant but low correlations between the individual attributes (r values between 0.21 and 0.492). The correlation between attribute 14 and 15 reached a moderate level ($r = 0.688$, $p < 0.05$). The inductive coding process shows clearly that the largest answer category relates to impact (192 answers). The vast majority of respondents, 87%, mentions one or more impact attributes.

It is remarkable to observe that only 43% of the respondents mentions the other main component of risk expressing uncertainty. Likelihood, probability, frequency or chance is mentioned only by 42 respondents. As almost all respondents mention impact criteria and less than half of them mentions likelihood there seems to be a predominant focus on impact/consequences.

Using the chi-squared tests there are no significant influences of individual characteristics observed. Education level, specific security trainings, age, and professional an security experience do not seem to influence the answers to the open ended questions.

4.2 Part two: Rating criteria

In the second part of this study the predefined list of risk attributes is presented to the respondents. Each of the presented attributes can be independently rated using a five point Likert scale: extremely important, very important, moderately important, slightly important,

Table 2: Descriptive analysis, inductive coding, sub categories of answers within the categories impact/consequence and likelihood/probability.

Question 1: When you **assess a security risk**, which criteria do you consider or take into account during your assessment?			
Free text entries, manually, inductive or grounded coding		Number of answers:	Percentage of respondents:
1	Perceived impact of the security risk	50	53%
2	Impact on health and safety of employees	23	24%
3	Impact on health and safety of customers, visitors	–	–
4	Impact on surroundings/community	10	10%
5	Impact on business process (including IT process/ downtime)	27	28%
6	Impact on supply chain	7	7%
7	Financial impact	13	14%
8	Legal impact/liability	3	3%
9	Environmental impact	4	4%
10	Damage to the reputation of the organisation	14	15%
11	Impact on public opinion	–	–
12	Physical damage to assets	8	8%
13	Data loss	11	11%
14	Data disclosure (including privacy sensitive data)	9	9%
15	Loss of data integrity	4	4%
16	Disclosure or loss of intellectual property	2	2%
17	Personal responsibility/accountability	2	2%
18	Damage to personal reputation	–	–
19	Management attention	2	2%
20	Personal liability	–	–
21	Personal financial loss (e.g. dismissal, loss of incentives)	–	–
22	Personal conscience	–	–
23	Regret of no or inadequate action	–	–
24	Probability/likelihood of risk	42	43%
25	Fear of a security risk	–	–
26	Involuntariness of risk taking	–	–
27	Uncontrollability of risk	–	–
28	Lack of knowledge about a risk	–	–
	Impact on trust	2	2%
	Impact on/for customers	5	5%
	Political/national impact	3	3%

not at all important. This list of attributes is considered an external stimulus to the respondents. It is analysed how this stimulus influences the priorities of the respondents.

The results show the influence of a stimulus: the respondents rate the majority of the attributes important even if they did not have them on top of mind at the first question. On average (the red graph in Fig. 2) the rating concentrates in the vicinity of "very important".

The reliability analysis (Cronbach's alpha) showed the rating to reach high internal reliability, $\alpha = 0.88$.

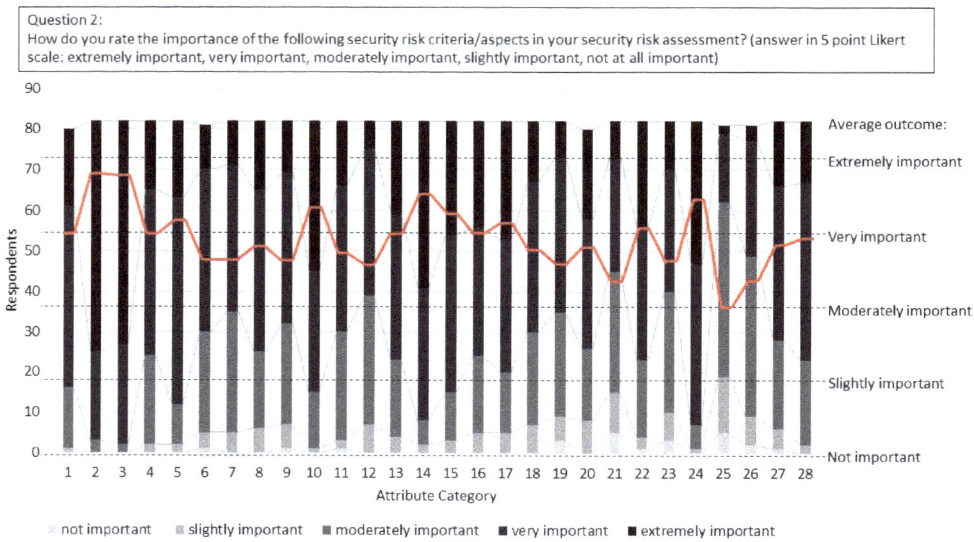

Figure 2: Descriptive analysis, rating of predefined risk attributes.

The absolute results of these answers are not considered of much value because the rating is assigned non-compensatory. The relative differences between the answers are considered of more value to be able to identify individual preferences.

The average answer is calculated by assigning a value to the Likert scale (extremely important is 5 points, etc.). The Likert scale is thus considered a continuous variable.

Correlation analysis showed various significant but low to medium correlations between the individual attributes (r values between 0.212 and 0.640). The only strong and significant correlation is identified between attribute 2 and 3 ($r = 0.900$, $p < 0.05$).

4.3 Part three: Ranking the attributes

When rating the attributes as analysed in the previous section respondents do not have to compare the attributes and can express their preferences without the need to make trade-offs. In the third part of the survey, however, the respondents are asked to rank the attributes and compose their individual top 10 of most important attributes. This is a form of compensatory decision making in which the aspects of and preferences to an attribute need to be weighed. This kind of decision making takes considerable cognitive effort and is considered a system 2 process. Each respondent can freely assign a rank (1 is most important etc.) to the 28 predefined attributes. To avoid order bias the attributes are presented in a random order to each respondent. 70 respondents completed this ranking task correctly. For overall comparison each top 10 listing is assigned a value (a number 1 listing 10 points etc.). The total value assigned to each attribute as well as the number of respondents listing an attribute in their top 10 is shown in Fig. 3.

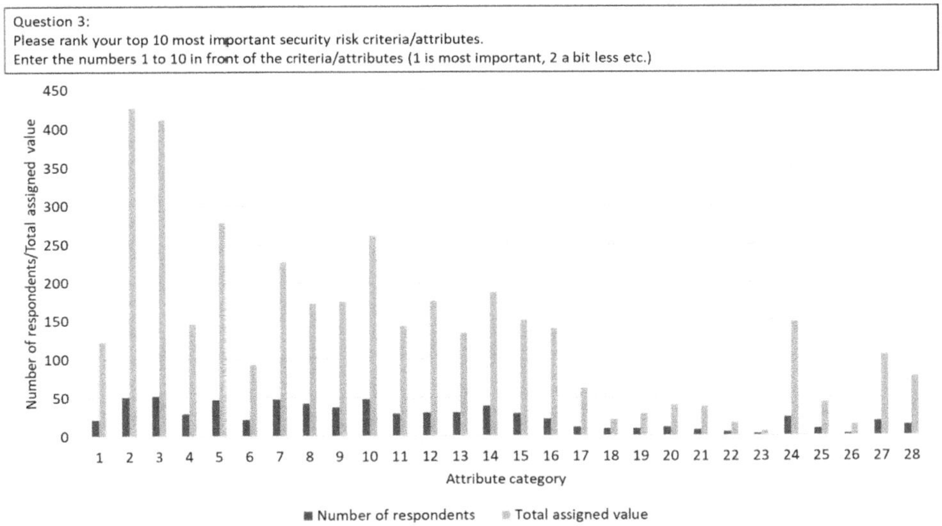

Figure 3: Descriptive analysis, ranking of predefined risk attributes.

It is clear that the impact on health and safety of employees (attribute 2) and of customers, clients and visitors (attribute 3) are overall considered the most prominent attributes in security risk assessments. This is in line with the results of the attribute rating (see previous section). Attribute 2 is listed by 73% of the respondents, attribute 3 by 74%. At the answers to the open ended questions only 24% mentioned health and safety.

The predefined list, in this experiment considered a stimulus, seem to have changed preferences of a large group of respondents.

The other main component of risk assessments: likelihood or probability shows a very different pattern. At the open ended questions 43% of the respondents state they take this attribute into account. At the rating question this attribute is rated on average between very important and extremely important. At the ranking question, however, only 34% of the respondents rank it in their top 10. This attribute received in total 148 points and rank at the 11 place of important attributes. As stated above the attribute likelihood is often listed at the open questions in combination with impact in general. These respondents (29%) seem to follow the, easily accessible, general definition of risk in their answers (risk = likelihood × impact). As they ranked the more detailed impact attributes in the third question they clearly choose impact over likelihood and might even ignore likelihood completely. These results comply to previous work, probabilities of events are not easy to define and people often disregard probability entirely [14], [15].

Table 3 finally presents an overview of the top 10 most prominent attributes over the three survey parts. The results show differences in priorities. The reaction of the respondents is clearly influenced by the list of predefined attributes that is inserted in the survey as a stimulus. A large group of respondents changes their priorities.

5 CONCLUSIONS AND DISCUSSION

The survey set up provided interesting information about the priorities of attributes in a security risk assessment. This section starts with a summary of the main results, followed by conclusions and discussion.

Table 3: Comparing the 10 most prominent attributes over three assessment processes.

	Question 1: When you **assess a security risk**, which criteria/attributes do you consider or take into account during your assessment? Please name the criterion and describe it briefly.		Question 2: How do you **rate the importance** of the following security risk criteria/attributes in your security risk assessment? (answer in 5 point Likert scale: extremely important, very important, moderately important, slightly important, not at all important)		Question 3: Please **rank your top 10** most important security risk criteria/attributes. Enter the numbers 1 to 10 in front of the criteria/attributes (1 is most important, 2 a bit less etc.)		
Top 10 rank	Top 10 is based on the proportion of the respondents listing attributes	% of resp.	Top 10 is based on the value the respondents assign to attributes (max. is 5: extremely important)	Avg. rating	Top 10 is based on the combination of the proportion of the respondents listing attributes **and** the value they assign to them (a top 1 listing is 10 points, a top 9 is 9 points etc.)	Total pts	% of resp.
1	1 Perceived impact of the security risk	53%	2 Impact on health and safety of employees	4.78	2 Impact on health and safety of employees	427	73%
2	24 Probability/likelihood of risk	43%	3 Impact on health and safety of customers, clients or visitors	4.76	3 Impact on health and safety of customers, clients or visitors	412	74%
3	5 Impact on business process (including IT downtime)	28%	14 Data disclosure (including privacy sensitive data)	4.49	5 Impact on business process (including IT downtime)	279	67%
4	2 Impact on health and safety of employees	24%	24 Probability/likelihood of risk	4.44	10 Damage to the reputation of the organisation	262	69%
5	10 Damage to the reputation of the organisation	15%	10 Damage to the reputation of the organisation	4.36	7 Financial impact	227	69%
6	7 Financial impact	14%	15 Loss of data integrity	4.23	14 Data disclosure (including privacy sensitive data)	187	56%
7	13 Data loss	11%	5 Impact on business process (including IT downtime)	4.16	12 Physical damage to assets	176	43%
8	14 Data disclosure (including privacy sensitive data)	9%	17 Personal responsibility/ accountability	4.15	9 Environmental impact	175	53%
9	12 Physical damage to assets	8%	22 Personal conscience	4.08	8 Legal impact/liability	173	60%
10	6 Impact on supply chain	7%	16 Disclosure or loss of intellectual property	4.06	15 Loss of data integrity	150	41%

The survey started with the open ended question: "When you assess a security risk, which criteria/attributes do you consider or take into account during your assessment? Please name the criterion and describe it briefly". This question allowed the respondents to answer based on their complete knowledge, perception and experience without restrictions and without any primer or influence from the researchers. The answers to these questions serve as an index of quickly or most available attributes. These are considered as most prominent by the security professionals at the point of time of answering the survey. The results show a predominant focus on impact attributes. Both in number of answers (192) as in the proportion of respondents mentioning one or more impact attributes (87%) this attribute category seems to be considered most relevant for security risk assessments. As a risk is often defined as a combination of uncertainty or likelihood and impact it is remarkable that less than half of the respondents (43%) mentions this second risk component. This might indicate that the likelihood of a security risk is not "on top of mind" and might be considered less important.

The survey continued with a set of rating questions. The respondents are confronted with a list of 16 context impact attributes, seven individual/personal impact attributes, a likelihood/probability attribute and four risk perception attributes. They are asked to rate the importance of each attribute using a five point Likert scale. The answers show a strong internal consistency and are, on average, centred around very important. The attributes rated highest are attribute 2: Impact on health and safety of employees (average 4.78 on a scale of 5) and attribute 3: Impact on health and safety of customers, clients or visitors (average 4.76 on a scale of 5). These two attributes have reach strong correlation ($r = 0.90$, $p < 0.05$). The majority of the respondents rate these attributes extremely important (attribute 2: 69.5% of the respondents, attribute 3: 67.1% of the respondents) and very important (attribute 2: 26.8%, attribute 3: 30.4%).

The third part of the survey consisted of a ranking question: "Please rank your top 10 most important security risk criteria/attributes". In this part of the survey the respondents are forced to make trade-offs between their favourite attributes. This is considered to be compensatory decision making (system 2). As in the previous rating question the health and safety attributes (attribute 2 and 3) are considered most important by the respondents. Overall the ten most highly ranked attributes are all context impact attributes. It is remarkable that the likelihood/probability attribute is ranked in their top 10 by only 34% of the respondents and ended overall at the 11th place. These results clearly indicate a predominant focus on impact attributes in accessing security risks by security professionals. Both the personal/individual impact attributes and the risk perception attributes (based on the SCM model) are not considered of much relevance by the respondents when confronted with the compensatory ranking.

This explorative study clearly shows the influence of stimuli on decision making by security professionals. Attributes that are not "on top of mind" and might even be, consciously or unconsciously, ignored in first instance, are considered very relevant after pointing to them. The most prevalent example are the two health and safety related attributes (attribute 2 and 3). They are only mentioned by 24% of the respondents in the first part of the survey. In the second part almost all respondents rate them extremely and very important while in the third part these attributes ended at the first and second place of the overall ranking. For real life daily praxis this could mean that without guidance the respondents take different attributes into account compared to if they are helped with tooling (in this case a predefined list). The consequence of this observed behaviour is that decisions made with or without tooling could be made on different grounds and define the outcome of the decision making process. A simple checklist could already help. Based on these results it can be concluded that attributes of security decisions that are considered extremely and very important by the majority of the respondents (see the rating question) are simply forgotten or ignored without help.

The second major finding is the lack of importance the security professionals in this study seem to appoint to likelihood/probability. At the open ended question less than half of the respondents (43%) mention likelihood or probability (87% of them mentions one or more impact attributes). At the rating, however, the majority rates it extremely important (43%) and very important (49%). When they are forced to compare the attributes in the third part of the survey only 34% of the respondents ranks likelihood/probability in their individual top 10. The assessment of likelihood or probability by people is based on their knowledge and beliefs and the assessments will thus vary over individuals. A subjective assessment of likelihood is hard for most people and they disregard likelihood entirely when confronted with risky choices [14].

Probability neglect is coined by Sunstein [15]. According to him this cognitive bias explains disregarding probability when assessing low-probability but high-impact threats. People tend to focus on the impact and ignore likelihood when strong emotions are involved. He also relates these emotions to the availability heuristic. Affect-rich decisions increase probability neglect [16]. This cognitive bias does not state that people neglect the likelihood, in situations where they can envision the impact (availability heuristic) and experience strong emotions the likelihood of occurring becomes less relevant or even irrelevant to them. Sieroń [17] added to this observation that, however the statistical likelihood of a high impact threat might be very small, people still want to avoid experiencing it. A small statistical likelihood does not mean this threat cannot affect the decisionmaker.

The respondents in this study might react according to these theories. The security risk domain is familiar to them so they can be expected to be able to envision the impact of security risks and threats. As it is there field of responsibility to decide upon or advice on managing these risks they can also be expected to feel affected by the possible impacts of these risks and threats. Finally, however small the statistical likelihood might be, the security risk or threat might materialize tomorrow and can affect their field of responsibility.

The important findings of this study might inspire other scholars to replicate them in other risk domains. They will raise awareness in both the academic as the professional security risk domain to the influence of cognitive biases on security risk decision making. This might lead to the development of de-biasing methods which can be added to existing security risk management processes enhancing security risk decision making. Managing security risks in organizations and society is of vital importance, understanding the decisions by individuals responsible for it is paramount.

REFERENCES

[1] Talbot, J. & Jakeman, M., *Security Risk Management Body of Knowledge* (Vol. 69), John Wiley & Sons, 2011.

[2] Hansson, S.O., A panorama of the philosophy of risk. *Handbook of Risk Theory*, Springer: New York, pp. 27–54, 2012.

[3] Möller, N., The concepts of risk and safety. *Handbook of Risk Theory*, Springer: New York, pp. 55–85, 2012.

[4] Cooke, R.M., *Experts in Uncertainty*, Oxford University Press: New York, 1991.

[5] Tversky, A. & Koehler, D.J., Support theory: A nonextensional representation of subjective probability. *Psychological Review*, **101**(4), p. 547, 1994.

[6] Yates, J.F., Veinott, E.S. & Patalano, A.L., Hard decisions, bad decisions: On decision quality and decision aiding. *Emerging Perspectives on Judgment and Decision Research*, Cambridge University Press, pp. 13–63, 2003.

[7] Svenson, O., Differentiation and consolidation theory of human decision making: A Frame of reference for the study of pre-and post-decision processes. *Acta Psychologica*, **80**(1–3), pp. 143–168, 1992.

[8] Tetlock, P.E., The impact of accountability on judgment and choice: Towards a social contingency model. *Advances in Experimental Social Psychology*, **25**(3), pp. 331–376, 1992.

[9] Kahneman, D., *Thinking, Fast and Slow*, Business Contact: Amsterdam, 2012.

[10] Van der Pligt, J. & Vliek, M., *The Psychology of Influence: Theory, Research and Practice*, Taylor & Francis: Abingdon, 2016.

[11] Gilovich, T., Griffin, D. & Kahneman, D., *Heuristics and Biases: The Psychology of Intuitive Judgment*, Cambridge University Press, 2002.

[12] Ribeiro, R.A., Fuzzy multiple attribute decision making: A review and new preference elicitation techniques. *Fuzzy Sets and Systems*, **78**(2), pp. 155–181, 1996.

[13] ANSI/ASIS/RIMS, *Risk Assessment RA1*, ASIS International: Alexandria, 2015.

[14] Evans, D., *Risk Intelligence: How to Live with Uncertainty*, Simon and Schuster: New York, 2015.

[15] Sunstein, C.R., Probability neglect: Emotions, worst cases, and law. *The Yale Law Journal*, **112**(1), pp. 61–107, 2002.

[16] Suter, R.S., Pachur, T. & Hertwig, R., How affect shapes risky choice: Distorted probability weighting versus probability neglect. *Journal of Behavioral Decision Making*, **29**(4), pp. 437–449, 2016.

[17] Sieroń, A., Does the COVID-19 pandemic refute probability neglect? *Journal of Risk Research*, **23**(7–8), pp. 855–861, 2020.

DEVELOPMENT OF A CONTINUOUS IMPROVEMENT APPROACH IN THE MANAGEMENT OF PRACTICAL PROBLEMS IN LOGISTICS AND TRANSPORT: CASE STUDY IN NUMILOG CONSTANTINE, ALGERIA

SAKER BESMA & CHAIB RACHID
University Fréres Mentouri Constantine 1, Algeria

ABSTRACT

This article describes the progress made in the development of a continuous improvement approach to managing the practical problems of logistics and transport in the Numilog-Est platform in the Constantine region of Algeria. This approach is built in agreement with those who are directly involved in the activity. Despite the benefits and services offered by these platforms, they are exposed to many types of risks and disruptions. This has a significant impact on its performance and efficiency. As a result, these companies need to manage these risks effectively, which is the goal of our research. These risks were identified from the analysis of the data and information collected on-site during the interview and the survey conducted with the administrators of this platform during our study and processed using the data processing program according to the prioritization series. Thus, it is possible to suggest the right ideas and actions to be taken. As a case study, we took the company Numilog-Est which has three logistics platforms among which those located in the city of Constantine. The latter is a company specializing in logistics and transport. It is the first of its kind in Algeria as a major player in the logistics chain.

Keywords: risk, continuous improvement, logistics and transport platform, 5M model, Pareto analysis, priority actions.

1 INTRODUCTION

For years, the problem of stocking and distributing goods has been a major problem for manufacturers, especially companies that do not have storage centers because of their high costs. Where the service of storage, exchange and distribution of goods depends on logistics platforms which are the real driver of the logistics system [1]. This has led to the use of transport and logistics companies due to the increasing need to store merchandise and manage deliveries interactively on long journeys across the national territory as well as to improve transport costs. However, despite the benefits gained from its services, these companies are exposed to many types of risks and disruptions that could jeopardize their work or even their sustainability. As a result, these companies need to manage the risks, to which they are exposed, objective of our research. As a case study, we took the company Numilog-Est. The latter is a company specializing in logistics and transport. It is the first of its kind in Algeria as a major player in the logistics chain. The company has three logistics platforms, among which the one located in the city of Constantine, our study zone.

The logistics platform can be defined as the place of receipt and storage of goods for a specific purpose to be reshipped in a very short time [2], it is also considered a tool for better managing time and space. Where It provides storage spaces in addition to several activities represented in the unloading and sorting of containers by destination, the loading and unloading of trucks, as well as the transport of goods [3]. Therefore, a slight performance failure can have significant consequences on operations where the slightest grain of sand in the logistics cogs can derail an entire company. As a reminder, risk can be defined as the chance of an uncertain risk, damage, loss, or other undesirable consequence that could

WIT Transactions on The Built Environment, Vol 206, © 2021 WIT Press
www.witpress.com, ISSN 1743-3509 (on-line)
doi:10.2495/SAFE210171

compromise the objectives of the organization [4], or even reduce the efficiency and effectiveness of the latter [5]. Consequently, the range of activities related to packaging, storage and transport activities are exposed to risk and must be overcome so that the company does not miss its objectives. From now on, to overcome these risks, it is necessary to measure them; "You can't improve what you can't measure" confirms [6]. In the past, the deadlines were long and not worrying, but today they are very short. It is calculated by the day, by the hour, even by the minute. Consequently, the authors try through this work and on the proposal of the manager of the company highlight the risks that cause delays in the loading, unloading and transport process at the Constantine logistics platform because of their impact on achieving the desired objectives and, therefore, customer dissatisfaction with delivery deadlines. These problems were established from the analysis of the data and information collected on this platform during our interview and survey carried out with the managers of this platform during our study stay. This paper is organized as follows: Section 2 describes the research methodology; Section 3 identifies the categories and sources of risks causing problems for the company; and Section 4 discusses and proposes corrective approaches by which the representatives of Numilog-Est design a framework to eliminate or mitigate these risks, or even engage in a process of continuous improvement.

2 METHODOLOGY

This article describes the progress accomplish to develop a continuous improvement approach in the management of practical problems of logistics and transport in the Numilog-Est platform in the Constantine region. This approach constructed in accordance with those who are directly implied in the activity. The quality of this work depends on the information collected about the system in several ways qualitative, semi-quantitative or quantitative [7], [8]. The envisaged approach is to use structured interviews with those in charge of the logistics and transport chain to understand the operating principle of this company and the problems associated with various on the site operations. This allows the authors to obtain real data from the field that allows the company to build a memory of the risks and develop an action plan that meets the company's expectations, while drawing on the experience gained in the event of accidents and incidents. Conclusions or even recall experience notes. It is clear from this experience and this observation that the prevailing problems in the company are the problems related to the delay in loading, unloading and transporting goods. To better understand and master these two problems a questionnaire was developed and distributed to Operations manager, QHSE Supervisor, Logistics manager and HR, This questionnaire was developed on the principle of Ishikawa making it possible to identify all the causes that have a more or less direct influence on the problems observed (Appendix), since the questionnaire technique is considered a measurable data collection tool. The next step is to represent the data which allowed the authors obtained using illustrative forms to facilitate reading and analysis, allowing us to get a clear picture of the issues raised by those responsible for transport and logistics. Two main problems were revealed, delays in loading and unloading of goods and delays in transporting goods.

3 PRESENTATION AND ANALYSIS OF DATA

For a better data analysis, the authors used the Ishikawa diagram (5M) to better understand the cause and effect. This method allows us to find the possible causes, even the root cause of the problem [9]. The first step was to highlight all the risks that cause these two problems (Figs 1 and 2).

From the first observation of Figs 1 and 2, we note that the two problems cite all risk categories (Material, Matter, Manpower, Method, Mother Nature "environment").

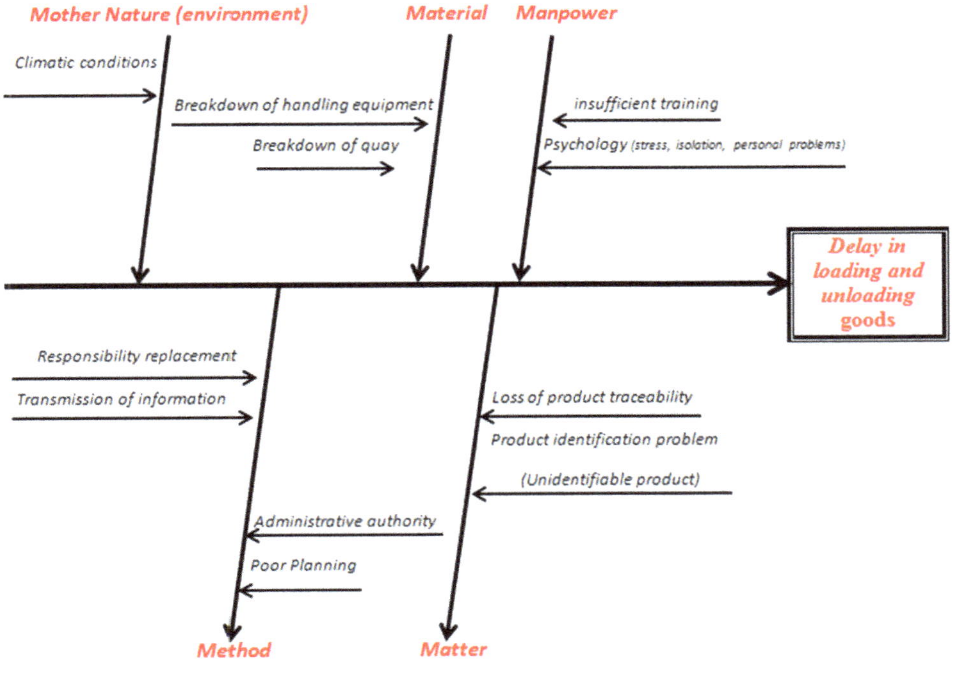

Figure 1: Cause and effect diagram of delays in loading and unloading goods.

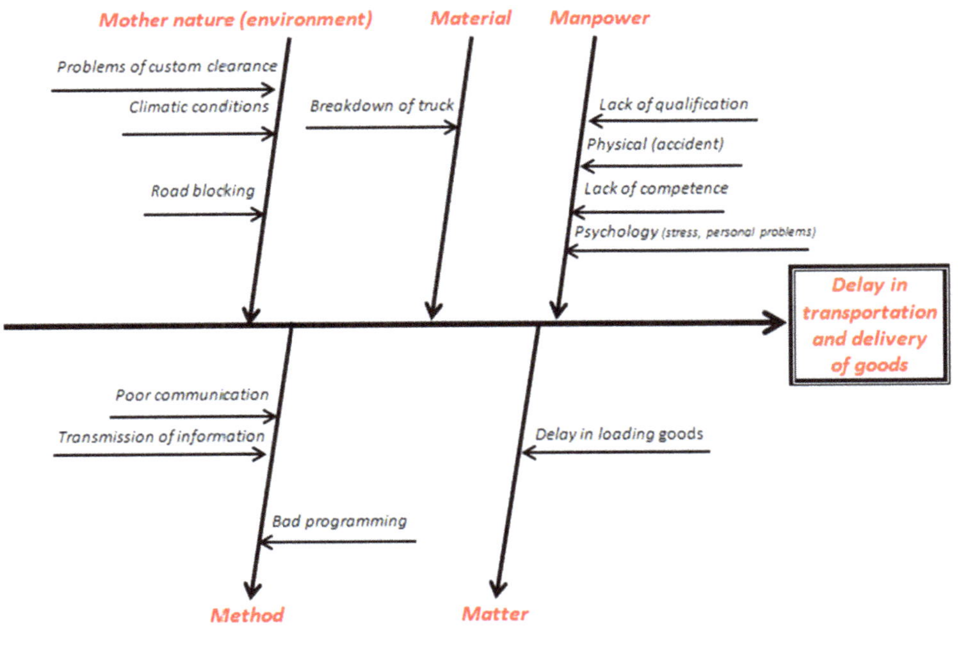

Figure 2: Cause and effect diagram of delays in transportation of goods.

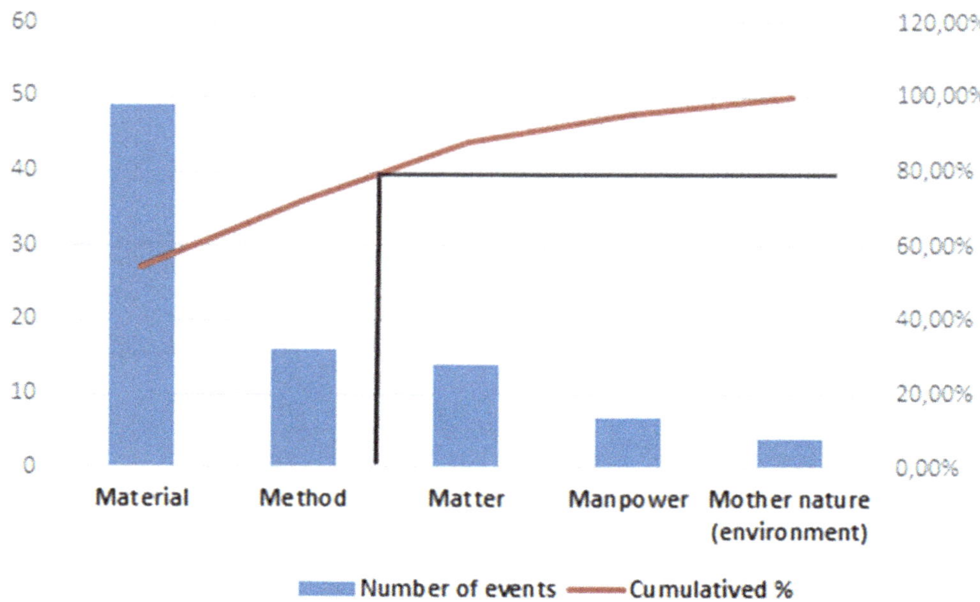

Figure 3: Presentation of the pre-prioritization of the problem of delay in loading and unloading of goods when treated by Pareto analysis.

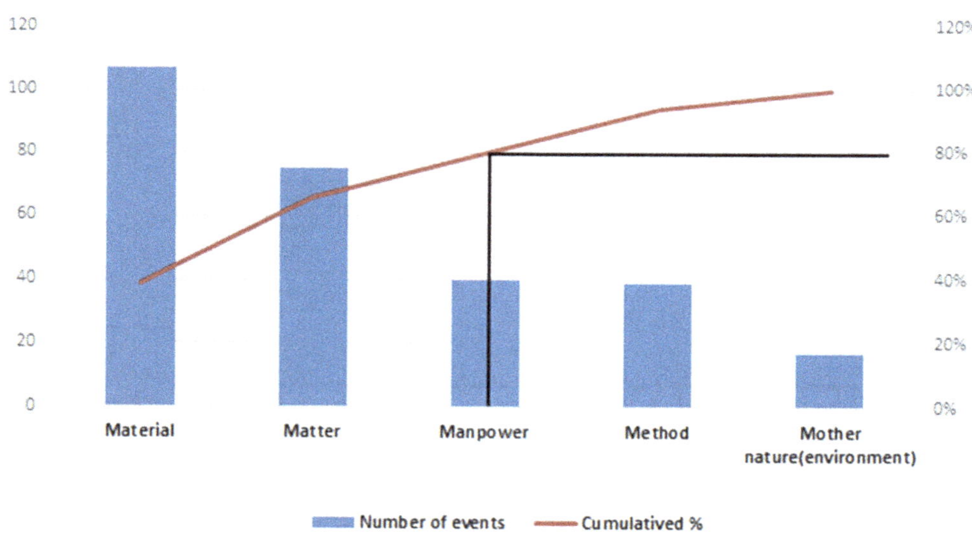

Figure 4: Presentation of the pre-prioritization of the delay problem in transportation of goods when treated by Pareto analysis.

To best ensure the operational safety of installations, processes and operations, a company must engage in the process of continuous improvement in its management and use of its observations. Thus, prioritization of the actions to be taken is therefore essential for their sustainability and efficiency. The approach used for prioritizing actions is the call to the Pareto chart (20/80%) [10], [11]. Thus, manage to prioritize the actions to be taken in the groups that have the most influence on the smooth running of the entity, or even conduct effective management guided by real field data.

The analysis of Fig. 3 shows that 80% of the risks that cause the delay in loading and unloading of goods are related, respectively, to the risks of method and material. As for the delay problem in transportation of goods, the results of Fig. 4 that 80% the risks are related to material, matter and manpower. For quality performance and to satisfy the customers at delivery time, users must pay attention to these risks.

4 DISCUSSION AND RECOMMENDATIONS

Following these results, the authors note that the risks raised are mainly related by method risks and material risks regarding the delay in loading and unloading the goods and, material, matter and manpower risks regarding the delay in the transportation and delivery of good. Special attention must be paid on these categories of risks, thus, to have a better and effective visualization in each category of risk, it was necessary to define another phase, the percentage of each risk in each category was used. In order to resolve these problems and engage in a continuous improvement process. The authors propose a set of tasks be assigned by priority of action. This solves many predominant risks caused the delay in loading, unloading and the transportation of goods and fulfills the needs of this system for better performance and company sustainability.

4.1 Essential action on risks related to the delay in the loading and unloading process

4.1.1 For material risks

Fig. 5 shows that breakdown of handling equipment is the dominant risk of 85.71%, with a significant effect on the one hand on delays in loading, unloading and transportation of goods and customer satisfaction on the other hand, on the saturation of the storage area and associated costs. As for the breakdown of quay, it is 14.28%. The latter is the main cause of the saturation of other quays leading to result in delayed activities and reduced quality of work.

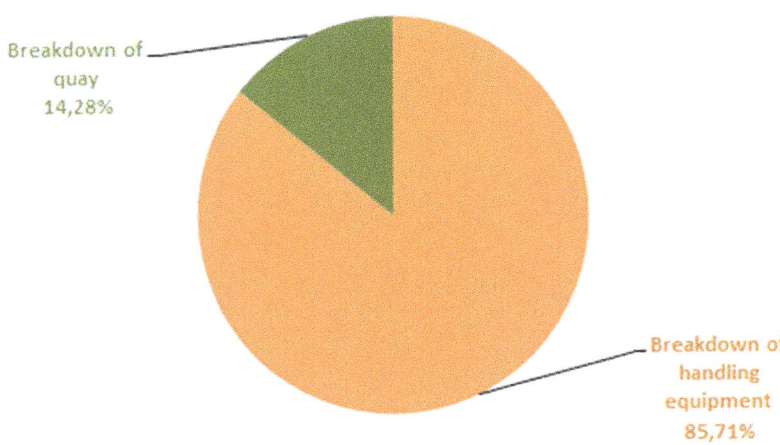

Figure 5: Representation of the material risks.

To minimize the impact of material failures and ensure good functioning during handling activities as well as achieving profitability objectives. Periodic maintenance is essential and must be performed out daily for the equipment after each use by making a check list for equipment monitoring. This list allows the maintenance department to know all the small and large repairs and even the replacement of equipment parts.

4.1.2 For method risks

According to our results in Fig. 6, it is shown that 43.75% of the risk is the transmission of information, this significantly affects in the loading process and therefore the delivery performance. The effect of information transmission on a company's performance cannot be hidden, most companies has become build all their management strategies on information transfer technology. 31.25% the authors note here that this risk could lead to the impossibility of achieving the desired objective, poor planning in this platform has led to a shortage of employees compared to forecasts so better results cannot be obtained with poor planning. Administrative authority and replacement responsibility takes 12.50%. Poor management by the administration and replacement of responsibility automatically leads to inconsistent and slow tasks. These method risks had a huge influence on the loading, unloading of goods and on performance and quality of their work.

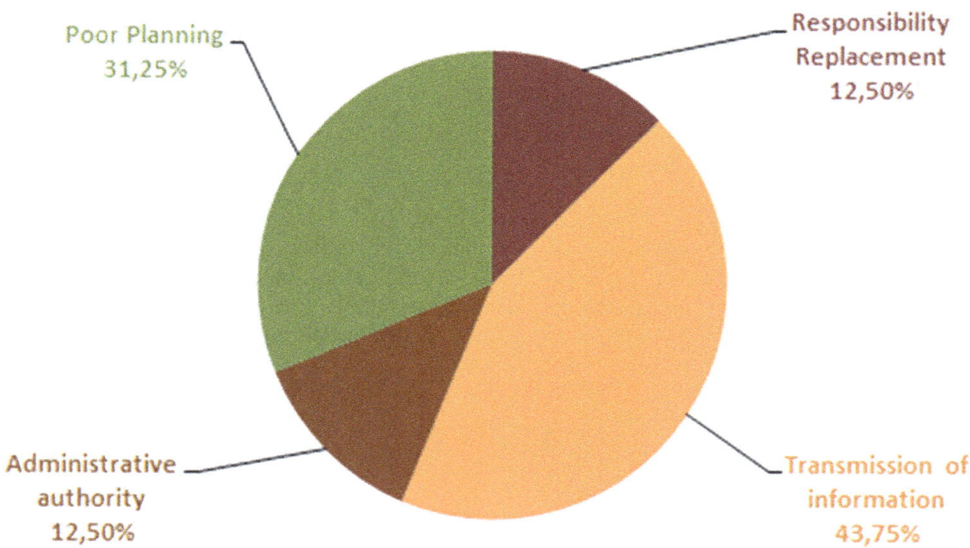

Figure 6: Representation of the method risks.

In our opinion, in order for the company to become more reliable in delivery times and more flexible in the process of loading and unloading goods, the manager must monitor and daily check the transfer and exchange of information to respond to customer requests, which positively affects the quality of work, as well a company must plan for a long time and in coordination with customers so that all changing forecast can be met without delay. Therefore, the forecast information shared with clients will enable them to determine their operational capacity. As well as focusing on cooperation between the different members of the company to coordinate the rapid response to any process.

4.2 Essential actions on risks related to delays in the transportation of goods

4.2.1 For material risks

Fig. 7 shows that the only risk related to material is the truck breakdown. In our opinion, Aged trucks and failure to held routine maintenance increases the occurrence of these risks.

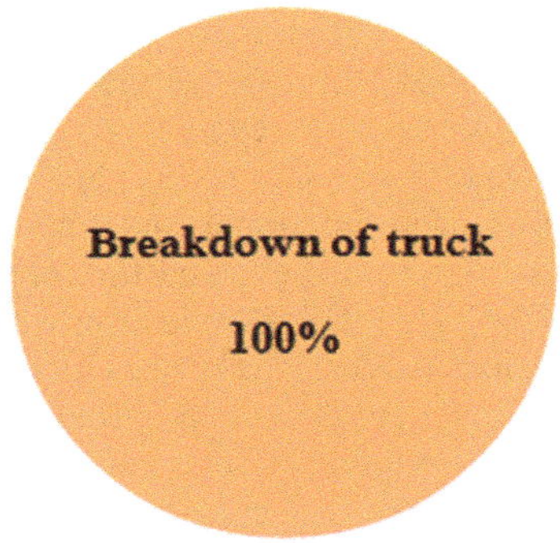

Figure 7: Representation of the material risks.

To reduce this risk, the authors suggest replacing and purchasing new trucks, also regular a maintenance service must be established at the company level; to allow for regular maintenance and create a record that includes all changes and repairs that have occurred on the truck so that it is possible to know which parts have been changed or repaired. Otherwise a contract should be entered with specialized maintenance workshops.

4.2.2 For matter risks

Fig. 8 shows that the only risk related with matter is the delay in loading goods. This risk is highly associated with delays in the transportation and delivery of goods. Whenever there is delay in the loading of goods, automatic there is delay in delivery of goods. This risk and the factors causing its occurrence and how to treat it have been studied in detail in the previous section.

Figure 8: Representation of the matter risks.

4.2.3 For manpower risks

Fig. 9 shows that 71.79% of manpower risks are psychological. Otherwise, all actions carried so far have proved to be wrong and disparate and staffs are often unable to manage their tasks without any training, information, motivation or guidance, often under the pressure of contradictory views of different responsible. Now, every organization is responsible for the occupational health and safety of its workers and others who may be affected by its activities. This responsibility includes promoting and preserving their physical and mental health [12]. These problems must be resolved to improve the work of drivers. Consequently, it is recommended to identify, upstream all the risk factors having a negative impact on the working climate and the health of the workers, even every deterioration in a worker's health situation affects his functional capacities or his work [13]. From now on, the consequences the consequences of these factors lead to functional constraints that have implications for the quality of life at work and the quality of work, even for the company's intended performance: sustainability and development [14], [15]. Despite the failure in the psychological level, only 2.56% of the risks related to physical accidents have been recorded. In our view, this is a very positive point for all of the company's employees. In terms of competence, the authors note that the majority of drivers do not exactly know the destination of customers, otherwise disruption in delivery times occurs. After analyzing the results, to mitigate this type of risk, can be used GPS. With regard to qualifications, it is very normal to be a new activity. Here, the authors propose the necessary training for new recruits to carry out this function in full compliance with the Traffic and Driver Rest Times Act (30 minutes rest after every 4 hours of driving, is obligatory).

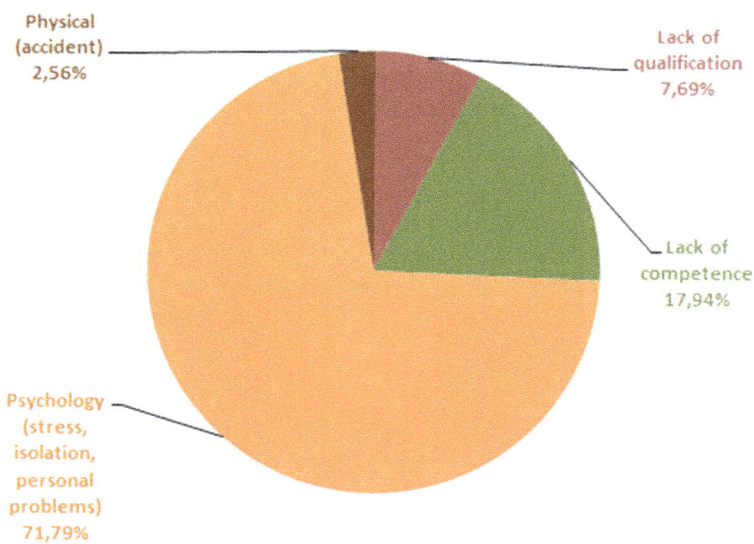

Figure 9: Representation of the manpower risks.

5 CONCLUSION

In conclusion, and based on the results of the analysis, it is clear that the management of the practice of the logistics organization is of paramount importance to attain a specific level of efficiency and quality of work. In addition, it can be concluded that the risks of delays in loading and transporting goods contribute to the emergence of a set of problems that prevent the achievement of programmed objectives. The use of the Pareto chart, allowed the authors to prioritize the actions to be carried, which are respectively the method, material risk for the problem of the delay of loading and material, matter and manpower for the delay of transport. These are the risks on which one the efforts of the manager of the organization should be concentrated without neglecting preventive measures for other risks. This work is part of a continuous improvement process. This work made it possible to categorize the problems according to its occurrence and importance and the actions to be taken according to the priorities. In addition, in order to identify and manage the risks and overcome the experience problems in an organization, a personnel cultural level must be part of the fundamental values of the actors of the work world in this field. Now, every organization is responsible for the occupational health and safety of its workers and others who may be affected by its activities. This responsibility includes the promotion and preservation of their physical and mental health. Thus, on the basis of data collected on the site, the authors propose a set of ideas that would certainly and effectively improve the management of the logistics and transport process. Subsequently, a promotion of preventive actions is also to be expected, namely:

- Promotion, information and training in occupational health and safety;
- Call on feedback. They can serve as a basis for undertaking future prevention actions;
- Develop training and welcoming programs for new employees, as well as continuing training for employees, without forgetting social dialogue (with employee representatives);
- In each action carried out, compare the results with the objectives set. To measure the effectiveness of actions, the establishment of criteria or indicators is necessary;
- Harmonize the health and safety policy with other company policies;
- Promote a multidisciplinary approach (technical, human and organizational);
- Analyze accidents and incidents, even problems by going back to the most upstream causes.

APPENDIX

Questionnaire established on the problems causing delays in loading, unloading and transport of goods in the logistics platform "Numilog-Est" in the region of Constantine "Algeria".

Age: .... **Years of experience:** ......

Risks that cause in the delay of loading, unloading and transport of goods

Material

1- Breakdown of quay
Yes No Number of cases observed ☐

2- Breakdown of handling equipment
Yes No Number of cases observed ☐

3- Breakdown of works
Yes No Number of cases observed ☐

4- Complexity of materials
Yes No Number of cases observed ☐

5- Truck breakdown
Yes No Number of cases observed ☐

Method

1 - Responsibility replacement
Yes No Number of cases observed ☐

2- Transmission of information
Yes No Number of cases observed ☐

3- Poor Planning
Yes No Number of cases observed ☐

4- Failure to validate the action
Yes No Number of cases observed ☐

5- Administrative authority
Yes No Number of cases observed ☐

6- Poor communication
Yes No Number of cases observed ☐

7- Check list incomplete
Yes No Number of cases observed ☐

Manpower

1- Lack of qualification
Yes No Number of cases observed ☐

2- Lack of competence
Yes No Number of cases observed ☐

3- Insufficient training in the use of handling equipment
Yes No Number of cases observed ☐

4- Physical problem (accident)
Yes No Number of cases observed ☐

5- Psychological problem (stress, isolation, personal problems)
Yes No Number of cases observed ☐

Mother Nature (environment)

1- Climatic conditions
Yes No Number of cases observed ☐

2- absent employee
Yes No Number of cases observed ☐

3- Problems of custom clearance
Yes No Number of cases observed ☐

4- Road blocking
Yes No Number of cases observed ☐

5- Accident while loading goods
Yes No Number of cases observed ☐

Matter

1-Loss of product traceability
Yes No Number of cases observed ☐

2- Updating of documents
Yes No Number of cases observed ☐

3 - Unidentifiable product
Yes No Number of cases observed ☐

4- Delay in loading and unloading of goods
Yes No Number of cases observed ☐

REFERENCES

[1] Raimbault, N., Douet, M. & Fremont, A., Les plateformes logistiques: Entre fluifité et fixité, Programme de recherche FLUIDE, 2010.

[2] Carrera, S., Planication et ordonnancement des plateformes logistiques. Thèse de Doctorat, Institut National Polytechnique de Lorraine, 2010.

[3] Direction général des infrastructures, Les entrepôts et les plates-formes logistique, p. 1, 2010.

[4] Abdel-Basset, M., Gunasekaran, M., Mohamed, M. & Chilamkurti, N., A framework for risk assessment, management and evaluation: Economic tool for quantifying risks in supply chain. **90**, pp. 489–502, 2019.

[5] Heckmann, I., Comes, T. & Nickel, S., A critical review on supply chain risk: Definition, measure and modeling. *Omega*, **52**, pp. 119–132, 2015.

[6] Harland, C., Brenchley, R. & Walker, H., Risk in supply networks. *Journal of Purchasing and Supply Management*, **9**, pp. 51–62, 2003.

[7] Yeboah, J. & Ewur, G.D.,. The impact of WhatsApp messenger usage on students' performance in tertiary institutions in Ghana. *Journal of Education and Practice*, **5**(6), pp. 157–164, 2014.

[8] Bolarinwa, O.A., Principles and methods of validity and reliability testing of questionnaires used in social and health science research. *Nigerian Postgraduate Medical Journal*, **22**(4), p. 195, 2015.

[9] Chaïb, R., Taleb, M., Benidir, M., Verzea, I. & Bellaouar, A., Failure: A source of progress in maintenance and design. *Physics Procedia*, **55**, pp. 185–191, 2014.

[10] Kharzi, R., Chaib, R. & Akni, A., Prioritizing the actions to be undertaken in health and safety at work: Case study region of Tiaret. *International Journal of Law and Management*, 2020. DOI: 10.1108/IJLMA-01-2018-0009.

[11] Mounira, K., Rachid, C. & Abdelaziz, K.A., Sensitive approach to identify black spots in urban transport, with application to cities in Algeria. *Journal of Transport Problems*, **16**(1), 2021. DOI: 10.21307/tp-2021-017.

[12] ISO 45001:2018.

[13] Glen, P., Kenny, J.E., Yardley, L. & Martineau, O.J., Physical work capacity in older adults: Implications for the aging worker. *American Journal of Industrial Medicine*, **51**(8) pp: 610–625, 2008.

[14] Aboubakr, M., Contribution à l'analyse des risques et à l'amélioration de la procédure d'identification, et la hiérarchisation des mesures de contrôle, mémoire de fin d'étude, 2017.

[15] Salmi, S. & Rachid, C., Feedback a learning tool for future organizations. *World Journal of Engineering*. DOI: 10.1108/WJE-12-2016-0169.

BEHAVIOR OF ACTIVE AND PASSIVE NOISE REDUCTION PERSONAL PROTECTIVE EQUIPMENT: A CASE STUDY IN A LIMESTONE QUARRY PLANT

GUIDO ALFARO DEGAN[1], ANDREA ANTONUCCI[1], GIANLUCA COLTRINARI[1], DIEGO ANNESI[2] & DARIO LIPPIELLO[1]
[1] University of Rome 3, Italy
[2] Italian Workers Compensation Authority (INAIL), Italy

ABSTRACT
In this study, an estimation of the behavior of three different passive and active personal ear protection equipment was done in order to evaluate their performances when used in a limestone quarry plant. Twenty measure points were chosen for the characterization of the noise emitted in the plant and for the calculation of the workers' noise exposition. Starting from the data we previously measured in laboratory tests, the noise reduction curves, characteristics of each protective equipment, were calculated for different configurations or ways of using. Results show that all the protection equipment are potentially able to reduce the noise level under the lower exposure limit in all the measured positions, but in most of the cases they produce hyper-protection. Differences were found for different ways of using or settings. When earplugs are not properly inserted, the noise could overcome 80 dB(A) in some of the points of our case study. When active earplugs are set to the maximum volume, the provided protection wasn't found to be sufficient in all of the noisier points of the plant. In some cases, depending on the spectra and amplitude of the external noise, the sound produced by active earplugs can also be higher than that measured in the external environment. On the contrary, when the earplugs are set on the middle volume, the protection was found to be always optimal. Since noise passing through active earplugs can be attenuated or amplified according to the amplitude of the external noise, the perceived sound is much more homogeneous, especially when passing across points at different noise levels, compared to that provided using passive earplugs. In conclusion, active earplugs, when properly set, could be the better choice, especially in those cases in which workers need to move across points at different noise levels. However, important is to correctly estimate the external noise and the properties of the active earplugs in order to give instruction to workers for their correct settings, so as to avoid both the hyper- than the under- protection.
Keywords: active earplugs, quarry noise, ear muffles, noise protection, noise risk, personal protective equipment.

1 INTRODUCTION

It's known that in quarry plants workers can be exposed to high levels of noise, due to the presence of noisy machines used for crushing and classifying stones, other than due to the massive movement of trucks and loaders [1]. These noises can represent other than a risk for workers [2], [3], also an environmental problem in those cases in which the plant is not far from urban centers [4], [5]. Solutions as encapsulation with sound-absorbing panels, can sometimes provide good results especially for fixed machines [6], but unfortunately, they are not always easy to apply, depending on the configuration of the plant and the spaces needed for their proper function and inspection. Otherwise, the use of personal protective equipment (PPE) inevitably remains the most common solution for workers' protection [7].

The market proposes many types of PPE, each one having different noise attenuation for different types of noisy sources, depending on both the amplitude of the sound than the frequency spectra [8]. The choice of the proper PPE requires, other than a careful analysis of the noisy sources, also the characterization of the different working tasks and areas in which

www.witpress.com, ISSN 1743-3509 (on-line)
doi:10.2495/SAFE210181

they take place, in order to guarantee good protection and avoiding possible interferences with other protective equipment [9].

In most of the cases in which the exposition to noisy sources varies a lot when moving across different places, a unique PPE it's often unable to satisfy all the properties it should have [10]. This can happen in quarry plants, in which workers don't operate in fixed positions, so the noise they are exposed to can frequently change. In these cases, when passive earplugs or ear muffles are chosen, a compromise of their properties in reducing the noise in all of the areas can't be avoided. What often happens in these cases, is that PPE is chosen in order to be able to guarantee good protection when workers are exposed to the noisiest point of the plant. Anyway, this compromise could result in a hyper-protection against noise in some other points. Hyper-protection can generate difficulties in comprehension of spoken words, annoyance, inconvenience, other than generating safety reduction, since could be difficult to hear alarms or horn signals [11], [12].

In recent years, there has been an increase of active noise protection devices on the market, such as electronic earplugs. These types of PPE, other than made by soft materials that provide the passive attenuation of the noise, are also equipped with an active device consisting of a microphone, a microprocessor, and an amplifier. The active part of the earplugs is able to modify the external sound, also providing an amplification at some frequencies in order to improve the perceived spoken language and reduce the hyper-protection [13].

The behavior of this type of active protection devices varies consequently with the variation of the amplitude of the external sound, following some pre-set curves of attenuation, which are pre-loaded in their software. Sometimes users can also be able to change the configuration of the chosen pre-set attenuation curves, simply by increasing or decreasing what is often indicated as the volume of the device, or simply turning it off. That means that when the device is switched off, the sound attenuation is only provided by the soft material they're composed by, while according to the setting volumes, workers can decide how much the speakers inside the devices will be able to increase or decrease the external sound. These devices would be consequently a good solution especially in those places in which the variation of the external noise does not allow the use of only one type of PPE.

Anyways, the pre-set attenuation curves are studied in laboratory tests at fixed noise emission, such as white or pink noise, and moreover at pre-set amplitudes [14]. That means that in real fields the attenuation of the external noise could not be that expected, and moreover, the speakers inside the device could produce a sound louder than that outside.

Due to the previous considerations, in this paper we aimed to evaluate the efficiency of this type of active earplugs when used in a quarry plant. Starting from some data we previously measured in laboratory tests, an estimation of the attenuation curves used by the software of the earplugs was made, so to be able to approximate their behavior in protecting workers against noise. We also used data from other types of passive protection equipment, in order to be able to compare their behaviors.

2 MATERIALS AND METHODS

The measures were taken in the production plant of a limestone quarry located in central Italy. A ½ inch free-field microphone connected to a 200 V–10 GΩ preamplifier from Larson and Davis and a four-channel data acquisition system (Soundbook ToughBook CF-19) were used for the measurements campaign. The data were analyzed using the software Samurai. Before and after the measures, the microphone was calibrated using CAL200 – class 1 – Larson and Davis calibrator.

Because the reduction of the production due to the restrictions caused by the pandemic, not all the plant was operative during the period of the present study, so the measure points were chosen only in those areas in which there were functioning machines or other activities.

Twenty measure points were chosen for the characterization of the noise emitted in the plant and for the calculation of the noise exposition in both cases of use and non-use of personal protective equipment (PPE). Most of the measures were taken proximally (between 5 and 15 meters, in the orthogonal projection) to the main production machines, such as rock breaking machines, impact crushers, vibrating sieves, and rotary drum mills. The other measures refer to the rest of the working areas, such as the loading area, the shop, the office and the mess hall. Fig. 1 reports the map of the plant with the measurement points and the machines or the working areas they refer to.

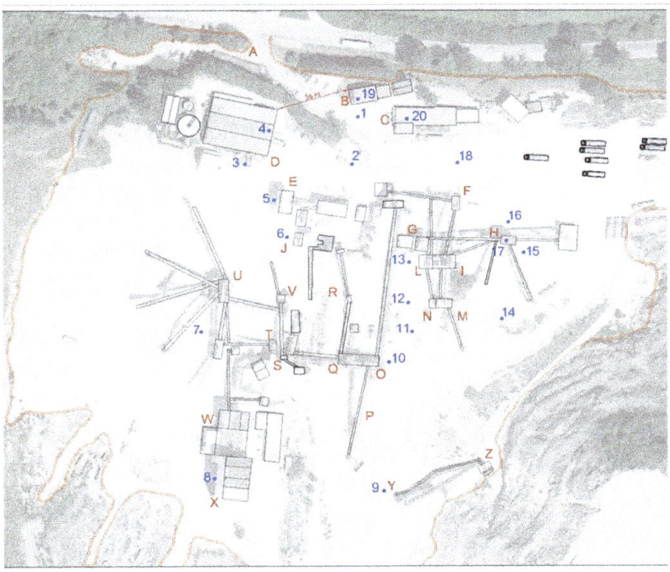

A: entrance
B: mess hall
C: office
D: shop
E: loading hoppers
F: vibrating sieve
G: impact crusher
H: vibrating sieve
I: impact crusher
L: rotary drum mill
M: vibrating sieve
N: impact crusher
O: vibrating sieve
P: primary conveyor belt
Q: vibrating sieve
R: impact crusher
S: impact crusher
T: mud remover
U: vibrating sieve
J: bag filters
V: cyclonic separator
W: rotary drum mill
X: loading hoppers
Y: primary storage
Z: feeding tunnel

Figure 1: Map of the plant with the numbered measure points and their locations.

Most of the activities in the plant are made using loader machines or other types of vehicles, so workers are partially protected against the external noise by their cabins, moreover in these cases it's not so frequent that workers operate very close to the plant's machines. On the contrary, in those cases requiring inspection activities and/or maintenance, workers operate outside, sometimes very close to the noise sources. Even when the required time for these activities doesn't last long, the amount of noise they are subjected to can be very high. For this reason, the calculation of the daily noise exposition was made supposing that only a part of the working day is spent at the production plant, while the rest of the time is spent in protected locations, such as the office/shop or inside the cabin of the vehicles. Two different scenarios were considered for the calculation, supposing both 5 and 10 minutes of permanence in each of the outside located eighteen positions, and fixing at 1 minute the required time for moving across two consecutive positions. The noise between two consecutive points was calculated as the average of the noise measured at the two fixed positions. Consequentially the two scenarios are as follow:

1st case (5' in each position): 109' at the plant; 311' at the office/shop; 60' at the mess hall; 2nd case (10' in each position): 199' at the plant; 221' at the office/shop; 60' at the mess hall.

The calculation of the $L_{ex,8h}$ was made according to the following equation:

$$L_{ex,8h} = 10 \, log \frac{1}{T_0} \sum_{i=1}^{n} T_i \, 10^{\frac{L_i}{10}}, \quad (1)$$

in which $L_{ex,8h}$ is the noise exposition of the worker for a working day of 8 hours; T_0 is the overall time of exposition; T_i is the exposition's time for the ith point of measure; L_i is the equivalent level of noise measured in the ith point.

A simulation of the noise protection provided by three different PPE was then made. They are as follow: 3M 1100 passive earplugs; Peltor X1A passive earmuffs; 3M Peltor LEP-100 EU electronic earplugs. The PPE were chosen among those able to protect workers at the noisiest point of the plant, other than among those we previously used in other studies. For the characterization of the performance of the active earplugs when used in the present context, we started using the data we previously measured in laboratory tests [13]. The measures were taken in an anechoic chamber, using a mannequin equipped with two microphones inside the head and positioning other two microphones at 10 cm outside the ears. The tests were made using different configurations of the hearing protection device (switched-off; volumes: 1, 2, 3) and measuring the noise attenuation referring to a speaker emitting pink noise at different amplitudes: 75; 80; 85; and 87 dB respectively.

Regarding the attenuation provided by the 3M 1100 earplugs, other than the data provided by the manufacturer, differences in terms of workers' protection in both the cases in which they are correctly inserted or not, were also estimated. In this case we used the results obtained in another our previous laboratory test [15], in which 100 subjects were asked to wear earplugs before and after the training on their correct use. For these laboratory tests, modified earplugs were used, in which a little plastic tube was inserted in the plugs in order to capture the noise inside the ears by the use of a micro-microphone connected to it. Another little microphone was positioned just outside the ears for the estimation of the noise attenuation. A white noise, generated by a speaker positioned at a fixed distance, was used as signal source.

The attenuation of the PPE in each frequency was calculated according to the equation:

$$L'_{Aeq} = 10 \, log \sum_{f=125}^{8000} 10^{0.1(L_f + A_f - APV_f)}, \quad (2)$$

in which L'_{Aeq} is the noise level resulting from the attenuation; f is the center frequency octave-band from 125 Hz to 8,000 Hz; L_f is the noise level at the considered frequency; A_f is the weighting-value (A) for the considered frequency; APV_f is the attenuation of the PPE at each frequency.

3 RESULTS

3.1 Characterization of the plant's noise

In Table 1 the noise levels measured in each of the eighteen external points are reported, other than the noise inside the office and the mess hall, in terms of linear, peak and weighted values (A; C), according to the numbers listed in Fig. 1. Note that the noise indicated as "office noise" is actually an estimation of the average noise coming from different measures: office, shop and inside the cabin of a truck. This approximation was necessary since we needed an estimation of the average noise exposition of workers when they stay in protected places.

Table 1: Measurement points and noise levels.

Measuring points	Location	Leq (dB)	Leq (dB(A))	Leq (dB(C))	Peak (dB(C))
1	Outside the mess hall (40 m from the plant)	86.9	74.9	85.4	102.3
2	Open area close to the gasoline station (25 m from the plant)	89.3	78.1	88.1	103.0
3	Outside the shop (20 m from the plant)	85.6	74.3	84.6	99.1
4	Inside the shop/entrance (18 m from the plant)	85.2	73.5	84.0	98.2
5	Close to loading hoppers (5 m)	85.3	73.1	84.4	98.5
6	Close to bag filters (5 m)	95.3	83.1	94.7	107.4
7	Open area not close to active machines (70 m from active plant)	80.5	70.6	79.0	95.1
8	Close to loading hoppers (5 m)	91.7	76.3	89.4	102.1
9	Open area nearby the primary storage	81.6	76.1	80.9	98.4
10	Close to vibrating sieve (5 m)	88.4	87.4	87.9	102.5
11	Between vibrating sieve and impact crusher (15 m)	89.5	88.5	89.1	112.7
12	Close to impact crusher (10 m)	93.6	92.4	93.2	112.4
13	Close to rotary drum mill (5 m) and impact crusher (7 m)	98.8	94.8	98.2	121.0
14	Open area, 25 m from vibrating sieve	84.5	78.4	83.3	98.1
15	Storage area between vibrating sieve and conveyor belts	85.8	82.2	85.1	99.1
16	Storage area between vibrating sieve and conveyor belts	87.7	83.1	87.1	100.0
17	On the gangway close to vibrating sieve (2 m)	100.5	99.7	100.0	115.0
18	Storage area close to vibrating sieve (15 m)	87.0	77.0	85.8	101.3
19	Inside the mess hall	76.2	63.5	76.0	86.1
20	Inside the office	76.8	65.5	76.3	88.8

Results show that the measured noise Leq(A) exceeds the lower exposure action values of 80 dB(A), as established by the European Directive 2003/10/EC [16], in eight points. Most of them are located close (less than 15 m, in the orthogonal projection) to the following machines: bag filters, vibrating sieves, impact crusher and rotary drum mills. In 5 of the previous points the noise also exceeds the exposure limit values fixed at 87 dB(A). Fig. 2 reports an estimation of the noise levels in the plant, considering those areas functioning during the period of the measurements campaign.

Figure 2: Noise levels in the plant. Red area: over 85 dB(A); green area: over 80dB(A).

It's not in the purpose of the present study to obtain a precise characterization of the plant's noise, so these data are intended to be just an approximation of the noise levels emitted in the plant, since the model used does not consider the sound power level emitted by each machine and its direction. In the modeling the noise sources are supposed to be punctual and in free field, without considering reflection surfaces and only partially considering the absorber panels or obstacles positioned around the machines. Dashed circles reported in Fig. 2 represent just an idea of the noise levels that could be in the other part of the plant, when the machines there located are supposed to be functioning. In this case, the same previous made hypotheses of punctual noisy source in free field were made, considering the same noise emitted by similar machines

As shown in Fig. 2, four of the previously indicated points, in which the noise levels overcome the exposure limit value, are located in the same area (points: 10,11,12,13), making it the most critical in terms of noise exposition. The spaces in which the noise could overcome 80 dB(A) and 85 dB(A) are indicated in green and red respectively. As can be noted, quite all the areas around the functioning plant were characterized by having high levels of noise, requiring workers to be protected.

3.2 Properties of the tested personal protection equipment

Fig. 3 reports the noise reduction properties of the passive PPE used in the present study, for each of the octave frequency bands ranging from 125 to 8,000 Hz. As shown in the figure, 3M earplugs are those performing the higher noise reduction according to the data reported by the manufacturer. The range of reduction is between 33.1 dB at 125 Hz to 48.8 dB at 4,000 Hz with a quite homogeneous increasing protection in the low and high frequencies, followed by a flat behavior in central frequencies. Our previous tests confirmed this behavior, even though at low frequencies, the protection measured when the earplugs are really used resulted to be a little lower compared with that declared. The difference is more marked starting from 2,000 Hz, where the measured protection was found to be quite 10 dB lower than that reported by the manufacturer.

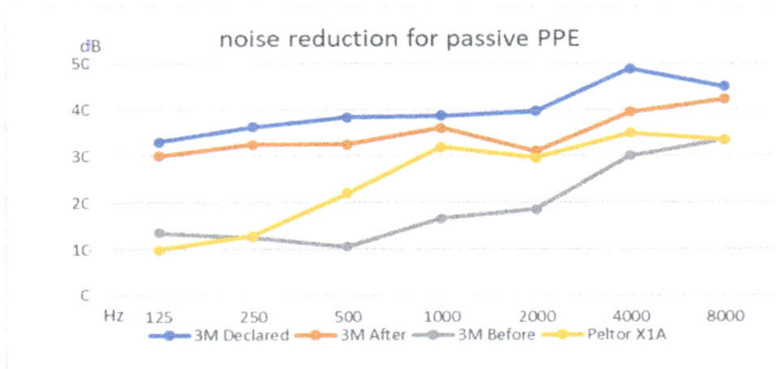

Figure 3: Noise reduction for Peltor X1A earmuff and for 3M 1100 ear plugs as declared by the manufacturer and measured before and after training.

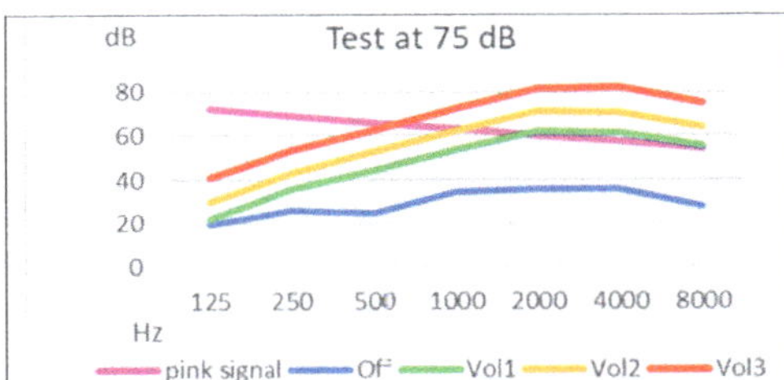

Mode	at ears [dB]	Reduction [dB]
Off	40.7	-34.3
Vol.1	65.2	-9.8
Vol.2	74.5	-0.5
Vol.3	85.3	+10.3

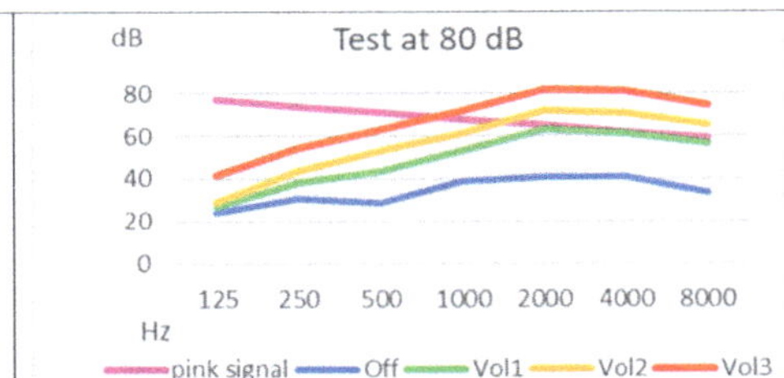

Mode	at ears [dB]	Reduction [dB]
Off	45.7	-34.3
Vol.1	65.8	-14.2
Vol.2	74.8	-5.2
Vol.3	85.0	+5.0

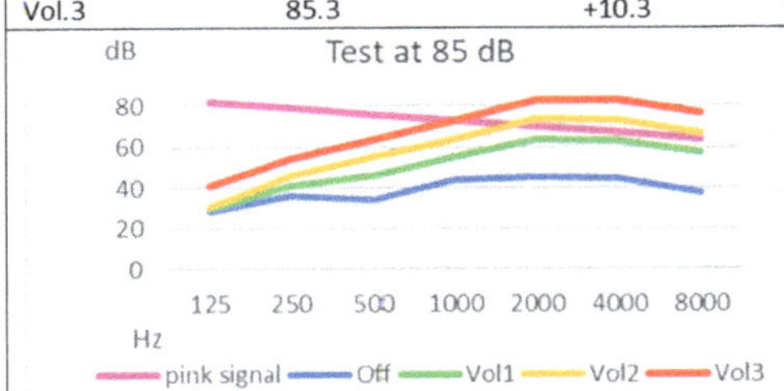

Mode	at ears [dB]	Reduction [dB]
Off	50.2	-34.8
Vol.1	67.4	-17.6
Vol.2	77.1	-7.9
Vol.3	86.6	+1.6

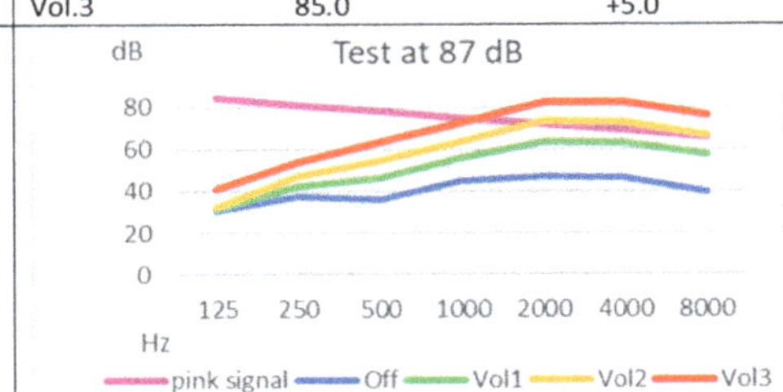

Mode	at ears [dB]	Reduction [dB]
Off	51.6	-35.4
Vol.1	67.3	-19.7
Vol.2	76.8	-10.2
Vol.3	86.0	-1.0

Figure 4: Behavior of the active earplugs at different noise levels.

When earplugs are not properly inserted, their performances are deeply affected in all the frequency ranges, with a reduction of the protection of about 20 dB in quite each frequency. For the Peltor X1A earmuffs, we only used the data reported by the manufacturer. These PPE are characterized by performing less noise reduction compared with the 3M earplugs, especially in the low frequencies range. The overall reduction is between 9.9 dB at 125 Hz, to 34.9 dB at 4,000 Hz.

Regarding the active noise control earplugs, in Fig. 4 are reported the data measured in our previous laboratory tests in terms of noise reduction, when pink noise is generated in the anechoic chamber at the different amplitudes of 75, 80, 85, and 87 dB. As shown in the graphs, the performance of active earplugs depends on the pre-set volume, other than on the amplitude of the noise at every single frequency. When the volume is set to 3, the overall noise at the ears could be also higher than that emitted by the noise source. This behavior can be useful in some circumstances in which the external noise really needs to be amplified, for example for earing spoken communications or alarm signals, but in some other circumstances can also be detrimental. Consequently, when this kind of active earplugs are used in the real field, it is important to carefully evaluate their behavior with respect to the amplitude and spectra of the environmental noise.

For the previous reasons, starting from the data reported above, in the present study we aimed to characterize the behavior of active earplugs when used in the real field, referring to the noise measured in the quarry plant. For this purpose, we first needed to characterize the attenuation or amplification provided by earplugs at every single frequency when varying the amplitude of the external sound. So, we analyzed, at every frequency and pre-set volume, both the amplitudes of the signals emitted by the source in the anechoic chamber, then the noise registered by the microphones inside the mannequin's ears. Results are reported in Fig. 5.

Since the noise emitted by the source in the anechoic chamber didn't reach high amplitude for all the frequencies, in order to be able to rebuild the attenuation curves, we needed to approximate the earplug's behavior at those missing amplitudes. In this case, we supposed that at 99 dB and for every setting volume, the amplification of the noise provided by earplugs is null. This hypothesis may have a sense, considering that when the noise is very high, it doesn't need to be amplified. In this way, the performances of the earplugs at 99 dB are supposed to be those provided by the average values, for the same frequency, measured when the pre-set volume is turned off. The noise reduction, starting from the highest measured amplitude in each frequency to the value of 99 dB, was supposed to be linear. Using the previously measured data and following our approximations, we were able to rebuild the attenuation curves that the electronic earplugs follow for reducing or amplifying the noise at each frequency and amplitude. Starting from the previously calculated curves, we estimated the noise reduction of the active earplugs in the hypothesis of their use in our case study. The PPE are supposed to be worn in each of the points in which the measured noise Leq(A) exceeds the lower exposure action values of 80 dB(A). Results are reported in Fig. 6.

When the amplification volume of the active earplugs is switched off, the overall reduction of the PPE is provided by the passive behavior of the materials they're composed to. In this case, the attenuation provided by the earplugs reaches its maximum value, that for the quarry plant noise it's estimated ranging between 24.2 to 28.9 decibels. Using the eqn (2), we calculated the performance of the other PPE when used in our context. For the 3M 1100 earplugs, in the hypothesis of their correct use, the noise reduction should be between 32 and 34.6 dB, providing the higher attenuation; while for the Peltor X1A (only data supplied by the manufacturer) it's supposed to be between 14.4 and 27.1 dB, providing the

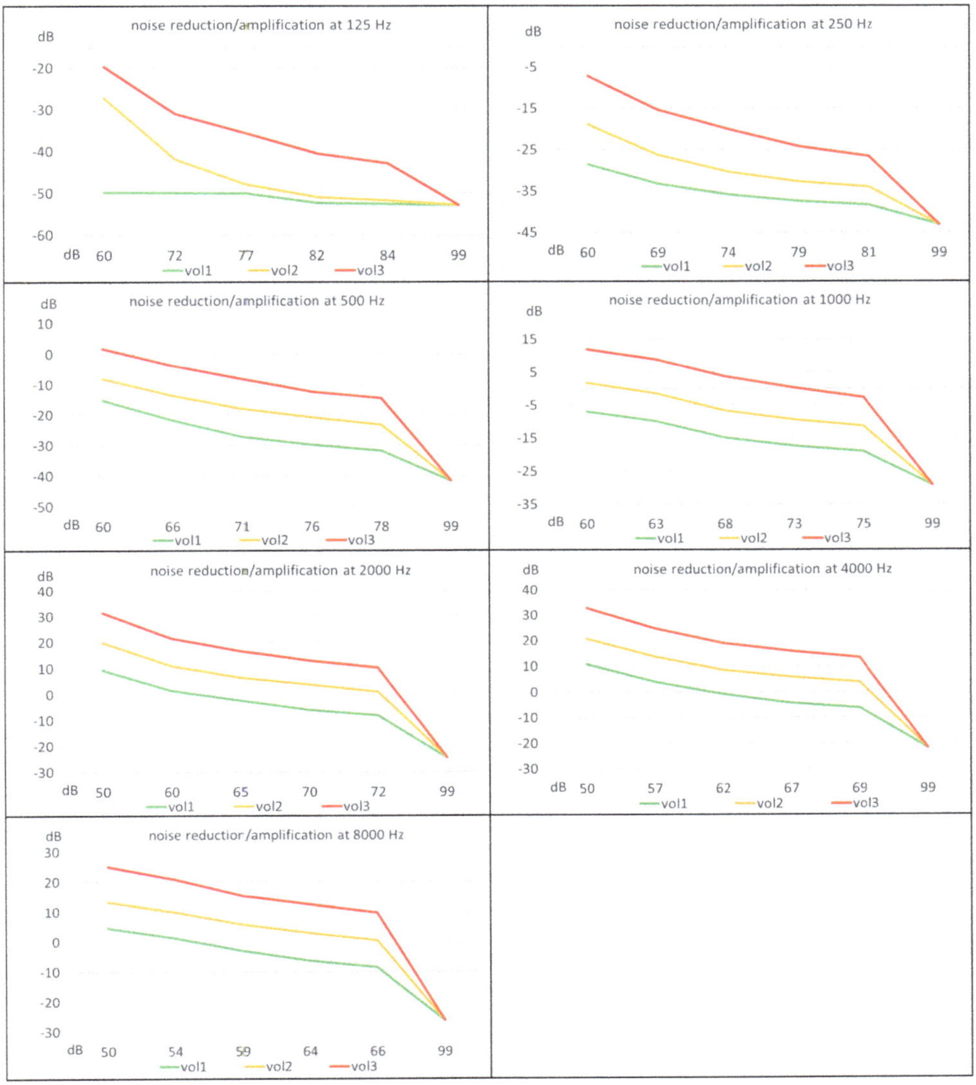

Figure 5: Each graph reports, at the specific frequency, the noise produced by the signal source in the anechoic chamber, measured just outside the ears of the mannequin (x axis), and the attenuation or amplification provided by the earplugs (y axis).

lower attenuation. Things change when the active earplugs are turned on. According to the curves previously calculated, at volume 1 they should perform a reduction, in our context, between 11.9 to 22 dB.

As can be seen in Fig. 6, the supposed reduction of the active earplugs is higher in the point 17 (22 dB), in which the noise emission is actually the higher (99.7 dB(A)), while they should perform the less reduction (11.9 dB) in the point n. 15, in which the noise measured

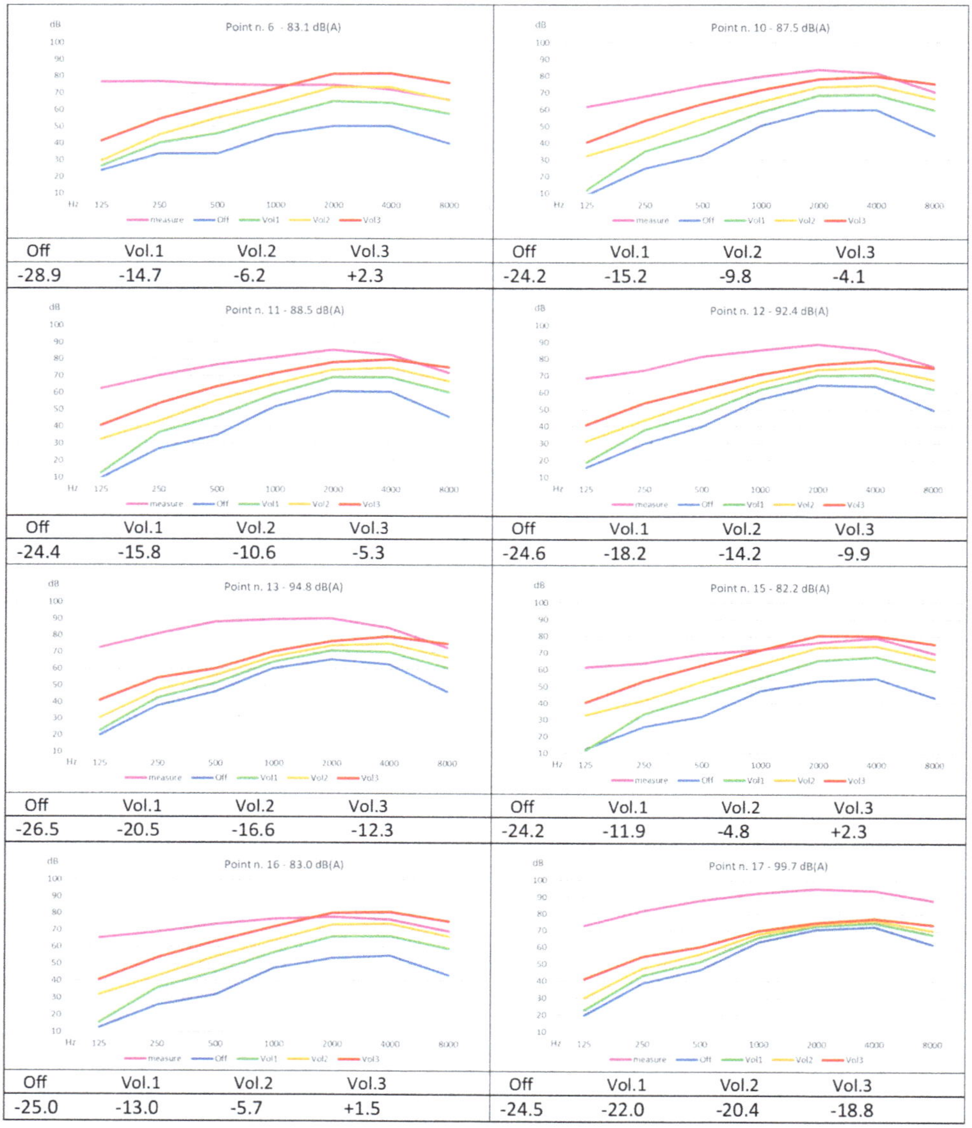

Off	Vol.1	Vol.2	Vol.3
-28.9	-14.7	-6.2	+2.3

Off	Vol.1	Vol.2	Vol.3
-24.2	-15.2	-9.8	-4.1

Off	Vol.1	Vol.2	Vol.3
-24.4	-15.8	-10.6	-5.3

Off	Vol.1	Vol.2	Vol.3
-24.6	-18.2	-14.2	-9.9

Off	Vol.1	Vol.2	Vol.3
-26.5	-20.5	-16.6	-12.3

Off	Vol.1	Vol.2	Vol.3
-24.2	-11.9	-4.8	+2.3

Off	Vol.1	Vol.2	Vol.3
-25.0	-13.0	-5.7	+1.5

Off	Vol.1	Vol.2	Vol.3
-24.5	-22.0	-20.4	-18.8

Figure 6: The graphs represent the noise perceived by workers at each point when active earplugs are used at different volumes. The numbers indicate the overall noise at each point and the overall reduction provided.

is the lowest (82.2 dB(A)). The same thing happens for the others pre-set volumes, confirming the good behavior of the active earplugs in reducing more noise when it is at high levels, and providing an amplification when the level is lower. Can also be noted that when the noise reaches high values the behavior of the device, for all of the configurations, is similar to what happens when it's turned off. Anyway, the overall protection doesn't always result in the more appropriate when the pre-set volume increases. As can be seen in Table 2, where are reported the attenuation/amplification data estimated for all the examined PPE in all the studied configurations, when the volume of the active earplugs is set to the maximum, they could perform a not enough reduction. In fact, according to our predictions, in this case, the noise perceived at the workers' ears should overcome the lower exposure action values of 80 dB(A), in all the noisiest points of the quarry plant.

Table 2: Noise at the ears level after the attenuation/amplification.

Points	Emitted noise Leq (dB(A))	3M 1100 Declared (dB(A))	3M 1100 After (dB(A))	3M 1100 Before (dB(A))	Peltor X1A (dB(A))	Active 100EU Off (dB(A))	Active 100EU Vol 1 (dB(A))	Active 100EU Vol 2 (dB(A))	Active 100EU Vol 3 (dB(A))
6	83.1	46.5	50.9	69.4	68.7	54.2	68.4	76.9	85.4
10	87.4	46.9	54.2	69.5	60.3	63.3	72.3	77.7	83.4
11	88.5	48.2	55.5	71.0	61.8	64.1	72.7	77.9	83.2
12	92.4	52.3	59.4	75.3	66.0	67.8	74.2	78.2	82.5
13	94.8	55.6	61.8	80.0	71.6	68.3	74.3	78.2	82.5
15	82.2	40.4	47.6	63.2	56.3	58.0	70.3	77.4	84.5
16	83.1	43.3	49.5	66.7	60.2	58.0	70.0	77.3	84.5
17	99.7	59.1	66.0	82.1	73.1	75.2	77.7	79.3	80.9

Another thing to note, is that in points number 6, 15 and 16, the noise produced by the active earplugs at vol. 3 should be higher compared to the noise measured in the plant, passing in the worst case from 83.1 to 85.4 dB(A). The reason is due to the particular spectra of these sounds which amplitudes result to be higher at the low frequencies, and, much more important, lower to that provided by the earplugs at the high frequencies.

If that estimated behavior should be confirmed in real field measures, the use of this kind of earplugs at volume 3, other than resulting to be not efficient in reducing the noise under the lower exposure limits, could also be detrimental in some of the quarry points (points: 6, 15, 16).

When the volume is set at 2, the active earplugs should generally perform a good protection, even if at some points the noise passing to the ears is close to the lower action limit.

Regarding the other examined earplugs, their performances seem to be often too high for the noise levels measured in the quarry plant. As can be seen in Table 2, the estimated value of the noise at the worker's ears is often below 65 dB(A), resulting in a possible over-protection. When the earplugs are not correctly inserted, there are no cases of over-protection, but on the other side, in the points n. 15 and 16, the noise reaching the worker's ears could be over 80 dB(A).

3.3 Simulation of the noise exposition

For the previous two hypothesized scenarios, the $L_{ex(8h)}$ noise exposition was calculated when workers spend 5 or 10 minutes in each of the external eighteen positions located in the quarry plant. Results are reported in Table 3. As can be seen, when workers aren't protected, the overall noise exposition exceeds the limits in both the scenarios, resulting over the lower limit value in the first case (82.7 dB(A)) and over the upper limit value in the second (85.5 dB(A)). When PPE are used, regardless of the model, settings, or proper use, the 8 hours noise exposition decreases under the limits. Anyway, as seen in Table 2, the noise reduction of 3M earplugs, can easily provide a hyper-protection in some points, while that provided by the active earplugs at volume 3 could be not enough. Moreover, the use of the passive earplugs generates a higher variability of the noise exposition compared to that provided by the active earplugs. In fact, looking at the differences between the maximum and the minimum noise perceived by workers while moving throw different positions, when passive earplugs are used or active earplugs are switched off, the delta is around 19 dB(A) (average value between 16.8 dB(A) for the Peltor X1A and 21 dB(A) for switched off earplugs); while at volume 1 it's around 9.3 dB(A) and at volume 2 only 2.4 dB(A). This is due to the fact that active earplugs tend to amplify the lowest noises and reduce the highest, providing a more homogeneous exposition, especially in those places, as the quarry plants, in which the exposition can varies a lot when moving across different points.

Table 3: $L_{ex(8h)}$ worker's exposition for both the hypothesized scenarios.

	No protection $L_{ex(8h)}$ (dB(A))	3M 1100 Declared $L_{ex(8h)}$ (dB(A))	3M 1100 After $L_{ex(8h)}$ (dB(A))	3M 1100 Before $L_{ex(8h)}$ (dB(A))	Peltor X1A $L_{ex(8h)}$ (dB(A))	Active 100EU Off $L_{ex(8h)}$ (dB(A))	Active 100EU Vol 1 $L_{ex(8h)}$ (dB(A))	Active 100EU Vol 2 $L_{ex(8h)}$ (dB(A))	Active 100EU Vol 3 $L_{ex(8h)}$ (dB(A))
1st case	82.7	66.5	66.6	69.4	67.2	67.2	68.4	70.4	74.3
2nd case	85.5	69.3	69.4	72.1	69.9	69.9	71.0	73.0	77.0

4 CONCLUSIONS

In the present study, different kinds of noise reduction personal protective equipment were analyzed in order to characterize their levels of protection when used in a limestone quarry plant. The activities made in the quarry are characterized by having high levels of noise in most of the areas close to the production's machines, other than a great variability of the noise when moving from a position to another. Even supposing an employee who spends less than 20 minutes in a day working in some of the noisy parts of the plant without any protection, the noise exposition ($L_{ex,8h}$) can easily overcome the lower limit value established by the European Directive 2003/10/EC.

Starting from the data of the noise reduction of some PPE measured in laboratory tests, and analyzing their behavior at different amplitudes and frequencies for characterizing their performances, a prevision of the possible protection for a worker that moves around the plant was estimated.

Results show that all the analyzed PPE are able to provide a good attenuation of the noise when used in our case study, even in the case in which earplugs are not correctly worn. Anyway, great variability was observed in terms of protection among those. 3M earplugs, Peltor X1A earmuffs and active earplugs when turned off, can provide hyper-protection on most of the noisy points of the plant, generating possible discomfort for workers, other than difficulties in perceived spoken words or horn signals. On the contrary, active earplugs, when set at the maximum volume, cannot protect properly the workers, sometimes generating more noise than that produced by the machines. Anyway, when the volume of the active earplugs is set on 1 or 2, we generally found a good response of this type of device, resulting the better choice for protect workers in our case study. Results of our simulations show that for the previous settings, the noise at the worker's ears does not overcome 80 dB(A) also in case of the external noise at 99.7 dB(A) and it never goes below 68.4 dB(A). Moreover, since the noise perceived when using active earplugs is attenuated or amplificated according to the amplitude of external noise, the variability of the perceived noise is lesser than that perceived using passive protection devices. So active earplugs seem to be preferred especially in those cases where workers need to move across positions at different noise levels, as often happen in quarry plants.

Anyway, the environmental noise, other than the properties of the active earplugs, must be carefully evaluated before allowing their use, since as shown in the present study, their behaviors are subjected to high variability when the external noise conditions change.

Since this study is modeling-based and moreover, we needed to introduce some approximations for the simulation of the behavior of active earplugs, some differences could be found in their real use. A possible follow-up of this study could be to measure the properties of these active earplugs in the same quarry plant, using the same equipment used in the anechoic chamber, in order to confirm the results here estimated.

REFERENCES

[1] Pathak, K., Durucan, S. & Kunimatsu, S., Activity accounting technique for environmental noise assessment in mechanized surface mining. *Environmental Monitoring and Assessment*, **56**(3), pp. 241–256, 1999. https://doi.org/10.1023/A:1005907720627.

[2] Fernandez, M.D., Quintana, S., Chavarria, N. & Ballesteros, J.A., Noise exposure of workers of the construction sector. *Applied Acoustics*, **70**, pp. 753–760, 2009.

[3] Abel, S.M., Kunov, H., Pichora-Fuller, M.K. & Alberti, P.W., Signal detection in industrial noise: effects of noise exposure history, hearing loss, and the use of ear protection. *Scandinavian Audiology* **14**(3), pp. 161–173, 1985. https://doi.org/10.3109/01050398509045937.

[4] Wichers, M., Iramina, W.S., Médici de Eston, S. & Marques Ayresda Silva, A.L., Using a noise monitoring station in a small quarry located in an urban area. *Environmental Monitoring and Assessment,* **190**(40), pp. 0–11, 2018. https://doi.org/10.1007/s10661-017-6404-6.

[5] Lippiello, D., Alfaro Degan, G. & Pinzari, M., Comparison of stochastic and deterministic methods for mapping environmental noise from opencast quarries. *American Journal of Environmental Sciences,* **12**(2), pp. 68–76, 2016. https://doi.org/10.3844/ajessp.2016.68.76.

[6] Yang, M. & Sheng, P., Sound absorption structures: from porous media to acoustic metamaterials. *Annual Review of Materials Research*, **47**, pp. 83–114, 2017. https://doi.org/10.1146/annurev-matsci-070616-124032.

[7] Arezes, P.M. & Miguel, A.S., Hearing protectors acceptability in noisy environments. *The Annals of Occupational Hygiene*, **46**(6), pp. 531–536, 2002. https://doi.org/10.1093/annhyg/mef067.

[8] Casali, J.G. & Gerges, S., Protection and enhancement of hearing in noise. *Reviews of Human Factors and Ergonomics*, ed. R.C. Williges, Vol. 2, Chapter 7, Human Factors and Ergonomics Society: Santa Monica, CA, pp. 195–240, 2006.

[9] Witt, B., Putting the personal back into PPE: hearing protector effectiveness. *Occupational Health and Safety (Waco, Tex.),* **76**(6), pp. 90–92, 2007.

[10] Acton, W.L., Problems associated with the use of hearing protection. *The Annals of Occupational Hygiene.* **20**(4), pp. 387–439, 1977. https://doi.org/10.1093/annhyg/20.4.387.

[11] Koskinen, H., Toppila, E. & Makinen, H., Selection of hearing protectors: How to ensure that the end user will get best possible protection? *Proceedings of 19th European Conference on Noise Control,* pp. 237–241, 2012.

[12] Byrne, D.C., Davis, R.R., Shaw, P.B., Specht, B.M. & Holland, A.N., Relationship between comfort and attenuation measurements for two types of earplugs. *Noise and Health,* **13**(51), pp. 86–92, 2011. https://doi.org/10.4103/1463-1741.77193.

[13] Alfaro Degan, G., Annesi, D., Antonucci, A., Coltrinari, G., Lippiello, D. & Nataletti, P., Hearing protection devices with active attenuation control for quarry employment. *XXth International Multidisciplinary Scientific Geo-Conference – SGEM,* pp. 135–143, 2020.

[14] International Organization for Standardization, Acoustics simplified method for a measurement of insertion loss of hearing protectors of ear-muff type for quality control purposes. Standard ISO/DIS 6290, Geneva, Switzerland, 1983.

[15] Alfaro Degan, G., Coltrinari, G., Lippiello, Nataletti, P. & Annesi, D., Noise exposure of workers and the correct use of the hearing protection device. *XIXth International Multidisciplinary Scientific Geo-Conference – SGEM,* pp. 649–657, 2019.

[16] European Parliament, European Directive 2003/10/EC on the Minimum Health and Safety Requirements Regarding the Exposure of Workers to the Risks Arising from Physical Agents (Noise), 2003.

SECTION 8
SAFETY ENGINEERING

CONCEPTS AND OPERATIONAL TOOLS FOR MULTIHAZARD RISK ANALYSIS IN ROAD TUNNELS: THE SAN BENEDETTO EVENT, ITALY

MARA LOMBARDI, ANGELO LIBERTÀ, DAVIDE BERARDI,
ELEONORA DI BENEDETTO & MASSIMO GUARASCIO
Sapienza University of Rome, Italy

ABSTRACT

Let's define "multihazard": a chain of sequential events intersection, each one conditioned by the previous. The present study affords an in-depth insight into road tunnel multihazard. The paper stems from interest in analyzing the conditions of occurrence and development of two important accident events: San Gotthard (Switzerland, 2001) and San Benedetto tunnel (Italy, 2016). The first originated from the improper human behavior of a driver who invaded the opposite lane colliding frontally with another vehicle. A fire developed involving the vehicles' fuel, cargo of goods, bituminous conglomerate of the pavement and collapse of the vault concrete with eleven victims and damage to the structure. The second which occurred during the seismic activity of 2016 in central Italy (Umbria), caused the break and relative 20–30 cm displacement of two sides of the tunnel pavement. The first part of the paper deals with the study of seismic hazard conditions around the San Benedetto tunnel and fragility of concrete lining and asphalt pavement. In the second part the deepening concerns the probability estimation of a potential intersection of a seismic event with forced swerve, collision, and fire. The seismic induced break plays the role of the human factor in the Gotthard multihazard accident. The overall length of TERN Italian tunnels is greater than the one of all other countries joint lengths and for this reason, Italy can be called the Country of Tunnels. The risk of fire in tunnels in Italy must be compliant with Alarp criteria of acceptability and tolerability according to a Legislative Decree of 2006 which in turn complies with the European Directive of 2004. The quantitative and interactive "Play Gu@larp" procedure is described, which allows and support in risk-based design, reducing risk under compliance conditions.

Keywords: multihazard, risk, earthquake, analysis, seismicity, fire, road tunnel.

1 INTRODUCTION

1.1 Major accidents in road tunnels

At the end of second millennium three major time accident occurred in three major Alps tunnels in Europe (Fig. 1). On 24 October 2001, in the St. Gotthard road tunnel (Switzerland) a long vehicle invaded the opposite lane colliding with another long vehicle (Figs 1 and 2). A fire developed involving other vehicles. The fire had been extinguished 24 hours later. There were eleven victims and the damage was around €12 million [1], [2].

In October 2016, an earthquake occurred in Norcia, a region of Central Italy. Here the San Benedetto tunnel reported severe damages to segments and pavement (Fig. 3).

In the present paper the results of the studies on seismicity potential role in a multihazard scenario for fire accident in tunnel are described and discussed. In particular, an in-depth insight of the San Benedetto tunnel case study has been developed using advanced geostatistical methods [3] adopted to the seismic territorial data. The geostatistical methodological tools used here are introduced and discussed in Guarascio et al. [4] and Guarascio [5].

WIT Transactions on The Built Environment, Vol 206, © 2021 WIT Press
www.witpress.com, ISSN 1743-3509 (on-line)
doi:10.2495/SAFE210191

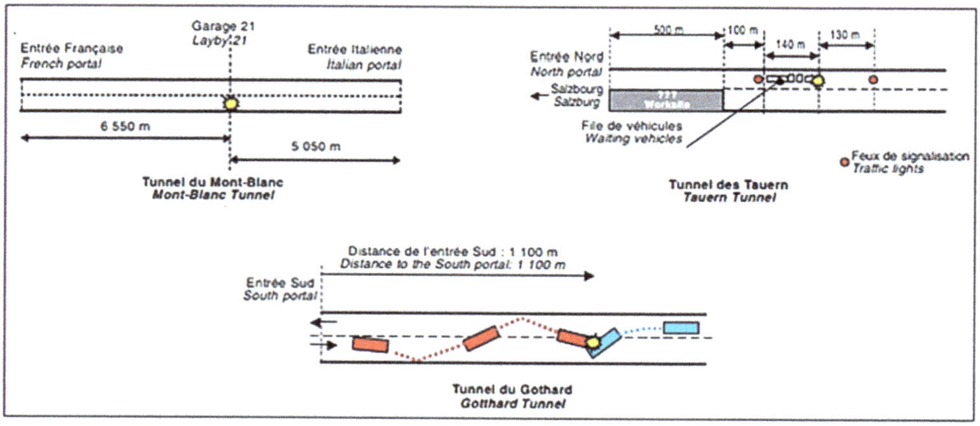

Figure 1: Locations of the three fire accidents (Mt. Blanc, Tauern, St. Gotthard, 2001).

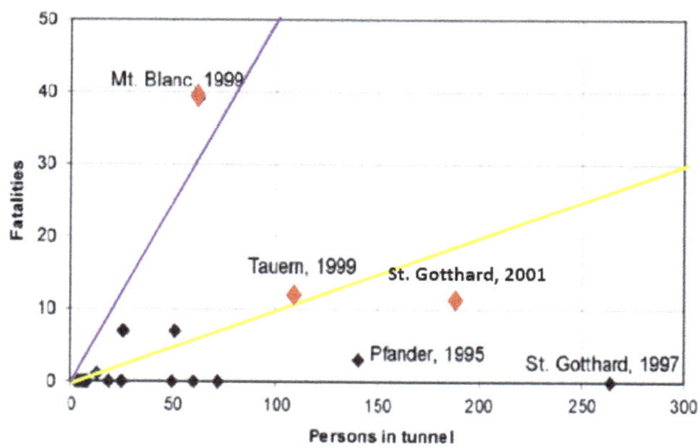

Figure 2: Mortality rates in the three fire accidents. Yellow line mortality rate 0.1; Purple line mortality rate 0.5.

Figure 3: Damage to the concrete segments and the pavement of the San Benedetto tunnel due to the Norcia earthquake.

1.2 Territorial seismicity in the Norcia area (Central Italy)

The observed and recorded historical seismic activity along 400 years until today is described in Fig. 4. The maps shown in Figs 5 and 6 give a quantitative idea of the spatial geographic relationship among the San Benedetto tunnel, the historical epicentre locations, the accelerometric measure stations and the grid points of the digital terrain model used for the geostatistical estimation of the seismic action component [4].

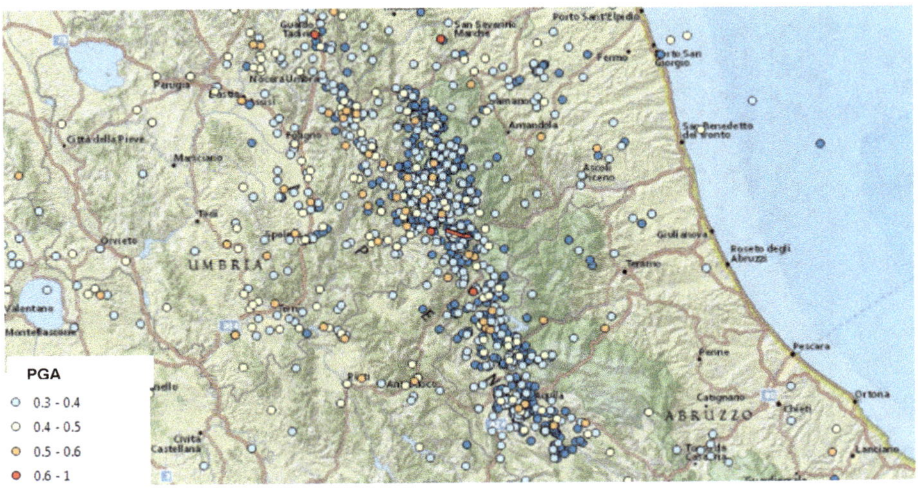

Figure 4: Observed and recorded historical seismic activity along 400 years in the Norcia area until today (Central Italy).

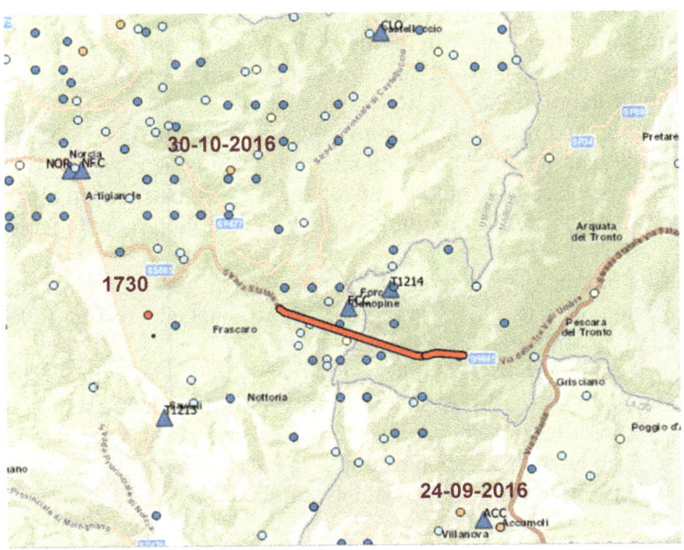

Figure 5: Historical seismic activity and location of the San Benedetto tunnel (red line). The dates written in red indicate major earthquakes (with PGA greater than 0.5).

www.witpress.com, ISSN 1743-3509 (on-line)

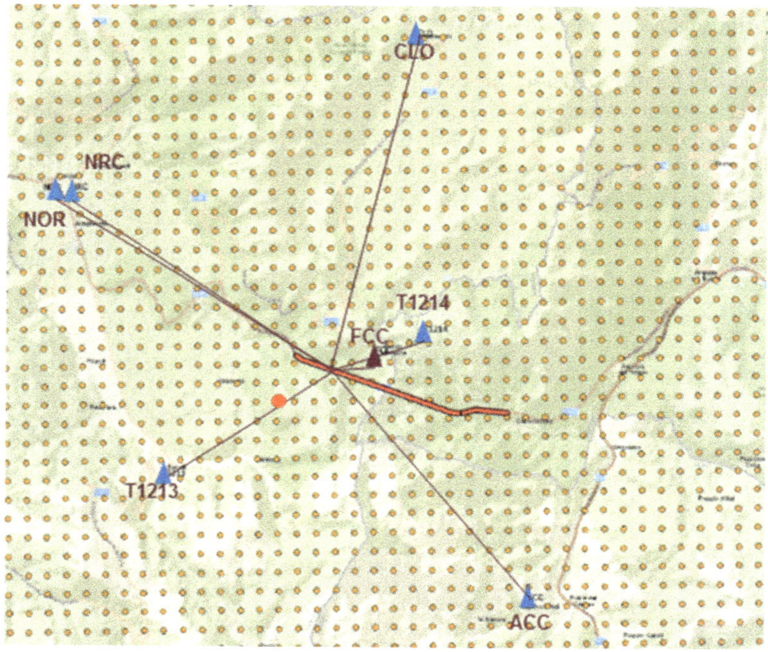

Figure 6: Grid points of the digital terrain model used for the geostatistical estimation of the seismic action component.

Examples of the contour maps of estimated useful variable are shown in Fig. 7.

2 CORRELATION BETWEEN STRUCTURAL DAMAGE AND SEISMIC ACTION

The cross sections locations shown in the Fig. 8 do describe the directional distance relationship between the area when the maximum values of seismic action have been estimated and the structure of the San Benedetto tunnel.

The cross sections of Figs 9 and 10 show the behaviour of PGA's vertical component value. In yellow the estimation confidence interval. In Fig. 11 is shown the site where the major structural damage occurred at 920 m from the western entrance in the tunnel, in which is possible to see changes and discontinuities. In Fig. 12 lithology and geology also shown [6].

The distribution of cracks in concrete segments (13.5 m each) of the tunnels is presented in the representation of Fig. 13, where the most damaged section (50 m around the progressive "920 m") mentioned are above.

A very significant comparative analysis between the behaviour of the vertical and longitudinal component of the PGA and also between their corresponding arrival times is shown in Figs 14 and 15. At the point of major observed damage, the vertical PGA component arrived about half a second before and its estimated value is higher than the epicentral one, at least 10% more. This is a relevant fact.

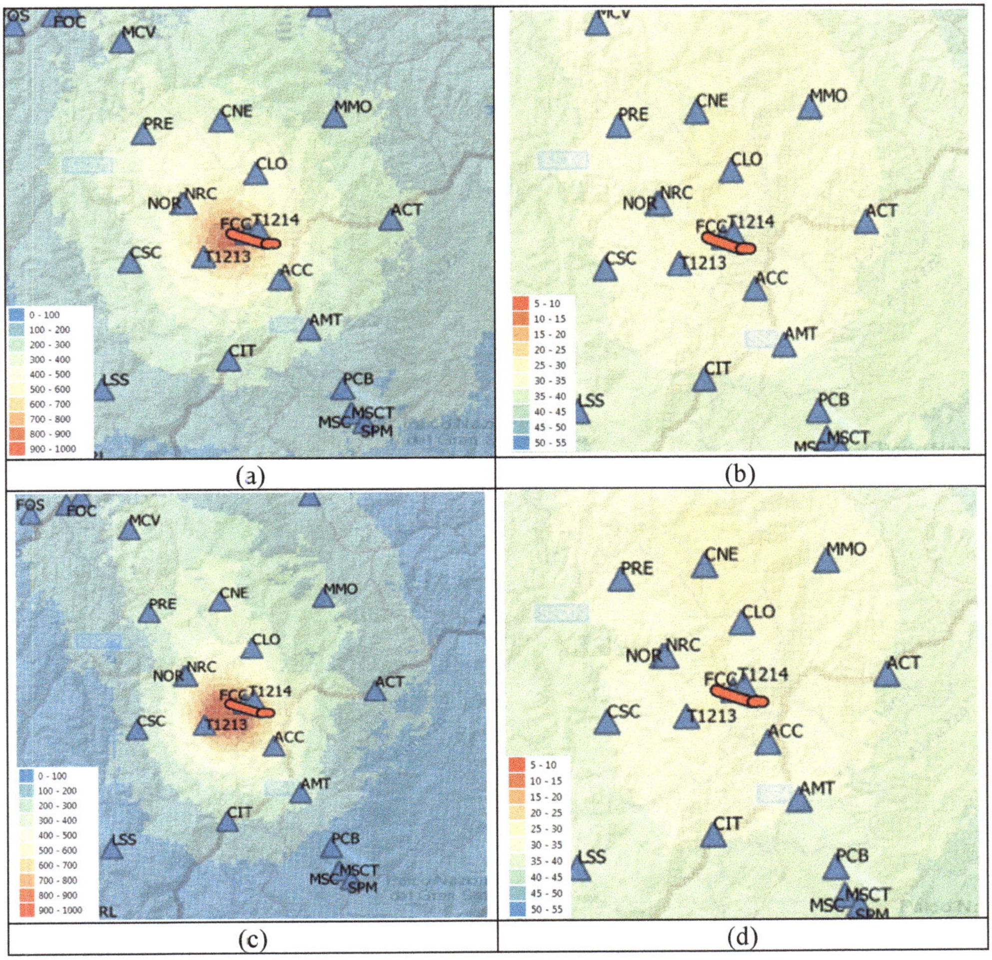

Figure 7: Contour maps of estimated variables. (a) Contour map of epicentral PGA; (b) Contour map of arrival time (epicentral PGA); (c) Contour map of vertical PGA; and (d) Contour map of arrival time (vertical PGA).

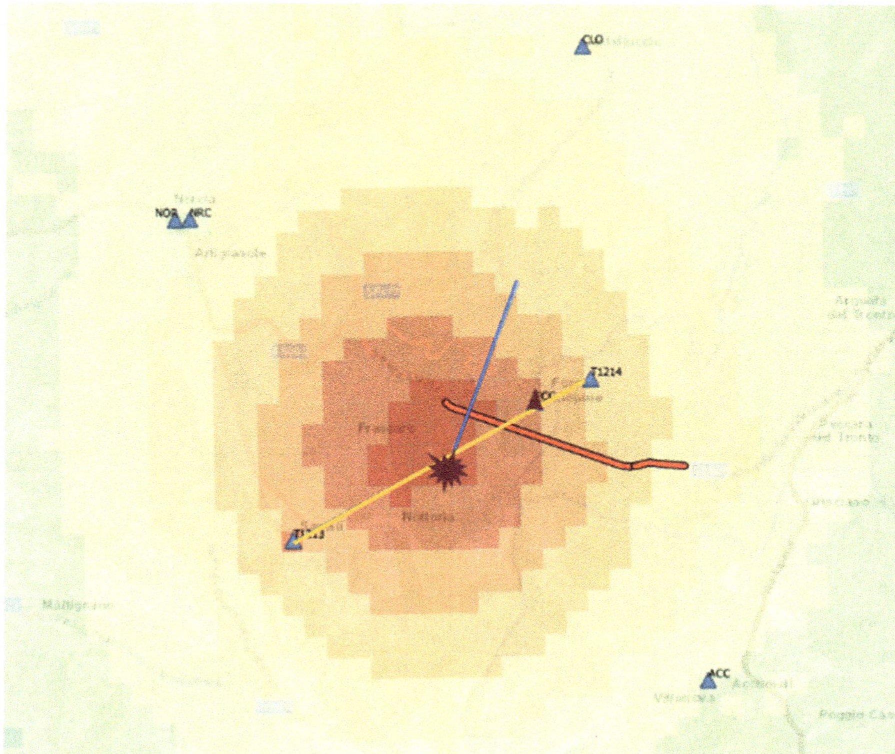

Figure 8: Yellow and blue line represent cross sections. The red symbol indicates point of maximum stress.

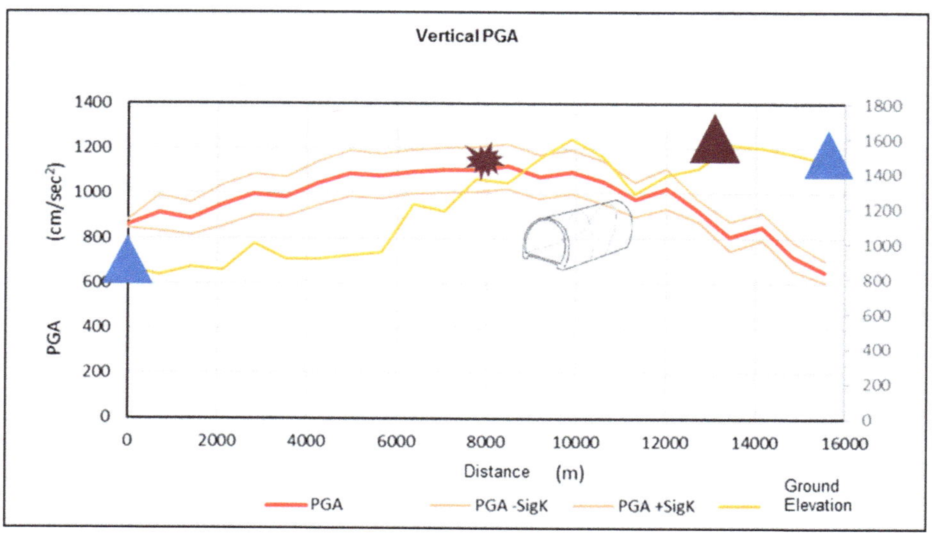

Figure 9: Behaviour of vertical PGA's value along the yellow cross section shown in Fig. 8.

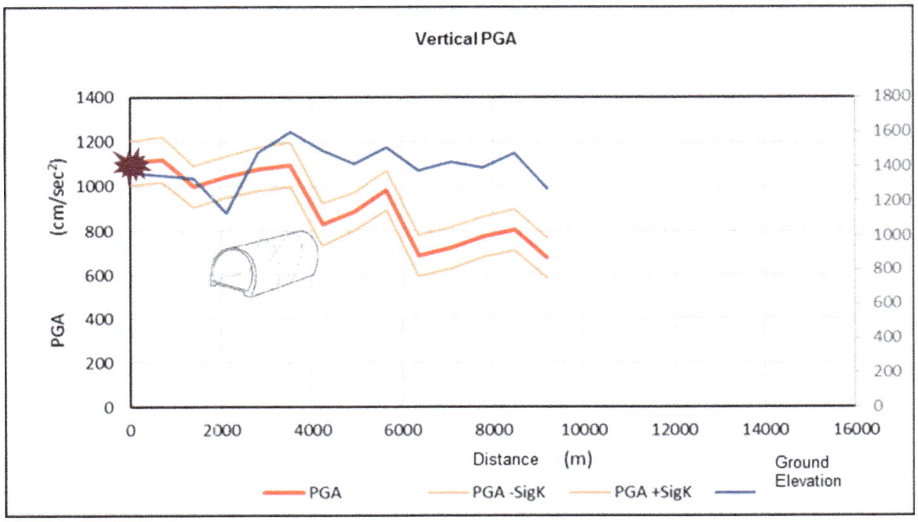

Figure 10: Behaviour of vertical PGA's value along the blue cross section shown in Fig. 8.

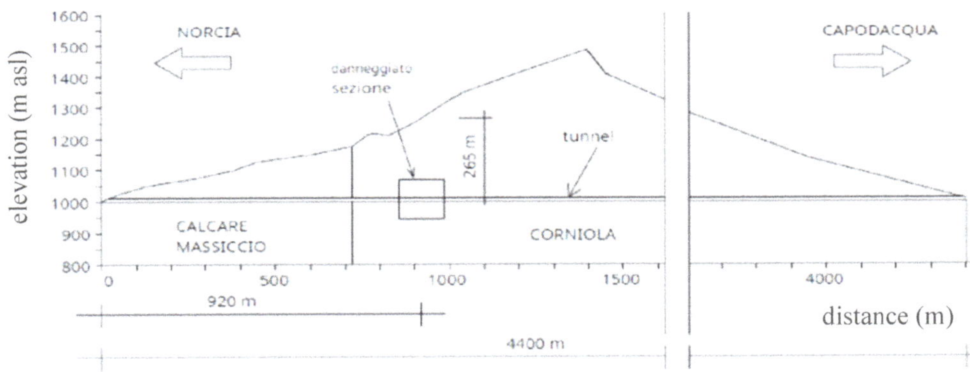

Figure 11: Site of San Benedetto tunnel where the major structural damage occurred.

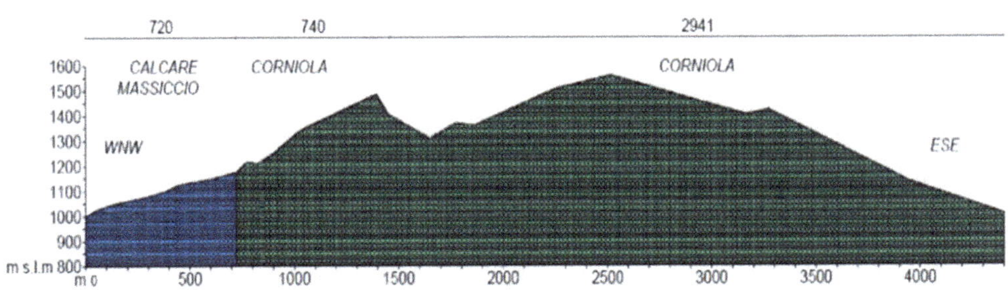

Figure 12: Lithology and geology of the site.

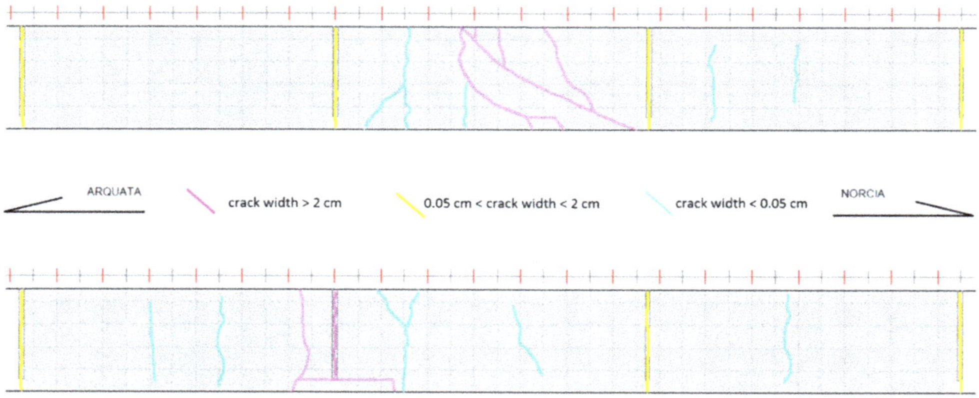

Figure 13: Distribution of cracks in the three most damaged concrete segments (50 m).

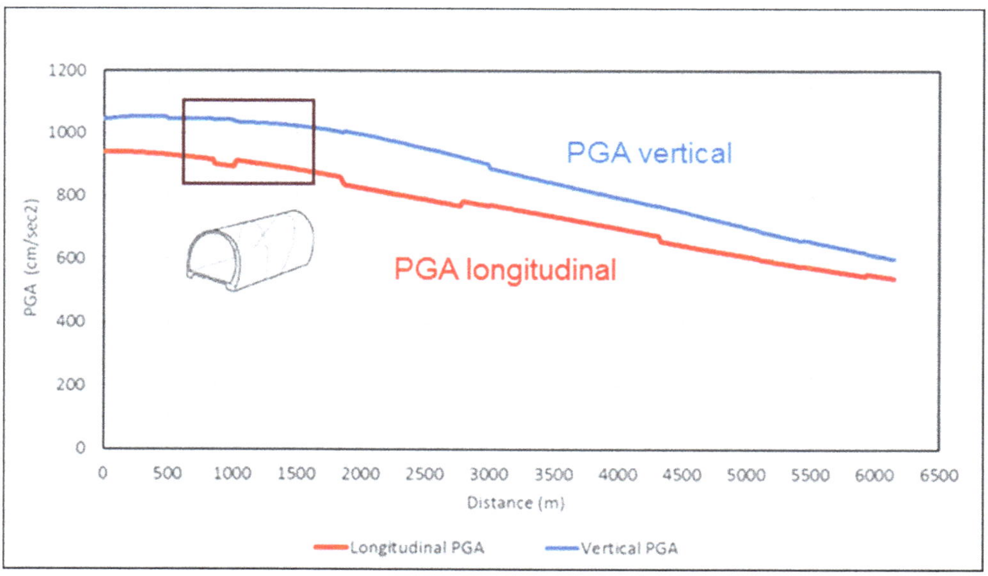

Figure 14: Behaviour of the values of the vertical and longitudinal PGA along San Benedetto tunnel.

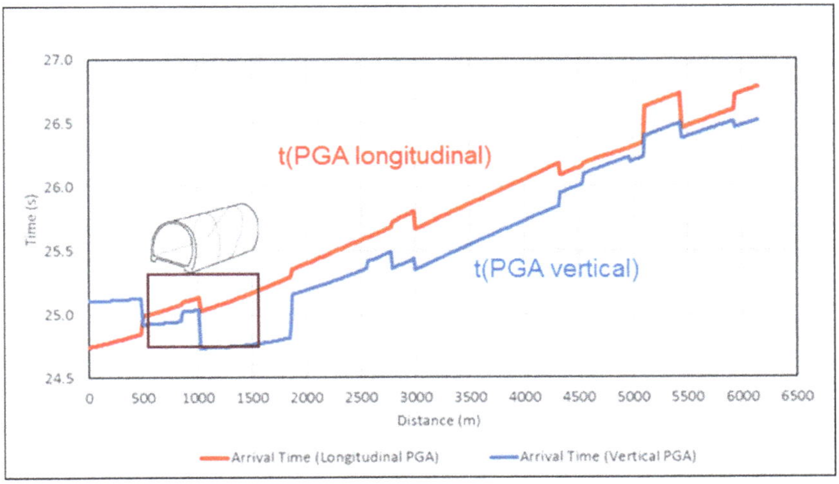

Figure 15: Behaviour of the values of the arrival times of vertical and longitudinal PGA along San Benedetto tunnel.

3 EARTHQUAKE INDUCED ACCIDENT IN ROAD TUNNEL

An event such as the one described in the paragraph, which occurs during traffic of heavy goods vehicles in both directions could generate Gotthard type accident. The last one together with that of Mont Blanc (France–Italy, 1999) and Tauern (Austria, 1999) were at the origin of the cultural debate inducing the European Commission to investigate the problems related to the safety of tunnels that ended with the enactment of the European Directive of 2004 for the tunnels of TERN. Here below we deal with the potential development of a chain of events in a bidirectional tunnel (like San Benedetto) induced by an earthquake that could generate a forced swerve followed by collision and fire in a crowded environment.

3.1 The point hazard curve

Let us consider the hazard curves describing seismic action in any given points of San Benedetto tunnel (Fig. 16) generated by potential future epicenters located in this south–east azimuthal sector. Looking at the profile and cross sections in Figs 9, 10, 14 and 15 we learn that component values, of the seismic action at the specific site of the "break" occurred, have been around 900 cm/s^2. From the above-mentioned hazard curves (Fig. 16), we are allowed to get the estimated probability of the seismic action corresponding to the type 30 October 2016 break event. Multihazard is an intersection of non-mutually exclusive events having, not only the probabilistic effect of conditioning the sequential events probabilities, but also a physical cause–effect relationship. An appropriated correlation analysis between the value of seismic actions and the cracking state produced allows us to estimate the so-called fragility curves that provide an estimate of the corresponding damage level probability.

3.2 Fire risk analysis and acceptability/tolerability criteria in Italian rules

In a tunnel like the San Benedetto the analysis of the potential hazard likely to be responsible of the occurrence of relevant fire accident, is then compulsory to take in seismic consideration

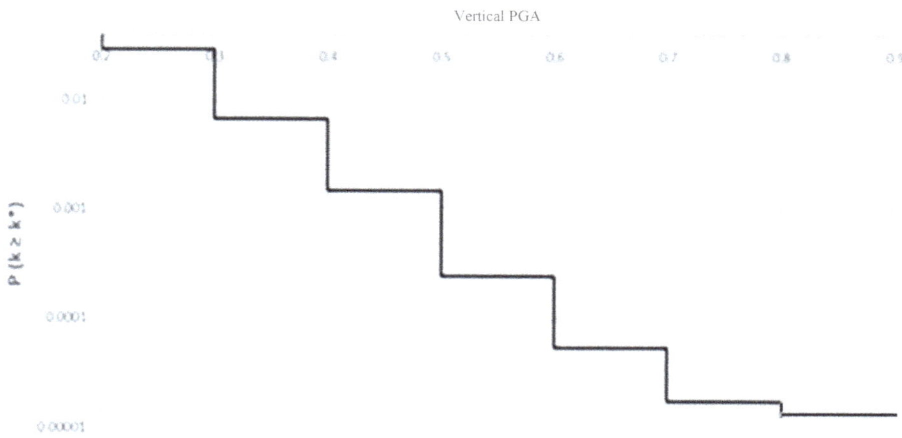

Figure 16: Hazard curve of San Benedetto tunnel.

the intersection of the possible different cause events of collisions and the possible consequent induced development of a fire in a relevant exposure scenario. So, both the seismic induced collisions and the ordinary traffic collisions analysis must be carried out and the corresponding initiating to event must be considered in the risk based design [7] relative compliance condition verification of three different design solutions and corresponding *F(N)-N* curves with compared to the risk acceptability and tolerability criteria (Fig. 17).

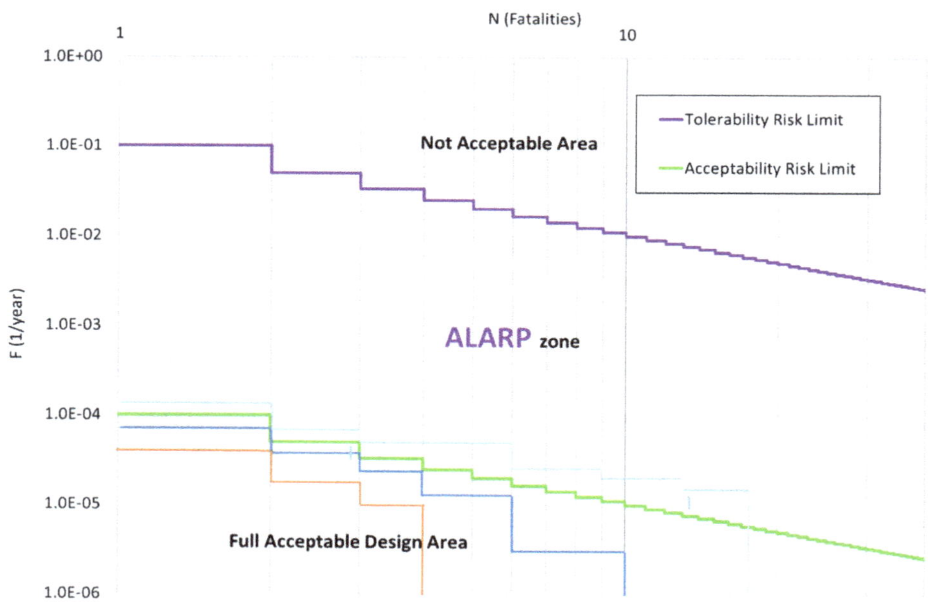

Figure 17: The Gu@larp broken lines and *F(N)-N* curves.

The calculation of the *F(N)-N* curve and related underlying event tree analysis must include the union of all the mutually exclusive design scenario events where the initiating trigger event is, at the time, the seismic of the ordinary traffic ones.

4 CONCLUSIONS

A country like Italy must consider multihazard events as an ordinary potential threat for its territorial infrastructure given the combination of factors: high seismicity, number of road tunnels and percentage of heavy vehicles in the very high volume of traffic.

It is imperative to handle property information regarding all the above hazard factors and identify the interconnection among them.

The result of the present study, preliminary developed in depth, demonstrate that geostatistical analysis is a very useful and powerful tool to achieve the demanded aims.

REFERENCES

[1] Voeltzel, A. & Dix, A., A comparative analysis of the Mont-Blanc, Tauern and Gothard tunnel fires. *Routes/Roads*, **324**, pp. 18–34, 2004.

[2] Jonkman, S.N., Lentz, A. & Vrijling, J., A general approach for the estimation of loss of life due to natural and technological disasters. *Reliability Engineering and System Safety*, **95**, pp. 1123–1133, 2010. DOI: 10.1016/j.ress.2010.06.019.

[3] Guarascio, M., Huybrechts, C.J. & David, M., Advanced geostatistics in the mining industry. *Proceedings of the NATO Advanced Study Institute,* Istituto di Geologia Applicata of the University of Rome, Italy, 13–25 October 1975.

[4] Guarascio, M., Libertà, A., Berardi, D., Di Benedetto, E. & Lombardi, M., Seismic geostatistics application to the territorial infrastructure resilience and sustainability. *WIT Press Transactions on The Built Environment*, vol. 206, WIT Press: Southampton and Boston, 2021.

[5] Guarascio, M., "As low as reasonably practicable": How does it work in the rail and road tunnels in Italian rules. Risk Acceptability/Tollerability criteria. The Gu@larp model. *WIT Press Transactions on The Built Environment*, vol. 206, WIT Press: Southampton and Boston, 2021.

[6] Marzi, V., Micheli, A., Cedrone, L. & Martino, M., Il dissesto della galleria San Benedetto in seguito al sisma del 30 ottobre 2017: messa in sicurezza provvisoria e attività di indagine funzionali al ripristino definitivo. Gallerie e grandi opere sotterranee n.123, 2017.

[7] Guarascio, M., Lombardi, M., Rossi, G. & Sciarra, G., Risk analysis and reliability based design in tunnel fire safety. *WIT Press Transactions on The Built Environment*, WIT Press: Southampton and Boston, 2021.pp. 575–584, 2009. DOI: 10.2495/SAFE090531.

USE OF EYE-TRACKING TO CONTROL A COBOT MOVEMENT: RESULTS COMING FROM THE APPLICATION OF TOBII PRO X2-30 TO AN OMRON TM5-700 ROBOT

MARCOS FAUNDEZ-ZANUY, ALEIX MORAL-VIÑALS & JOSEP LOPEZ-XARBAU
Tecnocampus, Universitat Pompeu Fabra, Spain

ABSTRACT
In this paper we present a connection setup to operate a collaborative robot (cobot) Omron TM5-700 by means of a Tobii Pro X2-30 eye-tracker. The application has been designed with the software OMRON TMFlow as well as the Tobii SDK (software development kit). The possibility to operate a collaborative robot by gaze can be used as a third arm, which permits to do more sophisticated activities by human beings but it can also facilitate the manipulation of dangerous or perilous substances, or in unsafe environments where the human being can stay far away of the dangerous area. It can be very useful for a plenty of safety and security applications. To demonstrate the correct operation of the cobot control application, we have implemented a practical demo and experimental setup. It consists of accuracy measurements, where the control of the position of the cobot is be defined by means of gaze, which defines a set of points in (X, Y) plane. Later on, the robot picks up an ink-pen and draw a graph in a piece of paper. This drawing consists of connecting these pre-defined dots by straights lines. To this aim, we have acquired a set of figures (parallelogram, pentagon, etc.), and compared the desired printed image on PC screen with the result produced by the cobot.
Keywords: collaborative robot, eye-tracker, gaze.

1 INTRODUCTION

Eye-tracking is a technology with applications in a large set of life areas. The eye tracking market is expected to grow from USD 368 million in 2020 to USD 1,098 million by 2025; this implies grows at a compound annual growth rate (CAGR) of 24.5% during the forecast period [1]. Scientific literature has also experimented an important grow in the last years. A Google scholar search of the terms "eye-tracking" and "safety" reveals 15,400 papers, with half of them produced in the last recent five years. Currently, those areas with the highest penetration of the use of this technology are health and retail, although the trend is that, in the coming years, automotive, neuromarketing and industry 4.0 will stand out. Eye-tracking technology can be very useful for a plenty of safety and security applications. Forecasts suggest that the most frequent applications of this technology in the coming years could be augmented reality and virtual reality, its incorporation into mobile devices, entertainment and gaming, or contactless biometric solutions. Especially after COVID-19, machines operated without touching any surface are more desirable in order to minimize the risk of virus contagion.

- Safety for construction workers. Some examples of research on safety improvement using eye-tracking on construction workers can be found in Jeelani, et al. [2], Hasanzadeh et al. [3] and Martinez-Marquez et al. [4]. Recent studies have also found that a strong correlation exists between viewing patterns of workers, captured using eye-tracking devices, and their hazard recognition performance. The analysis of the viewing patterns of workers can provide a better understanding of their hazard recognition performance.

WIT Transactions on The Built Environment, Vol 206, © 2021 WIT Press
www.witpress.com, ISSN 1743-3509 (on-line)
doi:10.2495/SAFE210201

- Automotive is one of the most important application areas of eye-tracking. The most direct application in this sector is in driver monitoring systems (DMS), where eye-tracking technology is being of vital importance. With the aim of creating safer and more advanced cars, the combination of facial recognition and eye-tracking enables information on the driver's attention, alertness or concentration at the driving task. So we can create warnings and notifications in driving. Some examples can be found in Carr and Grover [5], Vetturi et al. [6], Rjabovs and Palacin [7], Khan and Lee [8], and Le et al. [9]. Also the safe mobility of pedestrians has been analysed [10], [11].
- Neuromarketing is probably the first commercial field where eye-tracking technology entered. For advertising and marketing it is crucial to know the behaviour of customers, their preferences, their decisions and to predict their purchasing behaviour. For this they use neuromarketing techniques such as EEG, eye-tracking or implicit response tests. Regarding eye-tracking technology, it allows to identify, through an exhaustive eye-tracking, behaviour patterns during the visit to the point of sale. Therefore, the applications are endless, from software/hardware usability, advertising testing and product in stores, to analysis of the conception of the corporate image. See for instance the study about neuroergonomics [12].
- Robotics and industry in safe environments is a high potential area too. Eye-tracker widens the possibilities in collaborative robotics (human–robot). For example, with applications in camera inspection tasks, directing the camera with the eyes to the point where the inspection photograph is to be taken, and giving the order to shoot the photo by winking of one eye. Or the concept of the "third arm", which would allow people who in the production plant have to operate certain machinery or tools for which they need both hands, to instruct a robot with their eyes to carry out additional operations on the machinery. When manipulating perilous substances by means of a robot, eye-tracking can be a good solution to minimize the risk for human beings. See Chadalavada et al. [13] for instance. Another important application is the robot control by handicapped people in order to overcome their limitations and to enable an autonomous life [14].

In this paper, we setup the communication between Eye-tracker and cobot and we perform a set of experiments based on drawing performance based on gaze. This is a simple environment test that has already been explored by different authors using different setup [15]. Handwriting analysis has a wide range of applications in e-security and e-health [16], and a large set of tasks can be performed [17]. The possibilities of combining handwriting analysis and eye-tracking is large in e-health and e-security and will be probably developed in near future years.

We consider interesting to describe the setup of the system as a large amount of robots and eye-trackers exists and the interconnection is not trivial and must be ad hoc designed.

2 EXPERIMENTAL SETUP DESCRIPTION

In this paper, we have configured a setup based on the devices available in our lab: a cobot Omron TM5-700 and an eye-tracker Tobii Pro X2-30 attached to a laptop.

2.1 Elements description

Our system uses two different computers which communicate to each other by means of an Ethernet connection. Fig. 1 shows the experimental setup, which consists of the following eight parts:

1. OMRON Cobot-arm model TM5-700. Collaborative robots are designed to work safely with human operators thanks to technologies like force feedback, low-inertia servo motors, elastic actuators and collision detection technology that limit their power and force capabilities to levels suitable for contact. The safety standard ISO 10218-1, ISO 10218-2 and technical specification ISO TS-15066 define the safety functions and performance of the collaborative robot. We have used this robot because it was the unique collaborative robot available in our laboratory.
2. 3D grip attached to the cobot in order to hold/left the marker pen that performs the drawing on the surface area. The marker pen is a standard one, as the robot can use the grip to hold pens of different thick.
3. Drawing surface of standard DINA3 size, where the cobot performs the drawings by moving it arm.
4. Cobot controller screen, where the cobot is operated by a dedicated computer hidden in a box below the cobot. The computer executes a software programmed with the software OMRON TMFlow, which is a graphical human–machine interface (HMI). This facilitates the programming of a specific algorithm.
5. Cobot Remote control (robot stick) to start/stop the possibility to receive instructions by the robot.
6. Tobii Pro X2-30 Eye-tracker attached to the bottom part of the laptop screen.
7. Eye-tracker laptop, which operates the eye-tracker software programmed in phyton.
8. Ethernet connection in order to send instructions (communication) from eye-tracker computer to cobot computer.

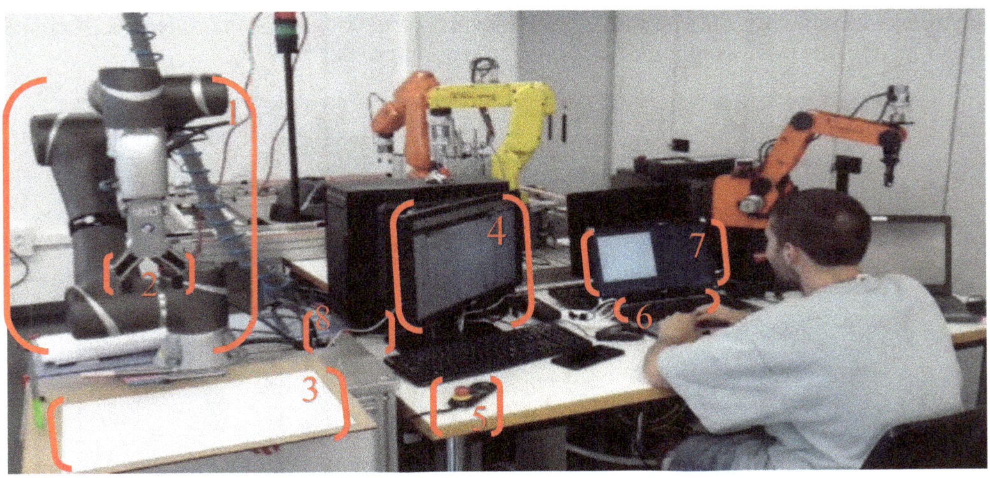

Figure 1: Experimental setup including: (1) Cobot-arm TM5-700; (2) 3D grip; (3) DIN A3 drawing surface; (4) Cobot controller screen; (5) Cobot remote control (3); (6) Tobii Pro X2-30; (7) Eye-tracker laptop; (8) Ethernet connection.

2.2 Technical block description

Fig. 2 shows the technical scheme of communication data flow between eye-tracker and cobot. This scheme evidences the server–client architecture. The server provides the data acquired by the eye-tracker while the client is the cobot.

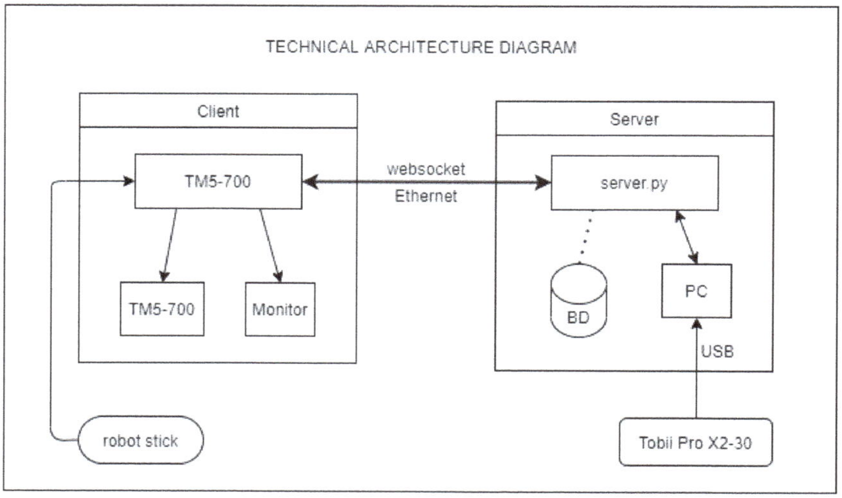

Figure 2: Technical scheme of communication data flow between eye-tracker and cobot.

Despite the fact the eye-tracker includes an automatic calibration tool, we decided to trim it by means of a manual calibration, as depicted in Fig. 3. This calibration has been done once the working area depicted in Fig. 1 was established. This calibration process permits an improved accuracy for the working area. It consists of looking at a predefined set of points (see Fig. 3). This process requires less than three minutes.

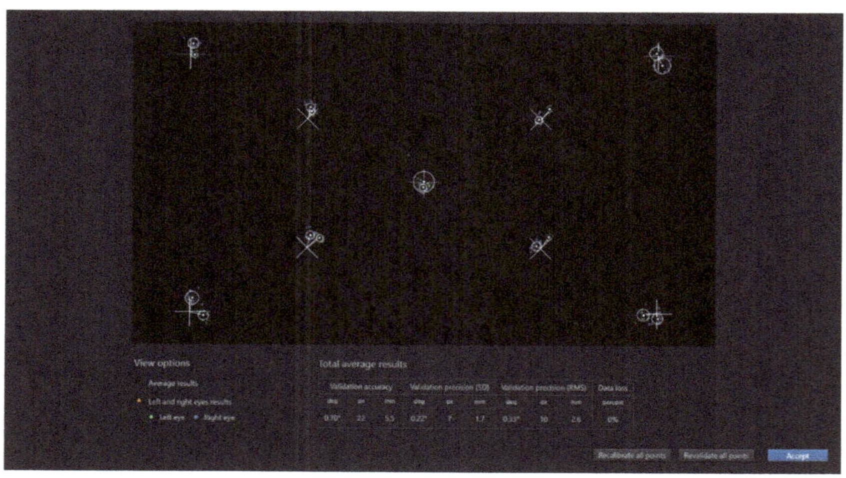

Figure 3: Eye-tracker manual calibration process.

2.3 Socket-host Python

Fig. 4 shows the flow chart of the algorithm programmed in python and executed on the eye-tracker computer. Basically, the application can be divided into two parts, the first part, the left branch, would correspond to the initialization, server establishment, connection between the client and the Tobii Pro X2-30 and finally the capture of the first point, called the reference point from now on.

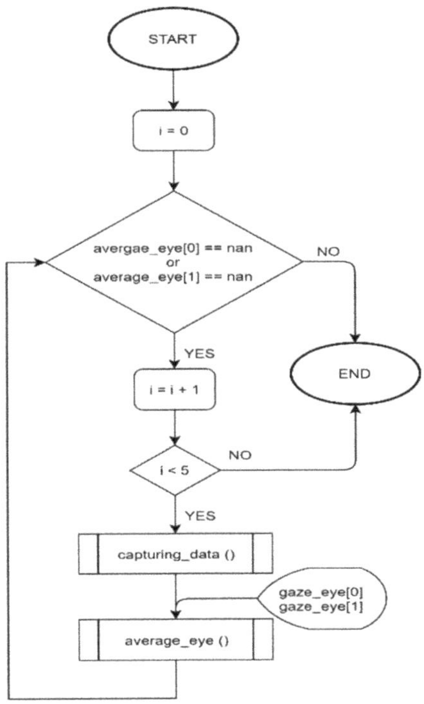

Figure 4: Flow chart of ant_nan () function.

Once the first part is finished, the second one begins, which is the branch on the right part in Fig. 4. This refers to the capture loop. This loop is performed indefinitely until the robot's message does not match what is expected, in which case the resulting dot graph is extracted and ends by disconnecting the server. We have used eight python libraries: Tobii_research, time, socket, pylab, numpy, PIL, winsound and func_aplication_CRAWE.

The most important phyton functions are described below:

- socket.socket (): In this function the server is initialized and the connection with the client is established (cobot). In case there is no established server the server remains opened waiting. The connection is established via TCP/IP by assigning the relevant IP and connection port.
- find_eyetracker (): This function is available in the Tobii_research library application. It is used to search for a Tobii device connected to the machine and establish a connection with it once located. This has been slightly modified to accommodate programming needs and to discard nonrelevant data from the device.
- capturing_data (): This is the Tobii capture function, available at the Tobii_research library application. Basically, this function performs the subscription to the Tobii Pro X2-30 to start data capture (left eye location and right eye location), giving a certain

amount of time for the user to have time to focus and locate the desired point to capture. At the end of the capture time, the last captures of each eye are stored, so that they can then be processed and sent to the cobot. In order to warn the user of the start of the data capture state, a warning noise is emitted thanks to the winsound library application. It should be noted that the data obtained with this function (left eye location and right eye location) have two components (x, y). These are called *gaze_left_eye* and *gaze_right_eye*.

- average_eye (): This function obtains the final lookout point captured by Tobii from the components (x, y) of the captured points (*gaze_left_eye* and *gaze_right_eye*).
- ant_nan (): When the eye-tracker does not directly record an acceptable value or basically does not detect the user's eyes, in the time set in the *capturing_data*() function, the extracted data is *nan* (not a number). This function is used to detect these situations and repeat the capture, that is, to run the *capturing_data()* function again, until either an acceptable capture is obtained or five capture attempts have been made and no one is acceptable. To do this, the loop of the Fig. 4 has been established. At the end of the software it continues as established in the flow chart of the application (Fig. 4).
- eyetracking_data_array (): This function is responsible for storing each capture made during the session in a two-dimensional array. The rows of the matrix correspond to the coordinates (x, y) of the captures, and the columns to the number of captures made, resulting in a [2 x *n*] matrix, where *n* is the number of captures taken.
- conv_to_robot (): This function is used to perform two actions: The first one is to convert the captured points to the work area established by the robot. In our case, this working area has been setup to DIN A3. The second one is to adapt the format of the data obtained to send them through the websocket protocol. To do this, this data must be converted to byte format, as required by the protocol.
- conv_to_host (): This function, like the previous one, serves to modify the format of the data. In this case the data sent by the robot to the server.
- plot_array (): Finally, the last executed function of the software, serves to depict in graph form the data matrix acquired previously. To do this, we set the x-axis of the graph to the x-components of the matrix and the same to the y-axis and the y-components of the matrix.

2.4 TMFLow

The software related to the operation of the cobot, programmed in TMFlow, can be summarized by the flow chart of Fig. 6.

The robot component of the designed control application starts from the resting point [PO], waits for the entry of the variable sent by the application, once obtained converts it into an array or array of data in string format.

Second, it checks if these variables are coordinates or [nan]. In the first case it continue with the software execution. In the second case, end the software returning to the resting point [PO].

Third, it extracts the component (x, y) from the previous variable, obtaining two independent variables, and then assigns these variables to the component (x, y) of the target point [PE], equivalent to the location captured by the eye-tracker but in the robot's work area.

The configuration of the points depends on the needs or desires of the application and the programmer. The components are, firstly, the coordinates (x, y, z) with respect to the base of the robot, and secondly, the rotation in (x, y, z) also with respect to the base. For example, for the developed application the point [PE] has been defined as follows:

$$[PE] = [var\_Eye0, va\_Eye1, z, Rx, Ry, Rz], \tag{1}$$

where *var_Eye0* and *var_Eye1*, are the variables which contain the (x, y) values acquired by the eye-tracker. Z is a fixed value, as the robot always draws on a 2D surface and *Rx*, *Ry* and *Rz* correspond to the robot axis rotation.

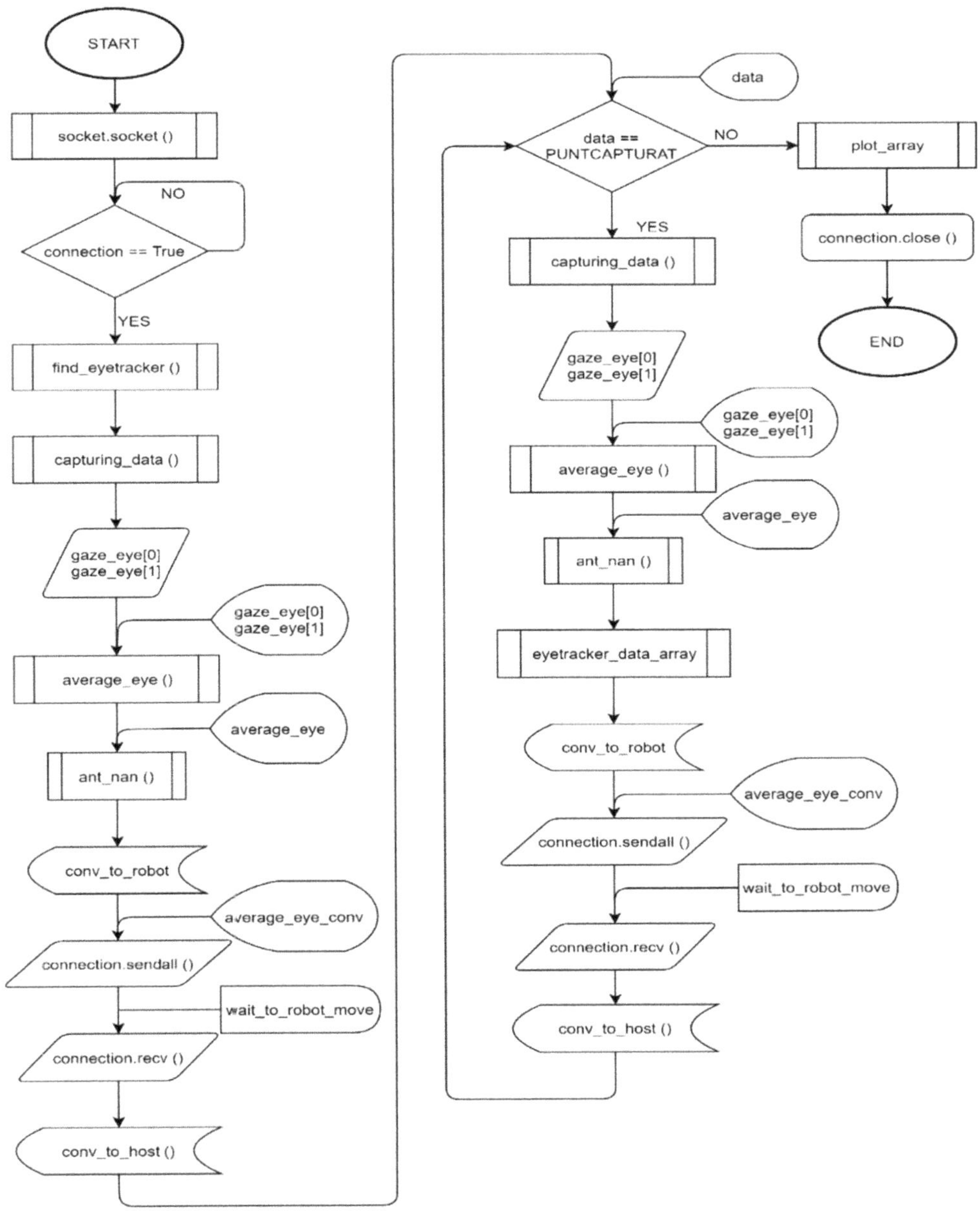

Figure 5: Eye-tracker flow chart.

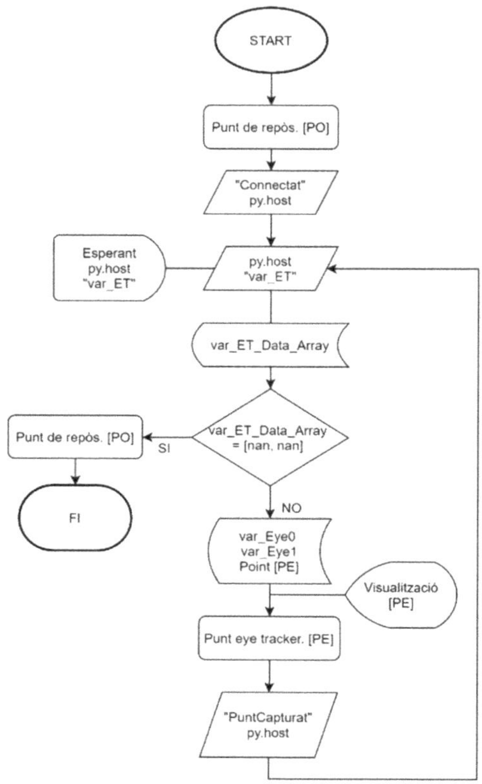

Figure 6: Cobot flow chart.

Finally, once it has reached the position determined by the [PE] point, a "CAPTURED POINT" message is sent to the server indicating that the robot has reached its goal and is waiting for another input. At this point, the data capture software comes back into action, and the robot waits.

3 EXPERIMENTS AND RESULTS

In order to check the system, we have stablished a set of simple exercises, consisting of drawing a simple set of figures (triangle, rectangle, pentagon and star. Fig. 7 represents these figures).The user must select first the desired image he wants to draw. Then the system shows the figure, the user looks at the corners of the figure (marked in red in Fig. 7), and the eye-tracker acquires the set of points. In this YouTube link we have included a demonstration of the system operation: https://youtu.be/U6KYuhel9Tk.

Once the acquisition/performing system has been setup, it is quite straight forward to adapt the system to a more realistic application such as robot control by handicapped people, in dangerous environments (toxic, hazardous, explosive substances, etc.) This is beyond the goal of this paper, as we do not have access to these scenarios.

Fig. 8 shows the experimental results performed by the cobot on a DINA3 paper. On the left column we can observe the result of the drawing performed by the cobot. On the right it

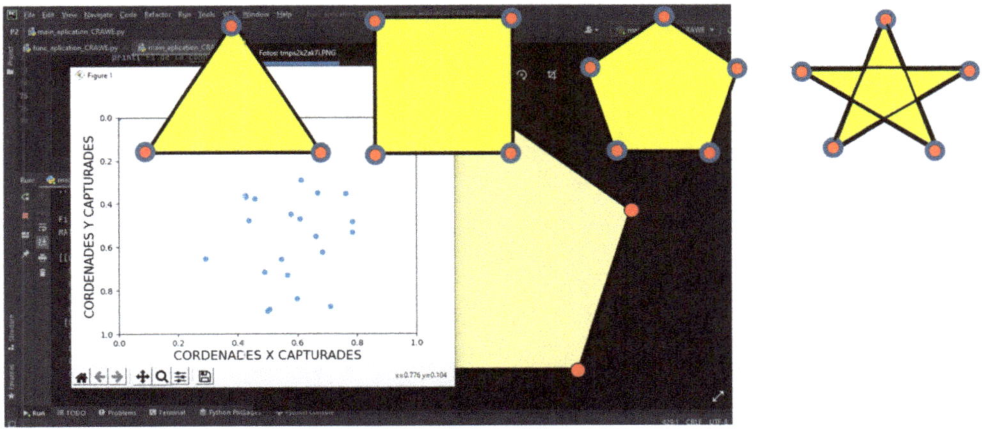

Figure 7: Top, from left to right: triangle, rectangle, pentagon and star figures presented on the screen of the eye-tracker. Bottom: user interface with several acquired dots.

is depicted the acquired points by the eye-tracker when the user looks to the corners of the desired image (from top to bottom: triangle, rectangle, pentagon and star inside a pentagon). In our software design the starting and ending point are marked separately by the user. For this reason, the figures look opened and not closed. In case of programming the application for performing a line connecting the last acquired point to the first one, the figure would be closed. Table 1 highlights the accuracy and precision of the system (accuracy refers to how close a measurement is to the true or accepted value. Precision refers to how close measurements of the same item are to each other).

Worth mentioning is that the user has performed the tasks without a lot of intensive training and it has been tested with several users. All of them agree that the system is easy-to-use.

4 CONCLUSIONS

In this paper we have presented an experimental setup to connect an eye-tracker Tobii PRO X2-30 with a collaborative robot (cobot) OMRON TM5-700. The most challenging part has been the communication between both elements. Once this has been solved, this opens the possibility to large amount of applications for robotic arm control by means of gaze.

The application has been designed with the software OMRON TMFlow for the robot side as well as the Tobii SDK for the eye-tracker. Although different robot and eye-tracker would require a new setup we think that this paper can be useful for the new system design involving different equipment. Eye-tracking technology can be very useful for a plenty of safety and security applications, including safety for construction workers, automotive and industry safe environments.

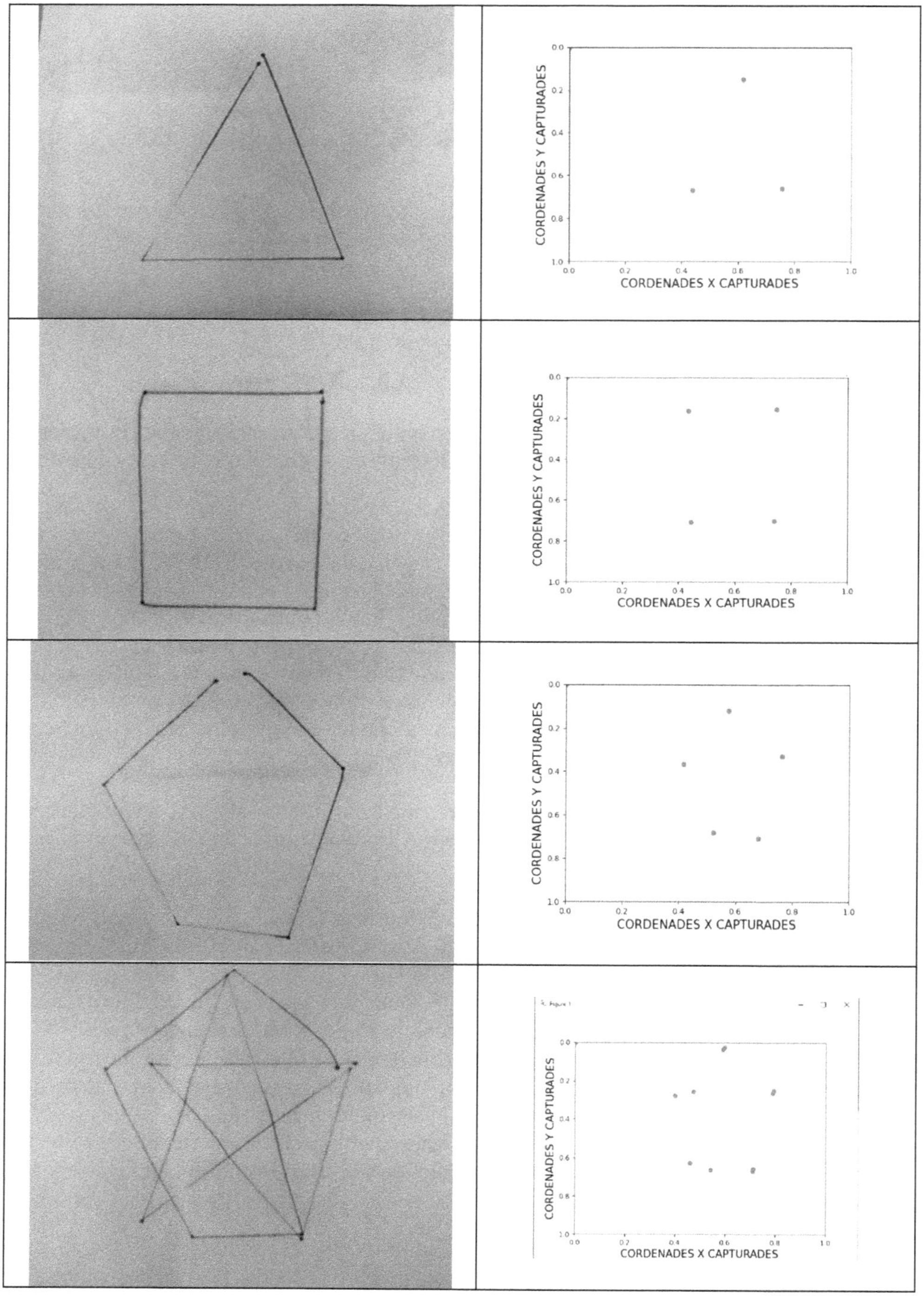

Figure 8: Triangle, rectangle, pentagon and star inside a pentagon. Drawing performed by the robot (left column) and points acquired by the eye-tracker (right column).

Table 1: Accuracy and precision of measurements. (SD = standard deviation; RMS = root mean square.)

Validation accuracy			Validation precision (SD)			Validation precision (RMS)		
degrees	pixels	mm	degrees	pixels	mm	degrees	pixels	mm
0.70°	22	5.5	0.22°	7	1.7mm	0.33°	10	2.6

ACKNOWLEDGEMENT

This work has been funded by Spanish grant Ministerio de ciencia e innovación PID2020-113242RB-I00.

REFERENCES

[1] Markets and Markets, Eye tracking market with COVID-19 impact analysis by offering (hardware, software, services), tracking type (remote and mobile), application (assistive communication, and human behavior and market research), vertical, and geography: Global Forecast to 2025. https://www.marketsandmarkets.com/Market-Reports/eye-tracking-market-144268378.html. Accessed on: 21 Sep. 2021.

[2] Jeelani, I., Han, K. & Albert, A., Automating and scaling personalized safety training using eye-tracking data. *Automation in Construction*, **93**, pp. 63–77, 2018. https://doi.org/10.1016/j.autcon.2018.05.006.

[3] Hasanzadeh, S., Esmaeili, B. & Dodd, M.D., Measuring the impacts of safety knowledge on construction workers' attentional allocation and hazard detection using remote eye-tracking technology. *Journal of Management in Engineering*. **33**(5), 2017.

[4] Martinez-Marquez, D., Pingali, S., Panuwatwanich, K., Stewart, R.A. & Mohamed, S., Application of eye tracking technology in aviation, maritime, and construction industries: A systematic review. *Sensors*, **21**(13), p. 4289, 2021. https://doi.org/10.3390/s21134289.

[5] Carr, D.B. & Grover, P., The role of eye tracking technology in assessing older driver safety. *Geriatrics,* **5**(2), pp.1–14, 2020. https://doi.org/10.3390/geriatrics5020036.

[6] Vetturi, D., Tiboni, M., Maternini, G. & Bonera, M., Use of eye tracking device to evaluate the driver's behaviour and the infrastructures quality in relation to road safety. *Transportation Research Procedia*, **45**, pp. 587–595, 2020. https://doi.org/10.1016/j.trpro.2020.03.053.

[7] Rjabovs, A. & Palacin, R., Investigation into effects of system design on metro drivers' safety-related performance: An eye-tracking study. *Urban Rail Transit*, **5**, pp. 267–277, 2019. https://doi.org/10.1007/s40864-019-00115-1.

[8] Khan, M.Q. & Lee, S., Gaze and eye tracking: Techniques and applications in ADAS. *Sensors*, **19**(24), p. 5540, 2019. https://doi.org/10.3390/s19245540.

[9] Le, A.S., Suzuki, T. & Aoki, H., Evaluating driver cognitive distraction by eye tracking: From simulator to driving. *Transportation Research Interdisciplinary Perspectives*, **4**, 100087, 2020. https://doi.org/10.1016/j.trip.2019.100087.

[10] Gruden, C., Otković, I.I. & Šraml, M., Safety analysis of young pedestrian behavior at signalized intersections: An eye-tracking study. *Sustainability*, **13**(8), p. 4419, 2021. https://doi.org/10.3390/su13084419.

[11] Babić, D., Babić, D., Fiolić, M. & Ferko, M., Factors affecting pedestrian conspicuity at night: Analysis based on driver eye tracking. *Safety Science*, **139**, 105257, 2021. https://doi.org/10.1016/j.ssci.2021.105257.

[12] Peysakhovich, V., Lefrançois, O., Dehais, F. & Causse, M., The neuroergonomics of aircraft cockpits: The four stages of eye-tracking integration to enhance flight safety. *Safety*, **4**(1), p. 8, 2018. https://doi.org/10.3390/safety4010008.
[13] Chadalavada, R.T., Andreasson, H., Schindler, M., Palm, R. & Lilienthal, A.J., Bi-directional navigation intent communication using spatial augmented reality and eye-tracking glasses for improved safety in human–robot interaction. *Robotics and Computer-Integrated Manufacturing*, **61**, 101830, 2020. https://doi.org/10.1016/j.rcim.2019.101830.
[14] Dahmani, M., Chowdhury, M.E.H., Khandakar, A., Rahman, T., Al-Jayyousi, K., Hefny, A. & Kiranyaz, S., An intelligent and low-cost eye-tracking system for motorized wheelchair control. *Sensors*, **20**(14), p. 3936, 2020. https://doi.org/10.3390/s20143936.
[15] Scalera, L., Seriani, S., Gallina, P., Lentini, M. & Gasparetto, A., Human–robot interaction through eye tracking for artistic drawing. *Robotics*, **10**, p. 54, 2021. https://doi.org/10.3390/robotics10020054.
[16] Faundez-Zanuy, M. et al., Handwriting biometrics: Applications and future trends in e-security and e-health. *Cogn. Comput.,* **12**, pp. 940–953, 2020. https://doi.org/10.1007/s12559-020-09755-z.
[17] Faundez-Zanuy, M., Mekyska J. & Impedovo, D., Online handwriting, signature and touch dynamics: Tasks and potential applications in the field of security and health. *Cogn. Comput*., 2021. https://doi.org/10.1007/s12559-021-09938-2.

SECTION 9
RISK REDUCTION IN TRANSPORT SYSTEM IN EMERGENCY CONDITIONS (SPECIAL SESSION ORGANISED BY F. RUSSO)

RISK REDUCTION IN TRANSPORT SYSTEM IN EMERGENCY CONDITIONS: A FRAMEWORK FOR DEMAND ANALYSIS

FRANCESCO RUSSO[1] & GIOVANNA CHILA[2]
[1]Dipartimento di Ingegneria dell'Informazione, delle Infrastrutture e dell'Energia Sostenibile (DIIES), Università Mediterranea di Reggio Calabria, Italy
[2]Environment and Public Works, Municipality of Motta San Giovanni, Italy

ABSTRACT
The aim of this work is to present a framework of the major contributions which consider demand in evacuation conditions. The production of works related to mobility simulation in evacuation conditions has significantly increased after Katrina event from one side and 9/11 event from the other, even if in the last years the attention is decreased. Several models have been developed, but, in most cases, these are implemented considering an isolated area with a non-system approach. In this work it is proposed a classification of literature models which simulate evacuation conditions and are demand-focused in relation to the effect in the transport system, considering mainly models that deal with natural disasters. The classification is presented, after a brief analysis of literature regarding demand models in the framework of behavioural-topological paradigm, core of transportation system model. The basic elements of the dynamic demand models developed in the literature are then recalled, classified according to the components that explain the dynamic aspects (attributes, parameters, residues). Sequential analysis is then proposed as a tool to highlight the presence of dynamic components.
Keywords: risk reduction, exposure, evacuation, dynamic demand, sequential analysis, transport systems, utility updating.

1 INTRODUCTION

Demand models are a fundamental tool for solving most problems in transport systems planning and management. Several mathematical models to simulate transport demand are proposed in the literature. These are based on various assumptions and can be subdivided in relation to different elements. In most cases, these models belong to discrete choice models [1], [2].

Discrete choice models are usually derived under the assumption of utility-maximizing behaviour by the decision maker and are applied to simulate several choices of transport and mobility, such as mode choice and path choice [2], [3]. Features that are common to discrete choice models are (Fig. 1):

- the choice set formation, which is the set of options that are available to the decision maker;
- the alternative choice probabilities inside the choice set, which derive from utility-maximizing behaviour.

Choice set formation can be simulated by the Manski approach [4], [5].

Alternative choice probabilities, deriving from utility-maximizing behaviour, depend on attributes and hypotheses on random residual distribution. The distribution implies a different specification of the model. Based on the assumptions on the distribution it is possible to define two main classes. The first class is made up of models for which it is possible to express the closed-form probability and it is therefore possible to directly maximize the likelihood function; the GEV (generalized extreme value) models [6] and their different specifications belong to this class: multinomial logit, nested logit and cross-nested logit.

WIT Transactions on The Built Environment, Vol 206, © 2021 WIT Press
www.witpress.com, ISSN 1743-3509 (on-line)
doi:10.2495/SAFE210211

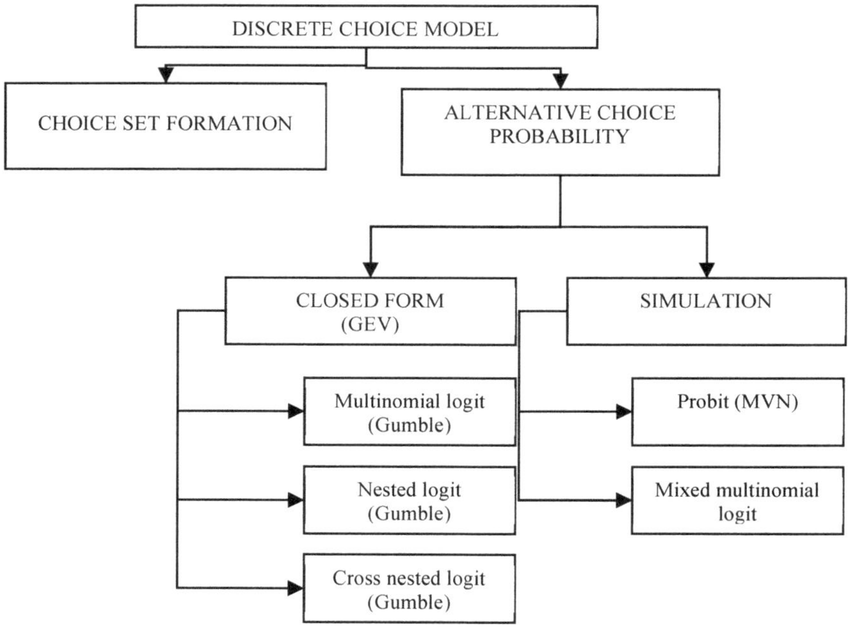

Figure 1: Main features of discrete demand models.

The second class includes models that cannot be solved in closed form and for which it is necessary to proceed with simulation methods to obtain the solutions; the multinomial probit [7] and mixed multinomial logit [8] models belong to this class.

In the ordinary situation, the demand for transport plays a central role in defining the conditions of the overall system. Demand interacts with supply, and the overall model, supply–demand-interaction, is studied within the topological-behavioural paradigm, modelled by the equations that define the so-called TSM (transportation system model) [9]. It is necessary to recall that the study of TSM in dynamic conditions implies the use for interaction, of a dynamic process in which both the within day and day to day components are considered. To define the dynamic assignment process, two models must be specified [9]: the choice updating model and the utility updating model. For the choice updating it is useful to recall [10], [11] and for the utility updating [12].

The risk situation is, by definition, a different situation from the ordinary one. The modelling of the risk situation in a transport system, while maintaining the same framework of the TSM formalized for the ordinary situation, implies deep modifications of the individual component models. In this note, the focus is placed on the evolution of demand models in risk conditions. Fig. 2 shows a picture of the field investigated here, placed at the intersection of the demand models with the models that simulate exposure inside the risk conditions, where the risk is considered in his main factors [13], [14]: occurrence, vulnerability, exposure.

The demand models developed for ordinary conditions cannot be directly used for extraordinary conditions, as evacuation to reduce exposure. In this case various problems

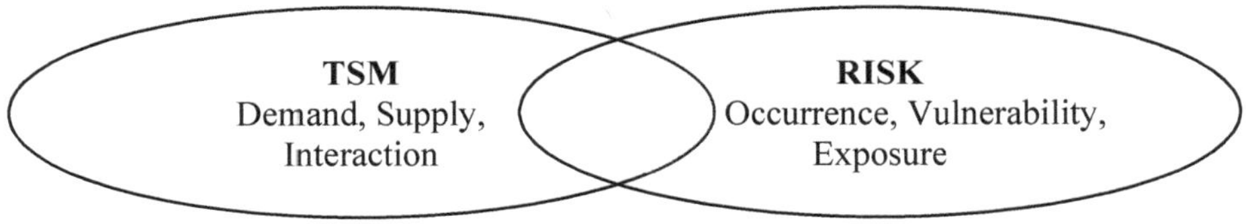

Figure 2: Intersections elements between TSM and risk models.

arise related to the type of event that induces evacuation. The different events and therefore the related effects can be classified with respect to: type of event; effect in time and space; effect in a component of the transport system (demand, supply or interaction).

On the basis of these four classes it is possible to identify an event and the related effect and it is possible to define the cases for which evacuation is necessary, the demand models estimated in the ordinary conditions cannot be applied due to: passage from a decision maker (user) to a multiplicity of decision makers (user, public administrator, security forces); modifications of the choice set of the alternatives, of the attributes of the single alternative, of the parameters of the attributes, of the random residuals.

In the literature, several specifications of dynamic models have been proposed, as regards different elements, such as attributes, parameters or random residuals. In this work are analysed only a set of these, belonging to discrete choice models. These models aren't characterized by a preliminary dynamic analysis of data used to model calibration. Dynamic analysis of these data could allow to:

- justify the use of dynamic relation to simulate user decisions, because dynamic analysis enables to individuate dynamic correlations among data in the time;
- suggest, given a specific kind of correlations among data in the time, the more appropriate dynamic approach to simulate user decisions.

It is important to develop the analysis of demand data relating to the single dynamic phenomenon considering on the one hand the large availability of data provided by the diffusion of the internet, and therefore available for public administrations and on the other hand the increase in internet use by users [15]. The increase of use implies the actualization of the choice and utility, as mentioned above.

A useful approach to analyse recorded data relative to demand in the time is sequential analysis, which explicitly considers the choice sequences. Sequential analysis arose as a science applied to observing social behaviour in psychology. It aims to demonstrate how to record observation data, related to user behaviour, in a way that preserves sequential information, and also how to analyse such data in a way that makes use of its sequential nature. In sequential analysis, data related to user behaviour are recorded and analysed, preserving sequential information and considering a statistical approach.

The paper is structured as follows: in Section 2 the literature on demand models under risk conditions (exposure) is summarized, in Section 3 the literature on dynamic probabilistic models is given and in Section 4 the basic elements of statistical sequential analysis are recalled. The conclusions summarize the main elements of the analysis carried out on the existing literature, proposing specific indications for the development of the research.

The work presented is of interest both to researchers of transport systems and to planning technicians, because on the one hand it summarizes the bases from which the TSM evolves towards the extraordinary conditions given by risk and on the other it identifies the main limits of use of TSM by signalling possible demand patterns that better represent extraordinary conditions in the commercial software.

2 DEMAND MODELS UNDER EXPOSURE CONDITIONS

On the basis of what has been said in the introduction, the first element of in-depth analysis, in a framework that wants to consider the analysis of the demand, is given by the different demand models that can be defined with respect to events, which are different for their effects in the space and time [13], [16].

In the international literature, the most analyzed natural event is the hurricane, and it can be referred to in order to rebuild the first methodological elements regarding the exposure reduction by means of evacuation. The assessment of demand in hurricane conditions is usually divided into two phases: estimating the total evacuation demand and estimating the departure time [17].

The models studied have very simple analytical structures. Consider that the most commonly used method to estimate demand is the use of evacuation participation rates [18]. A more refined model considers the response curves, in relation to the characteristics of the hurricane. The curves are calculated by relating the proportion of evacuation to the time elapsed since the issuance of an evacuation order [19].

With respect to the premise, in this section only some works are recalled that deal with dangerous events that have effects on the decisions to generate the movement, on the choice of destination and mode of transport, which are the models most influenced by the type of event and of effect. It is useful to suggest some general works relating to the choice of the path [20]–[22]. It is also useful to point out some works that consider the problem of the evacuation of transport systems and in this context also refer to the problems of demand. Daganzo and So [23] propose a strategy for the management of evacuation networks segmented into three classes: non-anticipative (not based on demand forecasting); adaptive (based on real-time traffic information on the whole network); decentralized (based on information available locally in each area of the overall network). Lindell and Prater [24] and Murray-Tuite and Wolshon [25], [26] highlight that there are two important lines of research, the one of social scientists, which studies the behavior of the population, and the other developed by transport engineers who study networks under evacuation; but the authors report that the two lines of research are poorly integrated. A particular attention to the analysis of natural and anthropogenic disasters that have affected millions of people, is the one supported by the Wessex Institute of Technology (WIT), which annually has two conferences on the various aspects of risk in territorial systems; you can refer to the wide-open source reading available on the WIT site starting from [27].

As mentioned above, the risk event most studied and reported in the literature is related to hurricanes. In Brebbia [28] the behavior during all phases of evacuation is analyzed considering some variables: level of risk of the area, decisions of public authorities, type of residence, preventive perception of personal risk, specific threat factors for hurricanes. In Dow and Cutter [29] the decision-making processes of families are analyzed on the basis of which the public decision-maker can plan subsequent evacuations; four issues are examined in the decision-making process: number of vehicles per household; departure times of evacuated families; distances traveled in the evacuation; importance of information in identifying the choice set of specific evacuation routes. Wilmot and Fu [18] assume that the two decisions relating to the execution of the evacuation and the time to carry it out are taken simultaneously; the authors hypothesize that each family continuously updates the information relating to the progress of the approaching hurricane, and in relation to the updating of the information update the choice, until a level of reduced utility is reached such to decide to evacuate in a specific time; the authors propose the use of a model called sequential logit. In Solis et al. [30] different models are compared highlighting the significant characteristics of the families' choice to evacuate due to the hurricane event, studying a sample of 1355 families in Florida.

It is interesting to recall the results achieved and published in the context of a research project financed with European funds, called SICURO, that is one of the few research projects related to anthropogenic delayed effects. As part of the project, several real evacuations of an urban center were carried out. The main results refer to a better understanding of: demand models, considering also SP surveys [31]; supply models with the calibration of the cost functions [32], [33]; design models [34] considering traffic junction regulation [35], [36] and design of the routes for emergency vehicles [37]; appropriate DSS [38], [39] considering multimodality [40], evidencing risk [41], [42]; planning methods [43], comprehensive of evacuation planning [44], [45] training [46], exercises and evaluation tools to reduce risk [47].

3 DYNAMIC DISCRETE CHOICE MODELS

In the previous section it has been recalled that in order to simulate some user choices dynamic models should be adopted. In this paper, choice models are classified as:

- non-dynamic, if they give the choice probability without considering system evolution;
- dynamic, if they give the choice probability while considering system evolution.

Under the assumption of random utility models (RUM) let be:

- t the generic current time interval;
- t^{-1} the generic previous time interval;
- n the generic user class;
- j^t the chosen alternative in time t;
- i^t the generic alternative in time t;
- k^{t-1} the chosen alternative in time t^{-1};
- r^{t-1} the generic alternative in time t^{-1};
- X^t, β^t attribute and parameter in t;
- X^{t-1}, β^{t-1} attribute and parameter in t^{-1};
- ε^{t-1}, ε^t random residuals in t^{-1} and in t;

with $j^t, i^t \in C^t$; $k^{t-1}, r^{t-1} \in C^{t-1}$.

In the evolution of the system various cases can happen; in the following are synthetized the more significant:

- evolution of user characteristics or level of service as perceived by the reference user, then $X^{t-1} \neq X^t$;
- evolution in user taste, hence in user evaluation and perception of attributes, then $\beta^{t-1} \neq \beta^t$;
- evolution of random residuals, then $\varepsilon^{t-1} \neq \varepsilon^t$;
- evolution of choice set, then $C^{t-1} \neq C^t$.

In the last case ($C^{t-1} \neq C^t$) an evident discontinuity is introduced in the system, since a new choice alternative arises from t^{-1} to t; this case is the opposite of the sequential case, and cannot be represented as a sequence, just because the process has no linear history, but from a general view it can be considered a point of discontinuity.

Given that $C^{t-1}=C^t\equiv C$, it is possible to assume that the choice set is fixed and then it can write, without loss of generality, $k^{t-1}=j^{t-1}$ and $r^{t-1}=i^{t-1}$, using for the same alternative the same representative variable in t and in t^{-1}; formally we can define:

- as a non-dynamic model, the model that gives the probability that user n chooses the generic alternative j independently of time interval as:

  $P^n[j^t]=prob(U^n[j^t]>U^n[i^t]\ \forall j^t,i^t\in C,j^t\neq i^t)=prob(V^n[j^t]+\varepsilon^n_j>V^n[i^t]+\varepsilon^n_i\forall j^t,i^t\in C,j^t\neq i^t)$
  with
  $P^n[j^t]=P^n[j^{t-1}]$

- as a dynamic model the model that gives the probability that user n chooses the generic alternative j^t in t, if $\beta^{t-1}\neq\beta^t$ and/or $X^{t-1}\neq X^t$, as:

  $P^{n,t}[j^t]=prob(U^{n,t}[j^t]>U^{n,t}[i^t]\forall j^t,i^t\in C,j^t\neq i^t)$
  with
  $P^{n,t}[j^t]\neq P^{n,t-1}[j^{t-1}]\ \forall j^t,j^{t-1}\in C,\ j^t\neq j^{t-1}$
  $P^{n,t-1}[j^{t-1}]=prob(U^{n,t-1}[j^{t-1}]>U^{n,t-1}[i^{t-1}]\ \forall j^{t-1},i^{t-1}\in C,j^{t-1}\neq i^{t-1})$

In the following it is analysed the main literature on dynamic models, proposing just some example of these in respect of dynamic element introduced (attributes, parameters, random residuals), without trying to consider all the produced literature. Dynamic models as regard choice set aren't considered in this note, because they cannot be represented as a sequence. It's important to highlight that, in some cases, several dynamic elements are introduced and it is difficult classify the model according to a specific element. Therefore, it is hard to introduce an unequivocal classification of literature models, saving the possibility to introduce blended classification (Table 1).

Table 1: Dynamic model: example.

Evolution in user taste ($\beta^{t-1}\neq\beta^t$)	Evolution of user characteristics or level of service ($X^{t-1}\neq X^t$)	Evolution of random residuals($\varepsilon^{t-1}\neq\varepsilon^t$)
	Manski and Sherman [48]	Russo [53]
Hensher and Le Plastrier [52]	Hensher and Le Plastrier [52]	
	Train [49]	
Nielsen and Sørensen [51]	Nuzzolo et al. [50]	Wilmot and Fu [18]
	Cascetta [9]	Cascetta [9]

Dynamic models as regard attributes are the most common in the literature. Among these, can be recalled models proposed in Manski and Sherman [48] and Train [49]. In these cases, a dynamic attribute is introduced in the systematic utility specification; this attribute is named as transaction search cost, a dummy variable which is equal to 0 if $k^{t-1}=j^t$ in the time t, equal to 1 otherwise. In Nuzzolo et al. [50] a dynamic path choice model for high frequency services introducing three kinds of attributes is proposed: constant attributes, which do not change from day to day and from one time to another; current and updated attributes, relative to within-day updating; attributes forecast by users on the day-to-day learning process.

In relation to distribution of parameters, in Nielsen and Sørensen [51] it is proposed a model considering a distribution of β parameters among individuals belonging to a given sample. It estimates a model for each data record and compares the estimates for all these models to assess the shape of the distribution. Estimations are performed on smaller subsets of data since it is not possible to perform model estimation for each individual except for very special data sets. The paper consider distribution inside the sample, but it doesn't analyse the evolution in the time.

A first rough hybrid model between these two classes (dynamic as regard attributes and parameters) can be considered the one proposed in Hensher and Le Plastrier [52]. They propose a discrete choice model simulating vehicle ownership, developing a series of linked choice models to explain household vehicle holdings and adjustments in the holdings over time. They suppose that the household periodically considers vehicles owned and decides on maintenance of the status quo, adjustment in the number of vehicles or adjustment in fleet composition.

Finally, in relation to dynamic models as regard random residuals, in Russo [53] it is proposed a dynamic discrete choice model applied to path choice simulation for high frequency services. In this case, if t^1, t^2, … is the run arrival time sequence and j^1, j^2, … the relative run sequence, the total unconditional probability of choosing a run arriving at t^w is defined through the probability of boarding and the probability that the users did not board the previous runs. This is equal to one minus the probability associated with the set of runs already passed.

Similarly, Wilmot and Fu [18] propose a sequential binary logit model to simulate the probability of a household evacuating at each time period before landfall. In the choice mechanism a temporal segmentation of the departure time is considered. In each time a household can choose whether or not to evacuate. Of course, if the household chooses not to evacuate, the choice is reconsidered at the next time and so on.

A hybrid model between dynamic models as regard attributes and random residuals is proposed in Cascetta [9], whit a model simulating day-to-day dynamic path choice behaviour, including two phenomena which are not highlighted in the equilibrium approach: user learning and forecasting mechanisms (utility updating model); user choice updating behaviour (choice updating model).

4 SEQUENTIAL DATA ANALYSIS

Literature models described in the previous section introduce dynamic evolutions of several phenomenon, but they aren't characterized by a statistical structure which allow to analyse the given dynamic phenomenon in the sample database and to highlight, if exists, a specific lag. In the literature which deal with data analysis, structure which allows to carried out dynamic analysis of phenomenon exists.

Sequential analysis arose as science applied itself to observing social behaviour in psychology. Sequential analysis aims to demonstrate how to record observation data, related to user behaviour, in a way that preserves sequential information, and also how to analyse such data in a way that makes use of its sequential nature [54], [55]. The main question to be analysed concerns whether or not specific transition frequencies from an antecedent to a consequent state differ from what would be expected if the two states were independent.

In sequential analysis, data related to user behaviour are recorded and analysed, preserving sequential information. Sequential analysis requires that data must be collected in a systematic way. The analyst must set the recording unit, which can be an interval or an event.

Depending on how data were recorded, the analyst can extract different representations from the same recorded data for different purposes.

Historically, there are at least four forms, termed the Sequential Data Interchange Standard:

- event sequences: a single stream of alternatives is presented without any information concerning time, whether onsets or offsets;
- state sequences: they are identical to event sequences, with the simple addition of timing information;
- timed-event sequences: this is a useful and general-purpose format, used if alternatives can co-occur and if their onset and offset times were recorded;
- interval sequences: this is designed to accommodate interval recording in a simple and straightforward way; alternatives are simply listed as they occur and interval boundaries are represented by commas.

Sequences are analysed in respect of transitional probabilities, so-called even if they recall the probability definition only as frequency. These probabilities are one kind of conditional probability. It may be recalled that an unconditional probability is just the probability of a particular target event occurring relative to a total set of events. A conditional probability is the probability of a particular target event occurring relative to another given event which happened. A transitional probability is defined as a kind of occurring conditional probability, which is distinguished from other conditional probabilities in that the target and the given events occur at different and sequent times. Therefore, by analysing realized data using transitional probabilities it can be obtained further information about their succession in time.

Data collected sequentially are summarized using a transition frequency matrix and a second matrix (so-called) transition probability matrix. The transition frequency matrix is a square matrix which has the same number of rows and columns as events; its generic element ϕ_{ij} represents the transition frequency from i to j (Table 2). These frequencies can also be converted by dividing each ϕ_{ij} by the row total of row i. The results can be arranged in a matrix so-called transition probability matrix (Table 3). The rows of this matrix sum to one.

Table 2: Transition frequency matrix.

		t		
		i	j	
t^{-1}	i	ϕ_{ii}	ϕ_{ij}	ϕ_{i+}
	j	ϕ_{ji}	ϕ_{jj}	ϕ_{j+}
		ϕ_{+i}	ϕ_{+j}	

Table 3: Transition probability matrix.

		t		
		i	j	
t^{-1}	i	π_{ii}	π_{ij}	π_{i+}
	j	π_{ji}	π_{jj}	π_{j+}
		π_{+i}	π_{+j}	

As regards transitional probabilities, in order to indicate the temporal displacement between target (in the current time) and given (in the previous time) event, the word lag is used.

5 CONCLUSIONS

In this paper a framework of demand models has been proposed, in relation to risk reduction in transport systems. The reduction considered is that deriving from the possible modification of exposure which is one of the crucial factors of risk together with occurrence and vulnerability. In the context of exposure, the role of evacuation was analysed.

A first point to highlight emerges from the available literature: mainly models dealing with natural disasters have been presented, while those of the anthropogenic type have a reduced presence in the literature despite some global focal events such as 9/11 or Chernobyl or Bhopal.

In the introduction, a synthetic structure of the literature related to the demand models developed in the context of the topological-behavioural paradigm, the basic nucleus of the transportation system model, has been proposed. The limits of the use, for evacuation, of the demand models developed for ordinary conditions were therefore highlighted. It was also noted that the diffusion of internet access profoundly modifies the situation in extraordinary conditions even more than in the past. From this another notable point (second) emerged: the need to develop the basic components (choice and utility updating) for the calibration and application of assignment models by means of a dynamic process.

The basic elements of the dynamic demand models currently available in the literature were then recalled, classified according to the components that explain the dynamic aspects: attributes, parameters, residues. A third point emerged from this survey: the lack of a statistical analysis of the dynamic phenomena significant for the demand. On this basis, it appears interesting to use sequential analysis as a tool to highlight the presence of dynamic components. The application of sequential analysis can reveal stochastic sequential patterns among the data collected over time, thus allowing to obtain additional information compared to traditional models.

On the basis of the points that have emerged, future objectives for the research are: (a) to study the evacuation related to anthropogenic event; (b) to analyse the data of evacuations carried out by means of sequential analysis; and (c) the development of models for updating utility and choice. The updating models should be specified and calibrated firstly for the ordinary conditions of path choice and then for the extraordinary conditions of evacuation.

REFERENCES

[1] Domencich, T.A. & McFadden, D., *Urban Travel Demand: A Behavioural Analysis*, American Elsevier: New York, 1975.

[2] Ben Akiva, M. & Lerman, S.R., *Discrete Choice Analysis: Theory and Application to Travel Demand*, MIT Press: Cambridge, MA, 1985.

[3] Train, K., *Discrete Choice Methods with Simulation*, MIT Press: Cambridge, MA, 2003.

[4] Manski, C.F., The structure of random utility models. *Theory and Decisions*, **8**, pp. 229–254, 1977.

[5] Morikawa, T., A hybrid probabilistic choice set model with compensatory and noncompensatory choice rules. *Proceedings of the 7th World Conference on Transport Research*, Avebury, 1995.

[6] McFadden, D.L., Modelling the choice of residential location. *Interaction Theory and Planning Models*, pp. 75–96, North Holland: Amsterdam, 1978.

[7] Daganzo, C.F., *Multinomial Probit: The Theory and its Application to Demand Forecasting*. Academic Press: New York, 1979.

[8] McFadden, D.L. & Train, K., Mixed MNL models for discrete response. *Journal of Applied Econometrics*, **15**, pp. 447–470, 1996.

[9] Cascetta, E., *Transportation Systems Engineering: Theory and Methods*, Springer Science and Business Media, 2013.

[10] Manski, C.F., Social learning from private experiences: The dynamics of the selection problem. *The Review of Economic Studies*, **71**(2), pp. 443–458, 2004. https://doi.org/10.1111/0034-6527.00291.

[11] Manski, C.F., *Social Learning and the Adoption of Innovations*, Blume and Durlauf, pp. 31–47, 2006.

[12] Russo, F. & Comi, A., Providing dynamic route advice for urban goods vehicles: The learning process enhanced by the emerging technologies. *Transportation Research Procedia – 24th EURO Working Group on Transportation Meeting, EWGT 2021*, 8–10 September, Aveiro, Portugal, Elsevier, 2021.

[13] Russo, F. & Vitetta, A., Urban road transportation analysis in emergency conditions: Models and algorithms. *6th International Conference on Urban Transport and the Environment for the 21st Century*, **6**, pp. 533–542. https://doi.org/10.2495/UT0005112000.

[14] Russo, F. & Vitetta, A., Safety of users in road evacuation: General methodology and main results. *Urban Transport XIII*, ed. C.A. Brebbia, WIT Press: Southampton and Boston, pp. 763–772, 2007.

[15] Campisi, T., Russo, A., Tesoriere, G., Bouhouras, E. & Basbas, S., COVID-19's effects over e-commerce: A preliminary statistical assessment for some European countries. *International Conference on Computational Science and Its Applications ICCSA 2021: Computational Science and Its Applications – ICCSA 2021*, pp. 370–385, 2021. https://doi.org/10.1007/978-3-030-86979-3_27.

[16] Russo, F. & Chilà, G., Safety of users in road evacuation: Modelling and DSS for demand. *WIT Transactions on Ecology and the Environment*, vol. 120, WIT Press: Southampton and Boston, pp. 465–474, 2009.

[17] Wilmot, C.G. & Fu, H., Static vs. dynamic and aggregate vs. disaggregate: A comparison between practice and research in hurricane evacuation travel demand modeling. *Transp. Research Board 86th Annual Meeting*, 2007.

[18] Wilmot, C.G. & Fu, H., A sequential logit dynamic travel demand model for hurricane evacuation. *Transp. Research Record*, **1882**, pp. 19–26, 2004.

[19] Wilmot, C.G., Fu, H. & Zhang, H., Modeling the hurricane evacuation response curve. *Transp. Research Record*, **2022**, pp. 94–102, 2007.

[20] Vitetta, A., Quattrone, A. & Polimeni, A., Safety of users in road evacuation: design of path choice models for emergency vehicles. *WIT Transaction on The Built Environment*, vol. 96, WIT Press: Southampton and Boston, pp. 41–50, 2007.

[21] Chen, Y. & Xiao, D., Emergency evacuation models and algorithms. *Journal of Transportation System Engineering and Information Technology*, **8**(6), pp. 96–100, 2006. http://doi.org/10.1016/S1570-6672(09)60008-8.

[22] Pel, A.G., Hoogendoorn, S.P. & Bliemer, M.C.J., Evacuation modelling including traveler information and compliance behaviour. *Proceedings of the 1st International Conference on Evacuation Modeling and Management*, eds S.P. Hoogendoorn, A.G. Pel & M.A.P. Mahmassani, *Procedia Engineering*, **3**, pp. 101–111, 2010.

[23] Daganzo, C.F. & So, S.K., Managing evacuation networks. *Transportation Research Part B*, **45**(9), pp. 1424–1432, 2011. http://doi.org/10.1016/j.trb.2011.05.015.

[24] Lindell, M.K. & Prater, C.S., Critical behavioral assumptions in evacuation time estimate analysis for private vehicles: Examples from hurricane research and planning. *Journal of Urban Planning and Development*, **133**(1), pp. 18–29, 2007. http://doi.org/10.1061/(ASCE)0733-9488(2007)133:1(18).

[25] Murray-Tuite, P. & Wolshon, B., Evacuation transportation modelling: an overview of research, development and practice. *Transportation Research Part C*, **25**, pp. 25–45, 2013. http://doi.org/10.1016/j.trc.2012.11.005.

[26] Murray-Tuite, P. & Wolshon, B., Assumptions and processes for the development of no-notice evacuation scenarios for transportation simulation. *International Journal of Mass Emergencies and Disasters*, **31**(1), p. 78, 2013.

[27] Brebbia, C.A. (ed.), *Disaster Management and Human Health Risk III*, WIT Press: Southampton and Boston, 2013.

[28] Baker, E.J., Hurricane evacuation behavior. *International Journal of Mass Emergencies and Disasters*, **6**, pp. 287–310, 1991.

[29] Dow, K. & Cutter, S.L., Emerging hurricane evacuation issue: Hurricane Floyd and South Carolina. *Natural Hazard Review*, **18**, 2002.

[30] Solis, D., Thomas, M. & Letson, D., Determinants of household hurricane evacuation choice in Florida. *Presentation at the Southern Agricultural Economics Association Annual Meeting*, Atlanta, Georgia, 31 January–3 February, 2009.

[31] Russo, F. & Chilà, G., Safety of users in road evacuation: RP vs. SP surveys in demand analysis. *14th International Conference on Urban Transport and the Environment in the 21st Century,* **101**, pp. 703–713, 2008.

[32] Musolino, G. & Vitetta, A., Short-term forecasting in road evacuation: Calibration of a travel time function. *WIT Transactions on the Built Environment*, vol. 116, WIT Press: Southampton and Boston, pp. 615–626, 2011.

[33] Musolino, G., Risk reduction in transport system in emergency conditions: A framework for supply analysis. *WIT Transactions on the Built Environment*, vol. 206, WIT Press: Southampton and Boston, submitted for publication.

[34] Vitetta, A., Risk reduction in transport system in emergency conditions: A framework for network design problem. *WIT Transactions on the Built Environment*, vol. 206, WIT Press: Southampton and Boston, submitted for publication.

[35] Marcianò, F.A., Musolino, G. & Vitetta, A., A system of models for signal setting design of a signalized road network in evacuation conditions. *WIT Transactions on the Built Environment*, vol. 111, WIT Press: Southampton and Boston, pp. 313–323, 2010.

[36] Marcianò, F.A, Musolino, G. & Vitetta, A., Signal setting optimization on urban road transport networks: The case of emergency evacuation. *Safety Science*, **72**, pp. 209–220, 2015.

[37] Vitetta, A., Polimeni, A. & Quattrone, A., Safety of users in road evacuation: Modelling and DSS for paths design of emergency vehicles. *WIT Transactions on Ecology and the Environment*, vol. 120, WIT Press: Southampton and Boston, pp. 485–495, 2009.

[38] Di Gangi, M. & Belcore, O.M., Risk reduction in transport system in emergency conditions: A framework for decision support systems. *WIT Transactions on the Built Environment,* vol. 206, WIT Press: Southampton and Boston, submitted for publication.

[39] Di Gangi, M., Safety of users in road evacuation: Modelling and DSS for pedestrian outflow. *WIT Transactions on Ecology and the Environment*, vol. 120, WIT Press: Southampton and Boston, pp. 497–508, 2009. http://doi.org/10.2495/SDP090461.

[40] Di Gangi, M., Planning evacuation by means of a multi-modal mesoscopic dynamic traffic simulation model. *Studies in Computational Intelligence*, **176**, pp. 99–115, http://doi.org/10.1007/978-3-540-89930-3_6, 2009.

[41] Di Gangi, M. & Polimeni, A., A mesoscopic approach to model route choice in emergency conditions. *Paper presented at the Springer Proceedings in Mathematics and Statistics*, **217**, pp. 547–555, 2017. http://doi.org/10.1007/978-3-319-67308-0_55.
[42] Di Gangi, M., Watling, D. & Di Salvo, R., Modeling evacuation risk using a stochastic process formulation of mesoscopic dynamic network loading. *IEEE Transactions on Intelligent Transportation Systems*, 2020. http://doi.org/10.1109/TITS.2020.3038478.
[43] Rindone, C. & Panuccio, P., Risk reduction in transport system in emergency conditions: A framework for evacuation planning. *WIT Transactions on the Built Environment*, vol. 206, WIT Press: Southampton and Boston, submitted for publication.
[44] Russo, F. & Rindone, C., Safety of users in road evacuation: Planning internal processes and guidelines. *WIT Transactions on The Built Environment*, vol. 96, WIT Press: Southampton and Boston, pp. 825–834, 2007.
[45] Russo, F. & Rindone, C., The planning process and logical framework approach in road evacuation: A coherent vision. *WIT Transactions on The Built Environment*, vol. 117, WIT Press: Southampton and Boston, pp. 415–425, 2011.
[46] Russo, F., Rindone, C. & Trecozzi, M.R., The role of training in evacuation. *WIT Transactions on Information and Communication Technologies*, vol. 44, WIT Press: Southampton and Boston, pp. 491–502, 2012.
[47] Russo, F. & Rindone, C., Urban exposure: Training activities and risk reduction. *WIT Transactions on Ecology and the Environment*, vol. 191, WIT Press: Southampton and Boston, pp. 991–1001, 2014.
[48] Manski, C.F. & Sherman, L., An empirical analysis of household choice among motor vehicles. *Transportation Research,* **14A**, pp. 349–366, 1980.
[49] Train, K., Auto ownership and use: an integrated system of disaggregate demand models. *Qualitative Choice Analysis: Theory, Econometrics and an Application to Automobile Demand*, ed. K. Train, pp. 134–170, MIT Press: Cambridge, MA, 1986.
[50] Nuzzolo, A., Russo, F. & Cristalli, U., *Transit Network Modelling: The Schedule Based Approach*, FrancoAngeli: Milan, 2003.
[51] Nielsen, O.A. & Sørensen, M.V., MSL for mixed logit model estimation: On shape of distributions. *Proceedings of the European Transport Conference 2003*, Strasbourg, France, 8–10 October, 2003.
[52] Hensher, D.A. & Le Plastrier, V., Towards a dynamic discrete-choice model of household automobile fleet size and composition. *Transportation Research,* **19B**, pp. 481–495, 1985.
[53] Russo, F., Modelli di scelta del percorso sequenziali per l'assegnazione dinamica a reti di servizi di trasporto collettivo urbano. *Metodi e Tecnologie dell'Ingegneria dei Trasporti*, eds. G.E. Cantarella & F. Russo, pp. 151–165, Franco Angeli: Milan, 1999.
[54] Bakeman, R. & Gottman, J.M., *Observing Interaction. An Introduction to Sequential Analysis*, Cambridge University Press: New York, 1997.
[55] Bakeman, R. & Quera, V., *Sequential Analysis and Observational Methods for the Behavioral Sciences*, Cambridge University Press: New York, 183 pp., 2011.

RISK REDUCTION IN TRANSPORT SYSTEM IN EMERGENCY CONDITIONS: A FRAMEWORK FOR NETWORK DESIGN PROBLEMS

ANTONINO VITETTA
Dipartimento di Ingegneria dell'Informazione, delle Infrastrutture e dell'Energia Sostenibile, Università degli Studi Mediterranea di Reggio Calabria, Italy

ABSTRACT
In emergency conditions, with dangerous events with effects delayed in the time, the transport supply requires configurations that allow the reduction of the negative effects on people. Therefore, it is necessary to define suitable design methodologies for risk reduction. In the context of the wider problem of the transport demand managing in all its components, this paper reports the problem for the road network design in transport systems. The problem analysed considers the design of the link directions and the traffic light regulation at the junction in the road networks for users and emergency vehicles. The paper: reports a state of the art of the problem; introduces a general framework; formulates the problem in terms of aggregate mathematical models. For further information on the topics reported, the literature cited can be considered.
Keywords: emergency condition, network design problem, vehicle routing problem, risk reduction.

1 INTRODUCTION

In the presence of natural or manmade disasters with effect on the people delayed in the time, the transport system must work at its best in order to allow the evacuation of users in the shortest possible time. Considering the three components of the risk, occurrence, vulnerability and exposure, with this type of events, the application of design methodologies for exposure reduction is fundamental (for more details see Fig. 1 cited in the Section 3 of this paper and the connected comments).

The design methodologies require the definition of different components of the transport system:

- the evacuation planning [1];
- the evacuation training [2], [3];
- the risk assessment and management [4];
- the analysis of the demand for mobility in terms of user behaviour [5];
- the study of the performance of the supply model in terms of disutility for users [6];
- the interactions between supply and demand models [7]–[9];
- the evaluation methods [10], [11].

The study of these components is beyond this paper scope.

This paper deals with some design methodologies with the assumption that the interaction between supply and demand is considered as a behaviour constraint of the optimum model. These methodologies can be used in the planning phase of the short or long term in the design of the best configuration of the supply system for risk reduction.

Among all the design models, the goal of this paper is to report a brief state of the art and to formulate the models commonly used to address the following problems: design of the link directions and the traffic light regulation at the junction in the road networks to be used by users and emergency vehicles (network design problem (NDP)).

Following this approach, the paper is divided into the following sections: Section 2 contains the state of the art for NDP; Section 3 reports a general framework of the method; Section 4 contains the formulations for the NDP. Finally, Section 5 reports some conclusions.

2 STATE OF THE ART

The NDP consists in defining the optimal configuration of the road and transit transport network in terms of topology and capacity. In the case of transport systems in emergency conditions, the NDP refers to the road transport system because transit transport vehicles are used and designed in emergency with VRP (vehicle routing problem) procedures reported in other papers, i.e. [12], [13].

The design of the topology consists in defining the optimal configuration of the directions and the allocation of the lanes in the road links. The allocation of the lanes involves the definition of the capacity of the links in each direction. One of the first papers concerning the road network topology design has been proposed in Chen and Alfa [14]. Other design methods have been proposed for the generation of optimal topological configurations under ordinary conditions, i.e. [15], [16]. The problem has been studied by numerous authors. Among all the papers proposed, some examples of methods for the design of the road network in ordinary conditions are cited, i.e. [17]–[20].

The road design problem has been extended to the emergency conditions. A method for vehicle assignment in emergency condition is reported in Kalafatas and Peeta [21]; a method for planning evacuation is reported in He et al. [22]; a method for contraflow operation is reported in Webster [23].

The design of the capacity consists in defining the optimal configuration of the regulation parameters at road intersections with methods valid for traffic light and priority junctions. The priority junctions, can be treated from a model point of view as the signalized junctions with appropriate mathematical analogies. Numerous methods have been proposed for the design of traffic light regulation in ordinary conditions. Some methods proposed are cited for single junctions, i.e. [24], [25] and interacting junctions, i.e. [26]–[29].

In ordinary conditions the topology and the capacity are studied inside the same model with the solution adopting heuristic procedures, i.e. [30].

In the SICURO project a general formulation for the risk has been proposed [4] the problem of the link topology design and the light regulation of the junctions has been studied under emergency conditions by experimentation, i.e. [31]–[34].

3 FRAMEWORK

The overall framework regarding NDP in emergency conditions is shown in Fig. 1.

3.1 A

The main part of the NDP is the risk reduction model (A in Fig. 1). This model takes as input:

- the occurrence values of the calamitous event and the vulnerability of the supply system (B in Fig. 1);
- the risk reduction objectives (C in Fig. 1);
- external and technical constraints (D in Fig. 1);
- the consistency between demand, flows and costs, which is also called behavioural constraint (from E to N in Fig. 1);

and it provides as output the configurations of the supply model in terms of topology of links and capacity at junctions (O in Fig. 1).

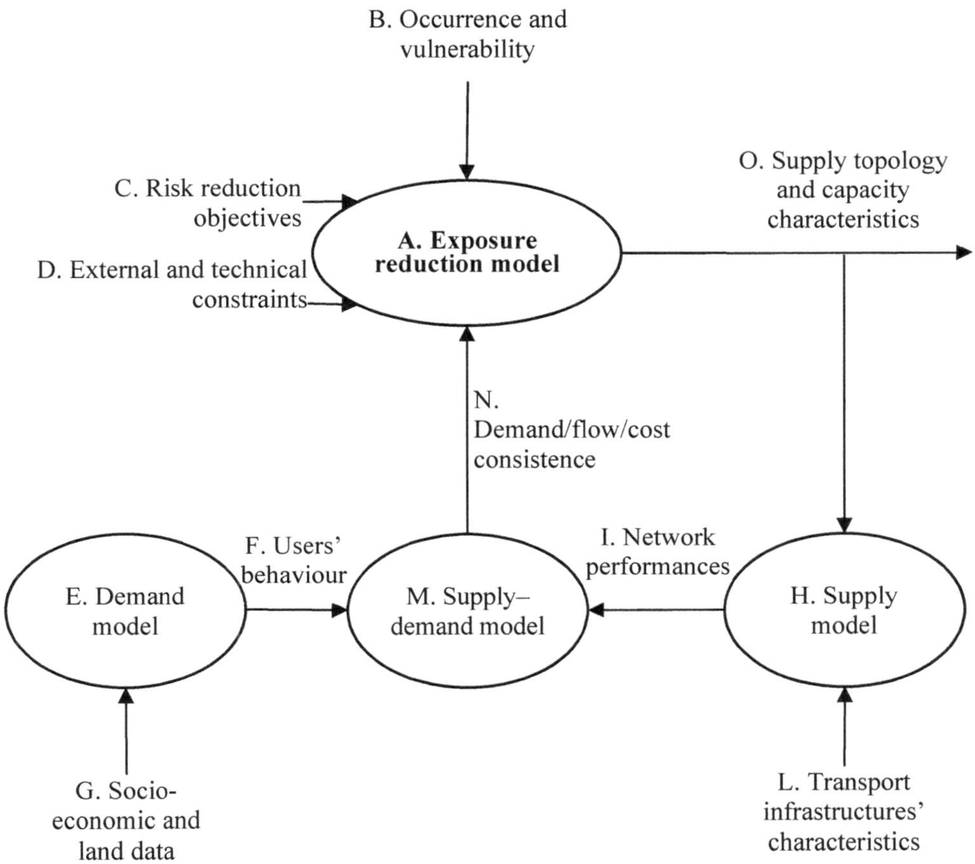

Figure 1: General framework.

3.2 B

The risk function is assumed to consist of three components: occurrence, vulnerability and exposure [4]. Each component is defined by a probability with the exception of the social risk case where the exposure is defined by a quantity in term of people. In the case of the transport network design, it is assumed that the first two components (B in Fig. 1) are defined a priori (it is the case of the occurrence of calamitous events in a transport system with predefined vulnerability). Therefore, the reduction of risk, in this case, coincides with the exposure reduction (A in Fig. 1).

3.3 C

The risk reduction in this specific case coincides with the reduction of exposure, which is implemented with evacuation measures. Evacuation has the purpose of making people move away and therefore reduces the number of people present in the system when the calamitous event generates effects in the system (C in Fig. 1). The objectives to be pursued with the reduction of exposure are mainly of two types: to minimize the evacuation time of the last user who leaves the system from the areas at risk; evacuate all people from the dangerous area (or, if it is not possible, maximizing the number of people evacuated before the dangerous effects).

3.4 D

The external and technical constraints must be respected and that cannot be modified with the project (D in Fig. 1). Some examples of technical constraints are for example the number of emergency vehicles available, the size of the safe areas. Some examples of external constraints are laws, regulations and guidelines.

3.5 E–N

The behavioural constraint models represents the users choice in the transport system on different levels and it is non-linear. The best choice to users can be recommended and it is difficult to force each user to make the choice. Therefore it is necessary:

- to build a demand model (E in Fig. 1) that models user behaviour (F in Fig. 1) starting from socio-economic data (G in Fig. 1);
- to build a supply model (H in Fig. 1); it models the performance of the system (I in Fig. 1) starting from transport infrastructure characteristics; it should be noted that the supply model has also input the supply topology and capacity characteristics (O in Fig. 1), output of the design model and which will be reported in the next paragraph (O);
- to build a model of interaction between supply and demand (M in Fig. 1) which provides as output the consistency between demand, flow and costs (N in Fig. 1) and takes as input the behaviour of users and network performances.

3.6 O

The model gives as outputs the topological and capacity configuration of the network (O in Fig. 1). It should be noted that the output of the design model is also the input of the behavioural constraint, that is of a non-linear type. For this reason, heuristic procedures [30] must be applied to solve the problem in real size transport systems.

4 MODEL

The NDP in a general form can be specified in the following form:

$$\begin{cases} Minimum_{y} \quad \phi(\mathbf{y}, \mathbf{f}) \quad (\mathbf{y}^* \text{ is the optimal value of } \mathbf{y}) & (1) \\ \text{Subject to:} & \\ \mathbf{y} \in \psi_{ETy}; \quad \mathbf{f} \in \psi_{ETf} \quad \text{(external and technical constraints)} & (2) \\ \mathbf{f} = \xi(\mathbf{f}, \mathbf{y}); \quad \mathbf{f} \in \psi_{Bf} \quad \text{(behavioural constraints)} & (3) \end{cases}$$

The output (O in Fig. 1) of the design model is given by the value **y*** of the vector **y** which minimizes the objective function $\phi(\bullet)$. The vector **y** contains the design variables that define the topological configuration of the network (discrete variables) and the light regulation strategy at junctions (continuous variables).

The objective function (*minimum* $\phi(\mathbf{y}, \mathbf{f})$) models risk reduction (C in Fig. 1). Considering that the occurrence and the vulnerability are predefined (B in Fig. 1) in this specific case of evaluation, the objective function (A in Fig. 1) is the reduction of the probability of presence of people (individual risk) or of the total number of people (social risk) in the areas at risk. Therefore the objective function can consider one of the following two criteria:

- minimum of the total evacuation time of all users;
- evacuate all people from the dangerous area or, if it is not possible, maximizing the number of people evacuated in a pre-established time and generally coinciding with the expected time of production of the negative effects on users.

The technical and external constraints (D in Fig. 1) can be indicated by belonging to the vectors **y** and **f** to the respective feasibility sets ψ_{ETy} and ψ_{ETf} ($\mathbf{y} \in \psi_{ETy}$; $\mathbf{f} \in \psi_{ETf}$). Very often these constraints are defined with linear inequalities (e.g. number of lanes lower than and equal to a predetermined value; number of vehicles available minimum or equal to a predetermined value).

The behavioral constraint ($\mathbf{f} = \Psi_f(\mathbf{f},\mathbf{y})$) considers the choice of users that follows the maximum of his utility (from E to N in Fig. 1). Therefore behavior must be modeled using established approaches in transport engineering [35]. The result of the users' choice is the flows vector **f** and the network performances. Each element of the vector **f** models the flow of users in each link of the network. Random utility models typically is applied.

In a static context (which does not well represent the system in emergency conditions), the behavioral constraint can be expressed as:

$$\mathbf{f} = \Delta(\bullet)\, P(\bullet)\, d(\bullet) \tag{4}$$

where

- $\Delta(\bullet) = \Delta(y)$ the link-path incidence matrix; it depends on y;
- $P(\bullet) = P(\Delta(\bullet)'\, \Gamma(\bullet), y)$ the path choice-origin destination pairs matrix; it depends on the path cost $\Delta(\bullet)'\, \Gamma(\bullet)$ and y;
- $d(\bullet) = d(\Delta(\bullet)'\, \Gamma(\bullet), y)$ the demand vector for the origin destination pairs; it depends on the path cost $\Delta(\bullet)'\, \Gamma(\bullet)$, and y;
- $\Delta(\bullet)'\, \Gamma(\bullet) = \Delta(y)'\, \Gamma(f,y)$ the path cost vector;
- $\Gamma(\bullet) = \Gamma(f, y)$ the link cost functions vector; it depends on f and y.

It is the results of the interaction between demand models (represented by means $P(\bullet)$ and $d(\bullet)$ and their relations) and supply models (represented by $\Delta(\bullet)$ and $\Gamma(\mathbf{f}, \mathbf{y})$ and their relations).

In emergency conditions the dynamic approach better represent the users' behavior. Two approaches can be adopted: within-day and/or day-to-day models [35].

In both cases, a compact representation can be adopted:

$$\mathbf{f} = \xi(\mathbf{f},\mathbf{y}) \tag{5}$$

In both cases the flow belongs to the flow feasible set ψ_{Bf}:

$$\mathbf{f} \in \psi_{Bf} \tag{6}$$

5 CONCLUSIONS

In this paper a model for the urban road transport network design has been reported. The model has been considered in the wider problem of risk reduction and, in particular, exposure reduction in presence of calamitous events. The quantitative design models is required in order to optimize the system in a preventive way and therefore be ready to allow optimal evacuation of users.

The main novelty reported in this paper concerns the specification of a model for the road transport networks design; it could be used for planning transport systems in emergency conditions. The design model has been reported in a unitary approach, starting from the state of the art. It is linked to the three components of risk: occurrence, vulnerability, exposure. The objective function was considered as a quantitative indicator of the risk exposure component in order to link network design and risk areas. The overall model has been specified also considering the behavioural constraint that models the choices of the users. The behavioural constraint is not linear and the design variables are continuous and integers. Therefore the problem requires the use of heuristic solution algorithms.

The methodologies proposed in this paper can be used for the medium or long-term design of the links layout and junctions regulation of a road network, aimed at reducing the risk. The optimal design of the transport network, especially in emergency conditions, would reduce the time required for the evacuation of users. Therefore, this methodology is a decision support that must be taken into account by operators in emergency conditions. The application of the methods reported cannot be implemented in real time. An emergency situation generates wide and differentiated management problems and scenarios. Therefore the design methods must be specified, calibrated and validated in advance using hypothetical risk scenarios. In real situations, the results and models obtained previously can be updated with changes in the real scenario compared to those assessed a priori; it could be considered as a decision support system. Therefore, for the application of these models, a preventive planning activity is required, including training.

In the future, the framework and the models reported require an extensive activity of specification–calibration–validation in real contexts followed by continuous training on the various components involved (system managers, operators, users, citizens).

ACKNOWLEDGEMENTS

This research is partially supported by the Dipartimento di ingegneria dell'Informazione, delle Infrastrutture e dell'Energia Sostenibile, Università Mediterranea di Reggio Calabria, and by the project "La Mobilità per i passeggeri come Servizio – MyPasS", Fondi PON R&I 2014–2020 e FSC "Avviso per la presentazione di Progetti di Ricerca Industriale e Sviluppo Sperimentale nelle 12 aree di Specializzazione individuate dal PNR 2015–2020", codice identificativo ARS01_01100.

REFERENCES

[1] Russo, F. & Rindone, C., Safety of users in road evacuation: Planning internal processes and guidelines. *WIT Transactions on The Built Environment*, vol. 96, WIT Press: Southampton and Boston, pp. 825–834, 2007.

[2] Russo, F., Rindone, C. & Trecozzi, M.R., The role of training in evacuation. *WIT Transactions on Information and Communication Technologies*, vol. 44, WIT Press: Southampton and Boston, pp. 491–502, 2012.

[3] Russo, F. & Rindone, C., Urban exposure: Training activities and risk reduction. *WIT Transactions on Ecology and the Environment*, vol. 191, WIT Press: Southampton and Boston, pp. 991–1001, 2014.

[4] Russo, F. & Vitetta, A., Safety of users in road evacuation: General methodology and main results. *WIT Transactions on the Built Environment*, vol. 96, WIT Press: Southampton and Boston, pp. 763–772, 2007.

[5] Russo, F. & Chilà, G., Safety of users in road evacuation: RP vs. SP surveys in demand analysis. *14th International Conference on Urban Transport and the Environment in the 21st Century*, pp. 703–713, 2008.

[6] Musolino, G. & Vitetta A., Short-term forecasting in road evacuation: Calibration of a travel time function. *WIT Transactions on the Built Environment*, vol. 116, WIT Press: Southampton and Boston, pp. 615–626, 2011.
Di Gangi, M., Planning evacuation by means of a multi-modal mesoscopic dynamic traffic simulation model. *Studies in Computational Intelligence*, **176**, pp. 99–115, 2009. DOI: 10.1007/978-3-540-89930-3_6.

[7] Di Gangi, M., Safety of users in road evacuation: Modelling and DSS for pedestrian outflow. *WIT Transactions on Ecology and the Environment*, vol. 120, WIT Press: Southampton and Boston, pp. 497–508, 2009. DOI: 10.2495/SDP090461.

[8] Di Gangi, M., Watling, D. & Di Salvo, R., Modeling evacuation risk using a stochastic process formulation of mesoscopic dynamic network loading. *IEEE Transactions on Intelligent Transportation Systems*, 2020.

[9] Russo, F. & Rindone, C., Safety of users in road evacuation: the logical framework approach in evacuation planning. *14th International Conference on Urban Transport and the Environment in the 21st Century*, pp. 751–760, 2008.

[10] Russo, F. & Rindone, C., The planning process and logical framework approach in road evacuation: A coherent vision. *WIT Transactions on The Built Environment*, vol. 117, WIT Press: Southampton and Boston, pp. 415–425, 2011.

[11] Vitetta, A., Polimeni, A. & Quattrone, A., Safety of users in road evacuation: Modelling and DSS for paths design of emergency vehicles. *WIT Transactions on Ecology and the Environment*, vol. 120, WIT Press: Southampton and Boston, pp. 485–495, 2009.

[12] Polimeni, A. & Vitetta A., The role of ITS in evacuation route optimization for emergency vehicles. *WIT Transactions on Information and Communication Technologies*, vol. 44, WIT Press: Southampton and Boston, pp. 517–529, 2012. DOI: 10.2495/RISK120431.

[13] Billheimer, J.W. & Gray, P., Network design with fixed and variable cost elements. *Transport. Sci.*, **7**, pp. 49–74, 1973.

[14] Chen, M. & Alfa, A.S., A network design algorithm using a stochastic incremental traffic assignment approach. *Transport. Sci.*, **25**, pp. 215–224, 1991.

[15] Foulds, L.R., A multi-commodity flow network design problem. *Transportation Research Part B*, **15**, pp. 273–283, 1981.

[16] Herrmann, J.W., Ioannou, G., Minis, I. & Proth, J.M., A dual ascent approach to the fixed-charge capacitated network design problem. *European Journal of Operational Research*, **95**, pp. 476–490, 1996.

[17] Solanky, R.S., Gorti, J.K. & Southworth F., Using decomposition in large-scale highway network design with a quasi-optimization heuristic. *Transportation Research Part B*, **32**(2), pp. 127–140, 1998.

[18] Gao, Z., Wu J. & Sun, H., Solution algorithm for the bi-level discrete network design problem. *Transportation Research Part B*, **39**, pp. 479–495, 2005.

[19] Poorzahedy, H. & Rouhani, O.M., Hybrid meta-heuristic algorithms for solving network design problem. *European Journal of Operational Research*, **182**(2), pp. 578–596, 2007.
[20] Xie, C. & Turnquist, M.A., Integrated evacuation network optimization and emergency vehicle assignment. *Transportation Research Part B*, 2009.
[21] Kalafatas, G. & Peeta, S., Planning for evacuation: Insights from an efficient network design model. *Journal of Infrastructure Systems*, **15**(1), pp. 21–30, 2009.
[22] He, X., Zheng H., Peeta S. & Li Y., Network design model to integrate shelter assignment with contraflow operations in emergency evacuation planning. *Networks and Spatial Economics*, **18**(4), pp. 1027–1050, 2018.
[23] Webster F.V., Traffic signal settings. Road Research Technical Paper No. 39, Road Research Laboratory: London, 1958.
[24] Allsop, R.E., SIGSET: A computer program for calculating traffic capacity of signal controlled road junctions. *Traffic Eng. Control*, **12**, pp. 58–60, 1971.
[25] Little, J.D.C., Kelson, M.D. & Gartner, N.H., MAXBAND: A program for setting signals on arteries and triangular networks. *Transportation Research Record*, **795**, pp. 40–46, 1981.
[26] Robertson D.I., TRANSYT method for area traffic control. *Traffic Engineering and Control*, **10**, pp. 276–281, 1969.
[27] Little, J.D.C., The synchronisation of traffic signals by mixed-integer-linear-programming. *Operations Research*, **14**, pp. 568–594, 1966.
[28] Cantarella, G.E., Improta, G. & Sforza, A., Road network signal setting: Equilibrium conditions. *Concise Encyclopaedia of Traffic and Transportation Systems*, ed. M. Papageorgiou, Pergamon Press, pp. 366–371, 1991.
[29] Cantarella, G.E., Pavone, G. & Vitetta, A., Heuristics for urban road network design: Lane layout and signal settings. *European Journal of Operational Research*, **175**(3), pp. 1682–1695, 2006.
[30] Marcianò, F.A, Musolino, G. & Vitetta A., Signal setting design on a road network: Application of a system of models in evacuation conditions. *WIT Transactions on Information and Communication Technologies*, vol. 43, WIT Press: Southampton and Boston, pp. I443–I454, 2010.
[31] Marcianò, F.A., Musolino, G. & Vitetta, A., A system of models for signal setting design of a signalized road network in evacuation conditions, *WIT Transactions on the Built Environment*, vol. 111, WIT Press: Southampton and Boston, pp. 313–323, 2010.
[32] Marcianò, F.A., Musolino, G. & Vitetta, A., Within-day traffic assignment and signal setting in road evacuation: A procedure with explicit path enumeration, *WIT Transactions on the Built Environment*, vol. 117, WIT Press: Southampton and Boston, pp. 403–414, 2011.
[33] Marcianò, F.A, Musolino, G. & Vitetta A., Signal setting optimization on urban road transport networks: The case of emergency evacuation. *Safety Science*, **72**, pp. 209–220, 2015.
[34] Cascetta, E., *Transportation Systems Engineering: Theory and Methods*, Springer: New York, 2009.

RISK REDUCTION IN TRANSPORT SYSTEM IN EMERGENCY CONDITIONS: A FRAMEWORK FOR SUPPLY ANALYSIS

GIUSEPPE MUSOLINO
DIIES – Dipartimento di Ingegneria dell'Informazione, delle Infrastrutture e dell'Energia Sostenibile
Università Mediterranea di Reggio Calabria, Italy

ABSTRACT

There is a rich literature, mainly developed in the last two decades, about risk reduction in transport system in emergency conditions. One of the challenges regards the development of ad-hoc transport system models (TSMs) in order to simulate emergency conditions. The paper focuses on the supply (network) component of the TSMs. The general objective of the research is the development of a framework capable to capture the relevant changes of the network from the ordinary conditions, due to an approaching calamitous event. The research will consider non-stationary conditions in transport supply models for continuous services. In relation to the consolidated static models, this evolution is necessary to capture the dynamics of the transportation network during the evacuation in terms of both topology (e.g. link closure, contraflow) and capacity of the links and of the network. The concept of learning process will be introduced to take into account how the transport costs (disutilities) in ordinary and emergency conditions will be perceived by the users.
Keywords: risk, transport system models, supply, evacuation, learning process.

1 INTRODUCTION

Risk is commonly defined as the product of three probabilities: occurrence, vulnerability, exposure. Transport system planning aims to reduce exposure, by means of evacuation procedures. The objectives of urban evacuations concern the minimization of the number of people and the quantity of goods involved in a calamitous event [1].

A fundamental element to support the evacuations planning and optimization is the transport system model (TSM) framework. TSMs in ordinary conditions are consolidated in the scientific literature. They are generally used by planners and technicians to identify the critical elements of the transport systems and to ex-ante assess alternative scenarios, or alternative transport system configurations, concerning both the demand (e.g. recurrent and no-recurrent demand peaks exceeding capacity), and the supply elements (e.g. new transport infrastructures or changes of existing services) [2]–[4].

However, TSMs in emergency conditions and/or when exogenous calamitous events affecting the transport system occur, have received minor attention in the literature. When an exogenous calamitous event starts its effects, both the demand and the supply components of the transportation system change in a relevant way. As far as concern demand, transport users travel on the network with different a behaviour (e.g. in terms of destination, mode, path choices), and they are not aware of network congestion, nor of reliability, of the transportation network. As far as concerns the supply, network in terms of both topology (e.g. link closure, contraflow) and capacity of the links and nodes changes, too.

Models and algorithms specified and calibrated in ordinary conditions for simulation and design transportation system cannot be directly applied in emergency conditions, when unexpected temporary over-saturation of some road links and nodes, queue formation and dispersion occur, and transport users are not aware of the current conditions of the network.

WIT Transactions on The Built Environment, Vol 206, © 2021 WIT Press
www.witpress.com, ISSN 1743-3509 (on-line)
doi:10.2495/SAFE210231

In order to capture and simulate the relevant changes that occur in the transport systems in emergency conditions, it is necessary to adapt TSMs developed for ordinary conditions. It is necessary to take into account two main elements: the introduction of dynamics in models and procedures and the consideration of the influence of emerging ICTs, both on the supply and on the demand modelling components.

The paper deals with the research area connected to transport supply modelling in the context of the TSM framework. The research novelty of the paper concerns the introduction of elements that allow to capture the dynamics of the transportation network in emerging conditions (e.g. during the evacuation), in relation to the consolidated static supply models. In particular, the paper focuses on the concept of learning process (model), that explain (simulates) how transport users (evacuees) forecast the costs (or disutilities) of the transport network that they will experience during evacuation, from previous experience of the network and/or from existing information systems (e.g. emerging ICTs). The learning process could take into account different types of information provided by emerging ICTs (e.g. internet of things, big data), to update within-day and day-to-day information.

The remaining part of the paper is articulated as follows. Section 2 deals with the current research areas connected to evacuation planning and optimization. Section 3 presents the main characteristics of the of TSMs, focusing on the supply–demand interaction (traffic assignment) models and of supply models. Section 4 introduces the elements concerning the dynamic extension of supply model consistently with within-day and day-to-day information, provided by emerging ICTs. The final section reports the research perspectives.

2 EVACUATION PLANNING AND OPTIMIZATION

The topic of evacuation planning and optimization has received little attention in the literature. Some studies were produced in the early 2000s [1], [5]–[7]. After the 9/11 and Katrina, the study of emergency conditions received great impulse (see, as example, [8], and the references included); however, according to the author, the systematic development of research on evacuation planning and optimizations, with the actualization to the new frontiers given by emergent ICTs is not present the literature.

In general, the following five research areas may be considered.

- *Emergency planning*, which is part of the broader disaster management cycle that, at international level, is identified with the following keywords: prevention; mitigation or prevention; preparedness; response; recovery [9]–[11].
- *Travel demand*, where the user evacuation is the main activity inside the exposure regarding risk reduction in transport systems. The evacuation problem in the TSM approach needs the development of different components; two of them are considered: the demand analysis considering also stated preference surveys and the planning process considering parametric and non-parametric methods [12]–[14].
- *Transport supply,* where the objective is the development of ad-hoc models capable to capture the relevant changes of the network in emergency conditions, due to an approaching calamitous event [15], [16].
- *Decision Support Systems (DSS),* where it is necessary to consider appropriate DSS taking into account multimodality for the evacuation design. It is also necessary to build TSMs able to simulate non stationary conditions and considering risk, taking into account both the availability of reliable paths and peculiarity of path choice in emergency conditions [17]–[19].

- *Network Design,* where it is necessary to build TSMs concerning: the calibration of the cost functions, the design of the traffic junction regulation, the design of the routes for emergency vehicles [20]–[22].

3 TRANSPORT SYSTEM MODEL (TSM)

TSMs are able to represent each single component of the transport system and the mutual interactions.

The main three modelling components of TSMs are the traffic assignment, the transport supply and the travel demand. The core of the TSMs is the transport network-demand interaction (traffic assignment) model, which simulates how travel demand and transport networks interact in transportation systems. The two modelling components are: the transport supply (network) model, which simulates performances and traffic flows, resulting from users and from transport infrastructures and services; and the travel demand model, which simulate users' choices resulting from activities and infrastructure and service performances.

The following parts of this section present a classification of supply–demand interaction (traffic assignment) models (Section 3.1) and of supply models (Section 3.2), according to the scheme depicted in Fig. 1. Travel demand models are described in Russo and Chilà [14].

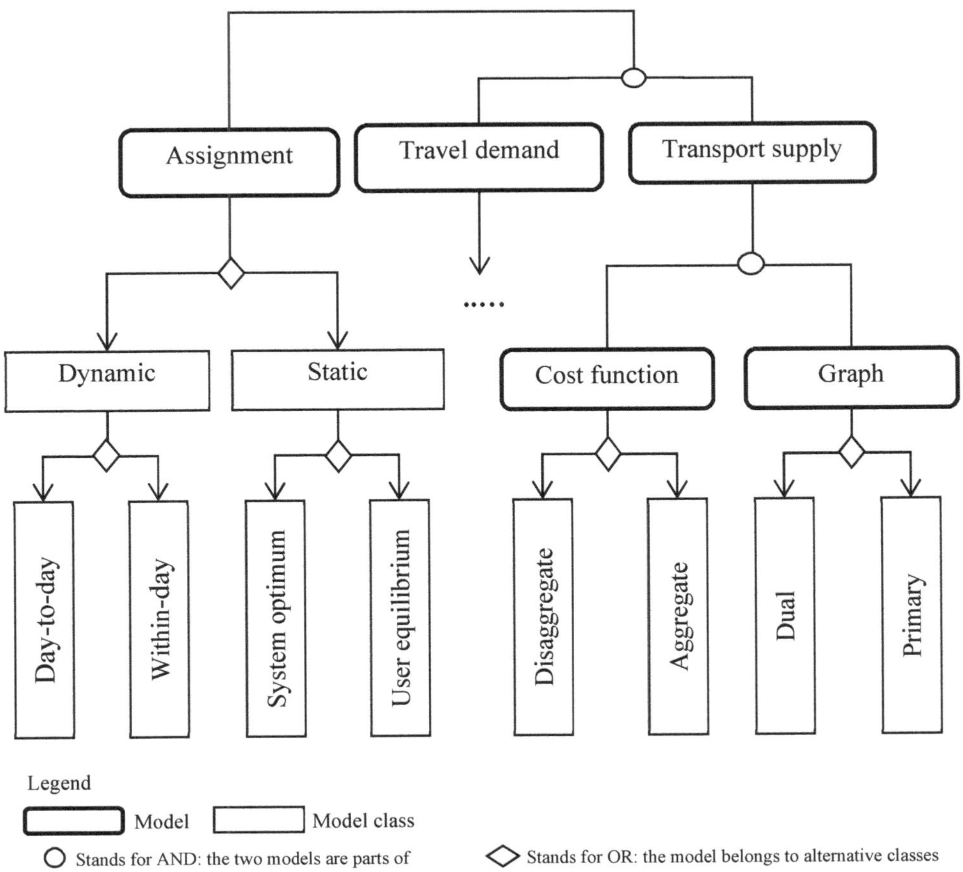

Figure 1: Classification of supply and of supply–demand interaction models.

3.1 Supply–demand interaction (traffic assignment) model

Traffic Assignment (TA) models allow to calculate performances (costs) and vehicular flows on each element (link and nodes) of the transport network, resulting from origin–destination demand flows, users' path choice behaviour and link costs.

Supply–demand interaction models may be classified according to different criteria: path choice, static vs. dynamic, equilibrium vs. process approach.

As far as concerns *path choice*, the specification of path choice models is based on the definition of the path choice set and the path choice models. The path choice set is behaviourally defined, by selecting admissible paths according to behavioural rules of the user. Paths belonging to the path choice set can be variable or fixed, if new paths can be periodically reconsidered from the users or not.

Path choice models can be subdivided into two categories: *pre-trip*, if the user chooses the path according to the origin, the destination and other characteristics, before starting the trip; *en-route*, if the users modifies, during the trip, the path according to the current traffic conditions on the network.

Two types of path choice models can be identified: *deterministi*c, if the users chooses the path of minimum cost (or disutility); *stochastic*, if the user associates a probability to each path belonging to the *path choice set*.

The *en-route* category of path choice models requires that information concerning the current and future traffic conditions of the network are available in some elements, such as diversion nodes. These information may be provided today by means of emerging ICTs, as briefly described below.

As far as concerns equilibrium vs. process approach, equilibrium is a (mathematical) condition, according to which path (link) flows are mutually consistent with the corresponding path (link) costs. This problem can be formulated with the so-called fixed-point models, variational inequalities or optimization models [3], [4]. Equilibrium traffic assignment models of the literature are based on two approaches:

- *User Equilibrium (UE)*, where path flows are estimated according to the First Wardrop principle [23], in order to minimize individual cost of users ;
- *System Optimum (SO)*, where path flow–path cost relationship is consistent with the Second Wardrop principle [23]; in order to achieve the minimum total cost on the network, although some users may follow a non-minimum cost path.

The process approach is associated with non-equilibrium, generated by the independence of link cost variables on vehicular flows in the static case. Fixed costs or uncongested networks result in non-equilibrium traffic assignment models, while variable costs or congested networks lead to equilibrium traffic assignment models.

As far as concerns *static vs. dynamic* approach, TA models may be classified into:

- *Static (STA) models*, when travel demand, path choices and the transport network are consistent in a reference time (stationary conditions);
- *Dynamic (DTA) models*, when stationary conditions are removed and transport system evolution is represented in a reference time, as in the case of emergency conditions.

Therefore, STA models achieve reasonable results when the time horizon simulation is long enough to assume stationary conditions. On the contrary, DTA models are necessary when frequent changes in network and demand occur, like in emergency conditions, and the stationary hypothesis cannot be assumed.

STA models belong to equilibrium category when the transport network is congested, in other words when link travel times depend on link flows (e.g. urban rods); while they belong to process category, such as network loading (NL) models, if link travel times do not depend on link flows (e.g. extra-urban roads). DTA models are generally associated to process category, even if the concept of dynamic equilibrium is available in the literature [24]. As matter of fact, it is reasonable to assume that users' choices in ordinary conditions may be based upon their experience of the network: when users have a certain experience of the network, equilibrium could be achieved and all users experience the same travel time. Emergency conditions are rare and, therefore, users have no experience of the network in these conditions. Hence, equilibrium, as defined in ordinary conditions, cannot be achieved.

DTA models may belong to two categories, that simulate two different temporal evolutions of the transport system [3], [4], [24], for an exhaustive description of the theoretical background):

- *within-day DTA models*, that simulate the within-day evolution of the transport system, that occurs over continuous time inside a given day; or inside a part of the day (e.g. morning peak hour);
- *day-to-day DTA models*, that simulate the day-to-day evolution of the transport system, that occurs over discrete time from (a part of) a given day and the next one.

3.2 Transport supply (network) model

According to the most common developed approach, transport supply models are represented by a network model. The network is composed by means of a graph, with links and nodes, and cost functions.

The graph can be primary versus dual. The cost functions can be classified according to different criteria: level of aggregation, element of the network.

According to the level of aggregation, cost functions range from a macroscopic to microscopic approach.

The *macroscopic approach* uses continuous flow representation and cost functions with aggregate variables (speed, density, flow) and with explicit capacity such as in static models. The representation of space and time can be continuous or discrete.

The *microscopic approach* is based on discrete flow representation allowing simulation of the behaviour of a single user. Cost functions are disaggregated in which the traffic flow is the result of individual choices, interactions among users and interactions between users and road facility.

An intermediate case is the *mesoscopic approach* where the behaviour of groups of users, termed packages, is simulated by means of a discrete flow representation [17], [18].

According to the element of the network, cost functions belong to two different classes:

- *link*, that simulate the (average) costs of users travelling on each link of the network;
- *node*, that simulate (average) costs of users at each node of the network (e.g. signalized and not-signalized intersections, merging and diverging areas, toll plazas).

Link models can be divided in two categories: macroscopic a microscopic.

Macroscopic link models are, generally, aggregate speed-density versus time-flow relationships, that estimate average values of traffic variables, as flow, density, speed.

Microscopic link models belong to two categories: *car-following* models, that simulate the movement of each vehicle and the interaction with vehicles travelling on the same lane;

lane-changing models, that simulate the interactions among vehicles travelling on different lanes.

Among *node* models, there are models for *signalized nodes* and models for *not-signalized nodes* [21]. Models for *signalized nodes* can be subdivided in: macroscopic, for an aggregate analysis of node performances and microscopic, for a disaggregate analysis of node performances, simulating explicitly every driver behaviour. Models *for not-signalized* nodes are based on the gap-acceptance model. Different gap-acceptance models have been developed, according to the assumptions of the critical gap distribution.

A general static formulation of a congested supply model consists of two equations:

- path costs versus link costs consistency equation

$$\mathbf{g} = \boldsymbol{\Delta}^{T}\, c\,(\mathbf{f}) \tag{1}$$

- path flows versus link flows consistency equation:

$$\mathbf{f} = \boldsymbol{\Delta}\, \mathbf{h} \tag{2}$$

where

- **g**, path costs vector;
- **Δ**, link-path incidence matrix;
- *c*(), link cost functions vector;
- **f**, link flows vector;
- **h,** path flows vector.

The mutual relationships among the input/output variables of eqns (1) and (2) are reported in Fig. 2.

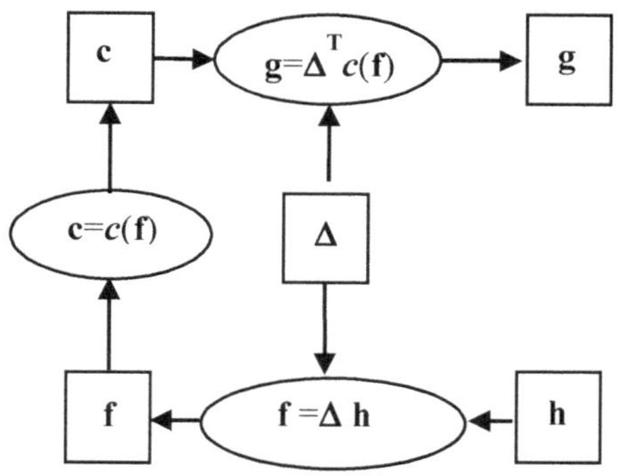

Figure 2: Mutual relationships among variable in a static supply model.

4 LEARNING PROCESS IN PRESENCE OF EMERGING ICT

The specification of a (dynamic) supply model consistent with a DTA model (see Section 3.1), according both to a within-day and a day-to-day pattern, requires an extension of the static supply model presented in eqns (1) and (2). In particular, it is necessary to include a modelling components related to the learning process of transport user, that could take into account also the different types of information (e.g. historical, real-time) provided by emerging ICT. Some elements are presented in the present section.

4.1 Emerging ICT

New, or emerging, forms of Information and Communication Technologies (ICTs) have been developed in the last decade. They quickly spread in many areas of economy and society (e.g. manufacturing, social media, etc.), influencing the behaviour of people at individual and collective way. Some applications are proposed in the field of city logistics [25] and of ports and maritime transport [26], [27].

In broad terms, the different emerging technologies present in the market may be classified into four groups (see [25]–[28]):

- BlockChain (BC), or "distributed ledger" is a "technology for keeping a chronologically ordered list of transactions (a ledger) on multiple, independent stakeholders (called nodes). Updates of the list should be identical on each of the nodes and each node checks the validity of the transactions before updating its list" [27];
- Internet of Things (IoT) is a "framework that leverages on the availability of heterogeneous devices and interconnection solutions, as well as augmented physical objects providing a shared information base on global scale, to support the design of applications involving at the same virtual level both people and representations of objects" [28];
- Artificial Intelligence and Machine Learning (AI&ML) aim to improving the prediction and decision making on transport operations, as well as to automatic repetitive (though sensitive) manual transactions;
- Big Data (BD) enrich information deriving from traditional data, in terms of volume and variety. ICT collects, manages and processes big data within a tolerable elapsed time, in order to improve the representation of traffic phenomena, and their modelling [25].

The 5G (or 6G) communication network is not considered because it is transversal to all other technologies.

The above classes of emerging ICTs are able to provide new opportunities to define and implement integrated and dynamic solutions to optimize the different components of the TSM.

It is worth noting that these emerging ICTs are mature today and they are, generally, part of the stand-alone digital platform of almost every transport user and/or decision-maker (e.g. transport system manager). They may be considered emerging, as their implementation inside the Intelligent Transport Systems (ITSs) is still *in fieri.*

4.2 Learning process

The modification of users' knowledge about the network may be described by means of the learning process, that explain how transport users (e.g. evacuees) forecast today the costs (or disutilities) of the transport network (e.g. during an evacuation), from previous experience of the network (e.g. in ordinary conditions) and/or from previous forecasts (e.g. by means of emerging ICTs).

In broad terms, the learning process of individual users may be simulated with the *(dis)utility updating* model, that allows to obtaining the forecasted/computed costs in a period t.

In the case of path costs, the vector of forecasted (or computed) path costs at time t, $\mathbf{g}^{fo,t}$, may be expressed as follows:

$$\mathbf{g}^{fo,t} = g(\boldsymbol{\beta}, \mathbf{g}^{ex,t-1}, \mathbf{g}^{fo,t-1}) \quad (3)$$

where

- $g()$, implicit updating function that may be differently specified;
- $\boldsymbol{\beta}$, vector of parameters to be calibrated;
- $\mathbf{g}^{ex,t-1}$, vector of experienced path costs at time t^{-1}, $\mathbf{g}^{ex,t-1}$;
- $\mathbf{g}^{fo,t-1}$, vector of forecasted (or computed) path costs at time t^{-1}, $\mathbf{g}^{fo,t-1}$.

The implicit updating function, *g()*, may be specified in different ways [4]. The simplest specification is the "yesterday" filter, where the forecasted path costs at time t are assumed to be equal only to the experienced path costs at time t^{-1} (yesterday). An extension of yesterday filter is the moving average filter, where the forecasted path costs at time t depend on the experienced path costs at time t^{-1}, t^{-2}, t^{-3}, ... In the exponential filter specification, they are defined by means of a strict convex combination of forecasted and experienced costs at time t^{-1}.

The learning process may evolve over time both according a within-day and day-to-day pattern. According to the literature, the within-day evolution is generally considered in continuous time, while the day-to-day evolution is generally considered in discrete time.
The new opportunities offered by emerging ICTs in providing historical, real-time and forecasted information can be exploited by introducing some modifications in the learning process, as traditionally specified in the classical problems.

Among the emerging ICTs classified in the previous paragraph, *IoT, BD* and *AI* may be considered the ones with the greatest impact on path choice [25]. According to their features, IoT allows real-time communication between vehicles (vehicle-to-vehicle communication) and between vehicles and infrastructure (vehicle-to-infrastructure communications), BD are able to provide an exhaustive and detailed representation of historical patterns of network costs, *AI* could offer the possibility to support short-term forecasting of network costs [28].

5 CONCLUSIONS

One of the main challenges of the literature about risk reduction in transport systems regards the development of ad-hoc TSMs able to simulate emergency conditions. In other words, models capable to capture the relevant changes of the network and of the users' behaviour, due to an approaching calamitous event.

The paper focuses on the supply (network) component of the TSMs. In relation to the consolidated static models, a modelling extension is necessary in order to capture two elements:

- the dynamics of the transportation network during the evacuation in terms of both topology (e.g. link closure, contraflow) and capacity of the links and of the network;
- the potentialities of emerging ICTs in providing valuable information according to both within-day and day-to-day patterns.

The concept of learning process has been introduced to take into account how the path costs (disutilities) in ordinary and emergency conditions will be perceived and updated by the users.

The new opportunities offered by the research on optimal path design in emergency conditions (e.g. during an evacuation) may be applied to incorporate inside the models the new potentialities offered by the emerging ICTs, such as the IoT to update within-day information, and the BD to update day-to-day information.

REFERENCES

[1] Russo, F. & Vitetta, A., Risk evaluation in a transportation system. *International Journal of Sustainable Development and Planning*, **1**, pp. 170–191, 2006.

[2] Ortuzar, J. & Willumsen, L.G., *Modelling Transport*, 3rd ed., Wiley: Chichester, 2001.

[3] Cascetta, E., *Transportation Systems Analysis: Models and Applications*, Springer-Verlag: New York, 2009.

[4] Cantarella, G.E., Watling, P.D., de Luca, S. & Di Pace, R., *Dynamics and Stochasticity in Transportation Systems: Tools for Transportation Network Modelling*, Elsevier, 2020.

[5] Murray-Tuite, P. & Mahmassani, H.S., Model of household trip-chain sequencing in emergency evacuation. *Transportation Research Record*, **1831**(1), pp. 21–29, 2003.

[6] Kindinger, J.P., *Programmatic Risk-Overview of Methods for Decision Making and Risk Management*, Los Alamos National Laboratory document LA-CP-04-0626, 2004.

[7] Goldblatt, R., Evacuation planning: A key part of emergency planning. *Proceedings of the Annual Meeting of Transportation Research Board*, Washington, DC, pp. 11–15, 2004.

[8] Lindell, M.K., Murray-Tuite, P., Wolshon, B. & Baker, E.J., *Large-Scale Evacuation: The Analysis, Modeling, and Management of Emergency Relocation from Hazardous Areas,* Routledge, 2019.

[9] Russo, F. & Rindone, C., The planning process and logical framework approach in road evacuation: A coherent vision. *WIT Transactions on The Built Environment*, vol. 117, WIT Press: Southampton and Boston, pp. 415–425, 2011.

[10] Russo, F. & Rindone, C., Urban exposure: Training activities and risk reduction. *WIT Transactions on Ecology and the Environment*, vol. 191, WIT Press: Southampton and Boston, pp. 991–1001, 2014.

[11] Rindone, C. & Panuccio, P., Risk reduction in transport system in emergency conditions: A framework for evacuation planning. *WIT Transactions on the Built Environment*, vol. 206, WIT Press: Southampton and Boston, submitted for publication.

[12] Russo, F. & Rindone, C., Safety of users in road evacuation: The logical framework approach in evacuation planning. *Proceedings of the 14th International Conference on Urban Transport and the Environment in the 21st Century*, pp. 751–760, 2008.

[13] Russo, F. & Rindone, C., Safety of users in road evacuation: Modelling and DSS for LFA in the planning process. *WIT Transactions on Ecology and the Environment*, vol. 120, WIT Press: Southampton and Boston, 2009.

[14] Russo, F. & Chilà, G., Risk reduction in transport system in emergency conditions: A framework for demand analysis. *WIT Transactions on the Built Environment*, vol. 206, WIT Press: Southampton and Boston, submitted for publication.

[15] Marcianò, F.A., Musolino, G. & Vitetta, A., A system of models for signal setting design of a signalized road network in evacuation conditions. *WIT Transactions on the Built Environment*, vol. 111, WIT Press: Southampton and Boston, pp. 313–323, 2010.

[16] Marcianò, F.A., Musolino, G. & Vitetta, A., Within-day traffic assignment and signal setting in road evacuation: A procedure with explicit path enumeration. *WIT Transactions on the Built Environment*, vol. 117, WIT Press: Southampton and Boston, pp. 403–414, 2011.

[17] Di Gangi, M., Watling, D. & Di Salvo, R., Modeling evacuation risk using a stochastic process formulation of mesoscopic dynamic network loading. *IEEE Transactions on Intelligent Transportation Systems*, 2020.

[18] Di Gangi, M. & Polimeni, A., A mesoscopic approach to model route choice in emergency conditions. *Paper presented at the Springer Proceedings in Mathematics and Statistics*, **217**, pp. 547–555. 2017.

[19] Di Gangi, M. & Belcore, O.M., Risk reduction in transport system in emergency conditions: A framework for decision support systems. *WIT Transactions on the Built Environment*, vol. 206, WIT Press: Southampton and Boston, submitted for publication.

[20] Musolino, G. & Vitetta, A., Short-term forecasting in road evacuation: Calibration of a travel time function. *WIT Transactions on the Built Environment*, vol. 116, WIT Press: Southampton and Boston, pp. 615–626, 2011.

[21] Marcianò, F.A., Musolino, G. & Vitetta A., Signal setting optimization on urban road transport networks: The case of emergency evacuation. *Safety Science*, **72**, pp. 209–220, 2015.

[22] Vitetta, A., Risk reduction in transport system in emergency conditions: A framework for network design problem. *WIT Transactions on the Built Environment*, vol. 206, WIT Press: Southampton and Boston, submitted for publication.

[23] Wardrop, J.G., Some theoretical aspects of road traffic research. *Proc. Inst. Civ. Eng.*, **2**(1), pp. 325–378, 1952.

[24] Cantarella, G.E. & Cascetta, E., Dynamic processes and equilibrium in transportation networks. *Transportation Science*, **29**, pp. 305–329, 1995.

[25] Russo, F. & Comi, A., Providing dynamic route advice for urban goods vehicles: The learning process enhanced by the emerging technologies. *Transportation Research Procedia – 24th EURO Working Group on Transportation Meeting, EWGT 2021*, 8–10 September, Aveiro, Portugal, 2021.

[26] Russo, F. & Musolino, G., The role of emerging ICT in the ports: Increasing utilities according to shared decisions. *Front. Future Transp.* **2**, 722812, 2021 DOI: 10.3389/ffutr.2021.722812.

[27] Carlan, V., Coppens, F., Sys, C., Vanelslander, T. & Van Gastel, G., Blockchain technology as key contributor to the integration of maritime supply chain? *Maritime Supply Chains*, eds T. Vanelslander & C. Sys, Elsevier, pp. 229–259, 2020.

[28] Atzori, L., Iera, A. & Morabito, G., Understanding the Internet of Things: Definition, potentials, and societal role of a fast evolving paradigm. *Ad Hoc Networks*, **56**, pp. 122–140, 2017.

RISK REDUCTION IN TRANSPORT SYSTEM IN EMERGENCY CONDITIONS: A FRAMEWORK FOR EVACUATION PLANNING

CORRADO RINDONE & PAOLA PANUCCIO
Dipartimento di ingegneria dell'Informazione, delle Infrastrutture e dell'Energia Sostenibile, Università degli studi Mediterranea di Reggio Calabria, Italy

ABSTRACT

Emergency planning is part of the broader disaster management cycle that, at the international level, is identified with the following keywords: prevention; mitigation or prevention; preparedness; response; recovery. The paper focuses on preparedness and in particular on the planning process for ordinary and extraordinary conditions. The specific focus is on the evacuation planning process. In the last years, the United Nations (UN) and European Commission (EC) dedicated a great effort to pursue disaster risk reduction. At aiming this goal, UN and EC promote frameworks to orient the urban emergency planning process. Despite the European addresses, there is no uniformity of approaches for evacuation planning adopted by the Countries. This paper proposes a framework for evacuation planning at the urban scale, with a focus on passenger mobility. The proposed framework integrates theories produced by science, rules produced by laws and implementation in real cases. Planning, training, exercises and evaluation tools are the main elements of the framework that work to reduce risk. Some elements of the Italy case study will be presented.

Keywords: emergency, risk, exposure, evacuation, urban planning.

1 INTRODUCTION

Emergency planning is part of the broader disaster management cycle that, at the international level, is identified with the following keywords: *prevention*, including the needed activities to avoid accidents; *mitigation* or *protection*, including activities to protect users by emergency's effects; *preparation*, including activities to increase the capability to cope with a disastrous event; *response*, including activities carried out immediately before and/or during and after the occurrence of a disastrous event; *recovery* including the activities necessary to return to ordinary conditions [1].

The *Sendai Framework for Disaster Risk Reduction* 2015–2030 is the adopted document by member states of the United Nations to pursue disaster risk reduction at global level. The framework indicates four priorities for action concerning: (i) *understanding disaster risk*; (ii) Strengthening disaster risk governance to manage disaster risk; (iii) *investing in disaster reduction for resilience*; (iv) *enhancing disaster preparedness*. The priorities are translated into seven objectives which correspond to specific quantitative targets [2].

European Commission supports Sendai Framework with a specific action plan. This integrates the *EU Civil Protection Mechanism (EUCPM)*, adopted since 2001 to promote the cooperation among national authorities for civil protection [3].

Emergency planning [4], [5] is strictly related to territorial planning that is possible to classify on (Fig. 1):

- planning for ordinary conditions [6], [7] where decisions concern the asset of a territory (urban, region, etc.) in terms of land uses [8] and quantitative standard allocation of infrastructures and services [9], [10] which respond to territorial needs of people, business and decision makers in usual situations [11], [12];

WIT Transactions on The Built Environment, Vol 206, © 2021 WIT Press
www.witpress.com, ISSN 1743-3509 (on-line)
doi:10.2495/SAFE210241

- planning for extraordinary conditions [13] where decisions concern external constraints for respecting rules [14] or specific regulations defined for risk reduction [15].

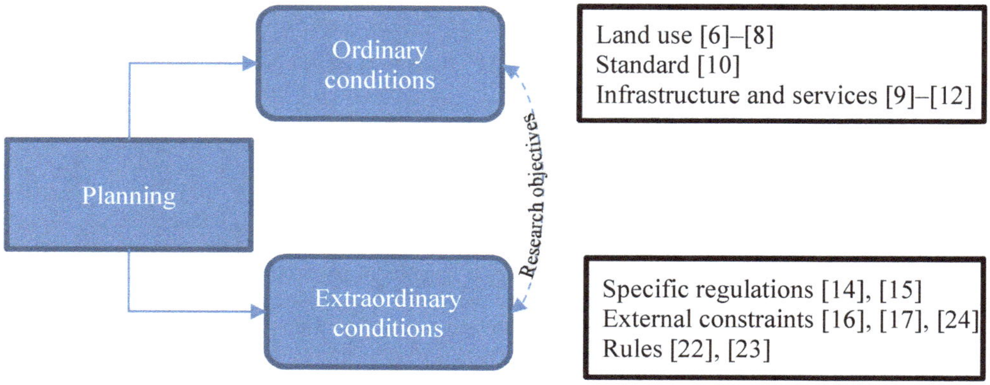

Figure 1: Classification of planning.

In literature and in real applications the two classes of planning are not interrelated [16], [17]. Despite UN or EU indications, there is no integration among territorial decisions that could produce effects for ordinary and/or extraordinary conditions. The present research has the objective to investigate the interrelation between the two forms of planning. The specific focus is on evacuation planning that has to be activated to reduce risk and in particular the exposure component. In particular, the paper proposes a framework for evacuation planning. The framework defines an integrated planning process that includes planned actions concerning training, experimentation and evaluation tools to evaluate progressive improvements and then corrective actions [18]. The framework is defined according to smart city concept where more urban systems are planned and managed in an integrated way, as system of systems where people and business needs are at the center of the decisions.

The paper is connected with other papers focused on risk reduction in the transport system. Planning needs of the support of Transport System Models (TSM) and in particular: models for network design problem [18]; transport demand analysis [19]; transport supply analysis [20]; integration of TSM in Decision Support System (DSS) [21].

After this introduction, the paper has four sections. Section 2 illustrates the main elements of the proposed emergency planning process. Section 3 illustrated the proposed framework for evacuation planning. Section 4 reports the main results of the Italian case study. The final section remarks some conclusive comments.

2 URBAN PLANNING FOR EMERGENCY CONDITIONS

2.1 Theories and laws for planning in ordinary conditions

Planning is a decision-making process of territorial organization designed by urban planning tools defined by specific laws [19], [20]. Urban planning tools are divided into ordinary and extraordinary, for the contents they analyze and the type of constraints they impose. For instance, in Italy, according to the national law published in 1942 [8], the urban local plan defines the structure of a city in an ordinary condition, respecting the constraints deriving from specific laws published at different territorial levels (e.g. hydrogeological plan [25], plan for seismic prevention seismic risk prevention plan [26], [27], landscape plan [28]). Each plan is characterized by the specific issue referred to the ordinary conditions. The knowledge, the interpretation of the data, the assignment and the consequent acknowledgement of the constraints, structure a general planning process that hardly makes the plan an integrated tool. The insertion of constraints within the urban planning tool is an activity that is processed in successive phases, overlapping superordinate indications. The plan for ordinary conditions contains indirect indications for the extraordinary conditions (Fig. 2).

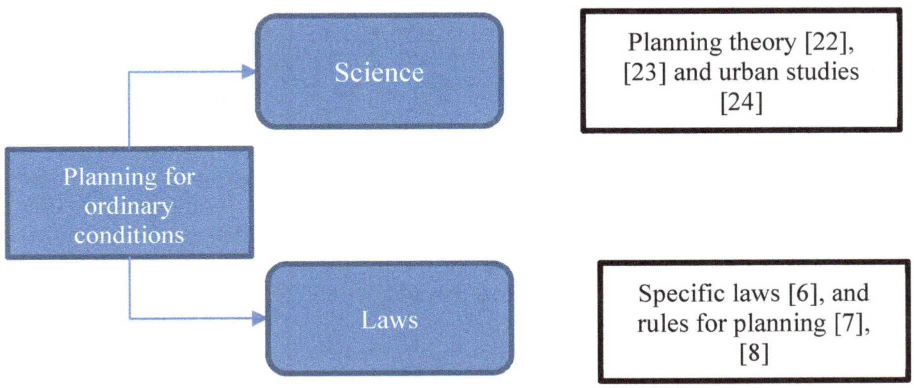

Figure 2: Planning for ordinary conditions.

As argued by Russo et al. [29], the integration of theory, rules and reality leads to the current city, defined as a smart city, a concrete expression of a system of interacting and implementing systems. For this type of city, the ordinary urban plan is inadequate, because it is based on insufficient tools for governing an inter-systemic complexity. Just as the city has adapted over the centuries, like the socio-economic processes that have characterized the different historical periods, characterizing new forms of cities, from New Towns to Smart Cities, similarly, urban planning tools, they should have made the same evolution. Planning, from a plan of regulation of fundamental norms to public life to Smart Planning to use innovative tools suitable for territorial governance, planned as an integrated smart system.

The urban plan has the task of governing the new urban complexity, proposing agile and immediate solutions for solving problems and mitigating risks. The strategic plan integrates rules and is a necessity of smart territory. It is a change of paradigm, the quantitative standards for ordinary conditions, give way to the implementation plan, based on integrated planning for ordinary and extraordinary conditions. Smart planning takes on this role as an integrated process; the scientific approach directs the design of an interconnected system, based on a holistic plan of priority actions and rigorous rules [30].

2.2 Theories and laws for planning in extraordinary conditions

Planning for extraordinary conditions regards decisions for defining activities aimed to reduce risk and its components (occurrence, vulnerability, exposure). These constitute the contents of:

- the safety science's sector that studies the asset of a territory during an emergency;
- the legislation and regulations that define urban development and land use management are finalised to prevent, protect and recovery community and environment.

Emergency planning aims to create safe, sustainable and resilient communities that can avoid or minimise the effects of emergencies and, at the same time, the ability to recover quickly and restore their socioeconomic vitality after an emergency event.

As far as concerns science, it is possible to recall planning theories and urban studies aimed to increase knowledge about emergencies. In the last twenty years, extraordinary events (from the 9/11 terrorist attack to pandemic related to COVID-19) generated disruptions in ordinary situations of territory producing different implications and new issues for research. New theoretical paradigms are needed to face extraordinary events. There are contributions from different disciplines that investigate emergency conditions from different scientific perspectives. Although many issues and functions are understudied, the knowledge base is increasing [31]. For instance, by focusing on mobility in emergency conditions, many scientific works concern the structure of transport planning [32] and the methods and models for supporting decisions in evacuation conditions.

The rules and laws are also adapting to new emergent issues and disruptions. For instance, at the global level, countries introduced new protocols to face pandemics related to COVID-19 [33]. Emergency planning needs responses from regulations and laws.

3 THE EVACUATION PLANNING FRAMEWORK

According to the general objectives of the research, this section presents the proposed framework to integrate planning for ordinary and extraordinary conditions (Section 3.1). To reach the integration, the smart city planning model is adopted. The two conditions are considered as two layers of a smart city that are connected for reaching a convergence among theories, rules and implementations. According to the specific objective of this paper, this section presents the proposed evacuation planning framework (Section 3.2). The integrated planning approach contributes to reducing risk, increasing the knowledge of exposure's component and, in particular, the phenomenon of the mobility in evacuation conditions; knowledge is translated into an evacuation plan that is experimented with exercises aimed to increase emergency's preparedness.

3.1 The proposed framework for planning

The smart city, as a set of integrated infrastructures and services that increase sustainability, can be a reference model for emergency planning and management.

In a smart city, social capital, citizen, operators and decision-makers use available resources to improve quality of life. Smart city's pillars like transport, Information and Communication Technologies (ICTs) are useful tools to understand community needs in ordinary and extraordinary conditions. The tools combination constitutes the equipment of a city control room for providing an overview of all activities in the city in ordinary and in extraordinary conditions. The collected data, information and quantitative simulation models are useful tools for planning, managing and monitoring infrastructures and optimal services [34].

Nevertheless, tools alone are not enough. A good capacity for planning and management is needed. Smart city planning use collected data and results of analyses to support decision-making processes. The city government authorities can analyze the data to prevent or to face future disasters. In this context, one of the main challenges for disaster risk management regards the use of spatial and temporal data deriving from technologies (e.g. weather and traffic conditions, the integrity of infrastructures, presence of people, etc.). Adaptation of decisions against disasters effects increases the efficiency and effectiveness of operators and decision-makers [33].

A smart urban planning model is needed to support collective actions relative to face natural and manmade disasters. Some applications are available in the literature. For instance, Hartama et al. [35] proposed a mobile-GIS platform that provides a real-time platform for spatial risk's visualization. The knowledge of available resources allows to planners in the definition of the evacuation routes to safe areas. A dynamic Decision Support System (DSS) transmits data to the population to mitigate risks, and to increase the resilience against the dangers of natural disasters [36]. Smart urban planning favorites the collaboration among agencies involved in the planning process, for example through integration, and sharing of information about space environment pre and post disasters.

For instance, the study proposed by Ullah et al. [37] identifies a multilayered *Technology–Organisation–Environment (TOE-based)* risk management framework for sustainable smart city governance and in particular for risk management. A set of risks are identified and grouped into TOE categories. The knowledge of risks and their categorization contribute to the smart city government that integrate the decision-making processes about environment, governance and risks.

Wang et al. [38] represents risk management as an iterative process constituted by five key steps ("identification, analysis, evaluation, treatment and monitoring and control of risks").

Following the smart city approach proposed by the authors of this paper themselves, to plan and manage the smart city in ordinary and extraordinary conditions, it is necessary to achieve a convergence between theory, rules and implementation (Fig. 3).

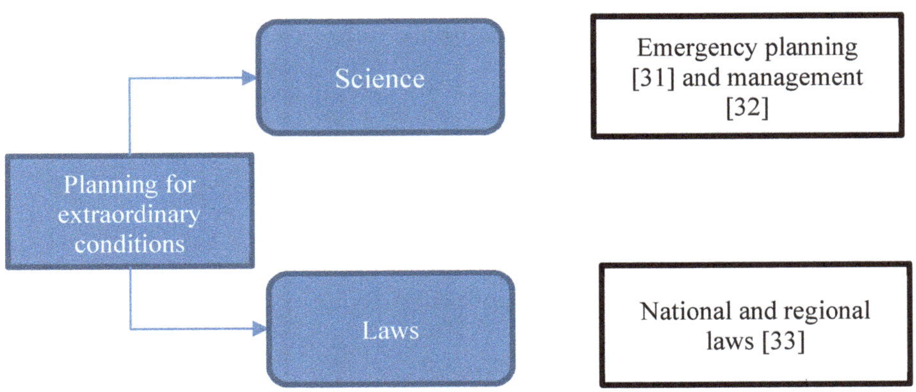

Figure 3: Planning for extraordinary conditions.

www.witpress.com, ISSN 1743-3509 (on-line)

3.2 The specific framework for evacuation planning

In the context of the disaster management cycle, planning is the core of the *preparedness* activities. Emergency preparedness is a continuous process for prioritizing actions, resource allocation for risk reduction. World Health Organization (WHO) defines emergency preparedness as "the knowledge and capacities and organizational systems developed by governments, response and recovery organizations, communities and individuals to effectively anticipate, respond to, and recover from the impacts of likely, imminent, emerging, or current emergencies". Planning constitutes the core of preparedness because it identifies, "at various levels, responsibility for a range of activities aimed at meeting specific objectives" [39].

By focusing on the problem of mobility in extraordinary conditions, and following a smart city approach (Fig. 4), it is necessary to reach a convergence among theories, rules and implementation to reduce exposure's risk component. With this aim, this paper proposes a methodology constituted by a circular process including:

- *planning*, that defines objectives and actions;
- *training*, in which, based on the results of the planning, the subjects and operators involved are trained;
- *exercises*, in which the planned actions are tested and the level of preparation of the operators, the subjects involved and the population is assessed;
- *improvement*, in which, based on the results of the previous phases, corrective actions are defined for risk reduction.

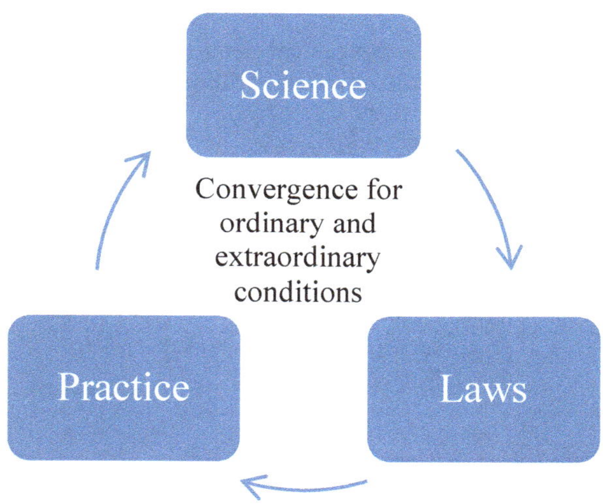

Figure 4: Integrated planning for ordinary and extraordinary condition. *(Source: Adapted from Russo et al. [29].)*

The proposed framework follows a circular process and it is supported by quantitative methods for simulation and monitoring of planned and implemented actions. The methodology allows to planners to define and control their decisions by means ex ante and ex post evaluations (Fig. 5). In the framework, TSMs play a central role because allow the analyst to simulate mobility in ordinary [41], [42] and evacuation conditions [43], [44] for supporting decisions road network setting [45], [46]. TSM fed Decision Support Systems (DSSs) [47], [48] constituting for the final goal is to produce a risk reduction [48] concerning each planned action [49] and then to increase response capacity and more in general the resilience [50].

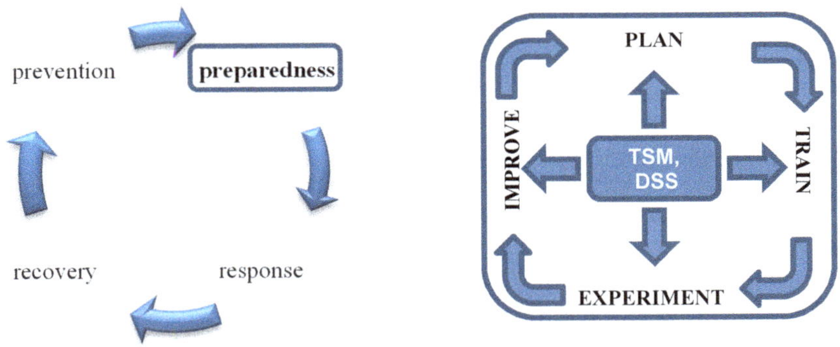

(a) (b)

Figure 5: General planning activities. (a) Disaster management cycle; and (b) Specification of preparedness activities.

4 EMERGENCY URBAN PLANNING IN ITALY

4.1 The emergency planning process

Italy has introduced national and regional laws on Civil Protection [51]. The laws constitute the principal national reference [52]. The Italian laws define:

- the types of events and relative competencies ("type a", "type b", "type c");
- in relation to type of event, the activities and task of civil protection (*prevision, prevention, rescue, recovery*);
- the components of the national service of civil protection (national, regional and local public administration, group of volunteers).

Prevention activities include alerting, emergency planning, training, dissemination of knowledge of civil protection as well as information to the population and the application of technical regulations, where necessary, and exercise activities. In the following the focus is on the planning activities. The Italian Law [53] imposed the obligation to coordinate the plans and programs for the management, protection and rehabilitation of the territories with the emergency civil protection plans, adopted at different territorial scales and in particular: at national scale, the prevention and rescue programs for reducing various risk hypotheses, with specific emergency measures; at regional scale, the regional prevention and protection programs, for following the national indications; at local scale, the provincial prevention programs, for following the national and regional programs, the Prefect's plan to deal with the emergency throughout the province, the Municipal Emergency Plan (PEC), or intermunicipal plan, for following the criteria and procedures set out in the operational guidelines adopted by the National Department of Civil Protection and by the regional councils.

In July 2021 the Italian Government published guidelines for producing the Civil Protection Territorial Plans [27]. The guidelines implement the "Codice della protezione civile" (Code) that introduced the necessity to regulate the modalities of organization and development of the civil protection planning ("monitoring, updating and evaluation"). The aim is to guarantee a coordinated framework at the national level for integrating civil protection systems at all territorial levels, respecting at the same time the autonomy of the Regions. The specific objective is to uniform the planning methodology at different territorial levels for the management of the activities connected to the calamitous events. The code establishes that territorial planning levels are: national, regional, sub-regional (provincial/Città Metropolitana; optimal territorial and organizational scope); local (towns).

As far as concerns the evacuation planning at the local level, it is important to underline the Code promotes the external planning integration: among near towns with the introduction of the "ambit plans"; inside the same local administration (urban planning, transport, technical sectors, …) with the coordination of the civil protection local service. Guidelines indicate the minimal strategical elements that represent the uniform contents of the emergency plans at different territorial levels, concerning:

- the uniform representation of territorial data with the "national catalogue of civil protection plans";
- the description of the organization of civil protection structures for ordinary and extraordinary conditions, to integrate planning and management;
- the territorial context of the occurrence territorial contexts;
- the individuation of the Coordination centres.

The local administrations have to define the safe areas and the routes for reaching them in the case of a tsunami. A specific focus of the guidelines is on the exercises. In particular, the guidelines introduce a classification of the exercises in relation to:

- a spatial criterion, comprehending:
  - international exercises involving national and international emergency organisations;
  - national exercises involving national service of civil protection aimed to test national plans for emergency conditions;
  - regional, local and scope exercises involving regions, provinces, metropolitan and town administrations.
- a modality of execution criterion, comprehending:
  - Command post exercise – CPX, for testing operative centres and their capacity of interrelations;
  - Field exercise – FX, for testing specific operative modules in real contexts;
  - Full-scale exercise – FSX, for testing (prevention, alerting, management) a planned scenario involving operative modules and population;
  - Table top (TTX) for testing emergency plans and in particular specific scenarios in an artificial environment;
  - Discussion-based exercise (DBX) for discussing and comparing hypothetic risk's scenarios.

Note that the guidelines do not report operative indications for network design, simulation tools of the transport system in emergency conditions.

4.2 The process at the local level

The final goal of the PEC is to increase safety and security in emergency conditions. To this end, the PEC defines the operational procedures to be implemented when an emergency event occurs. Starting from the analysis of the current asset of the territory and of the potential risk scenarios, the PEC must report the indications about the procedures (e.g. assistance to the population) and the monitoring activities in case of an emergency event. This has three parts:

- *general* that contains information about physical and socioeconomics characteristics of the territory;
- *planning* that indicates objectives to pursue employing, civil protection's responses for each emergency and the involved operators and their roles;
- *intervention model* that assigns responsibilities to the command and control levels, uses resources rationally, defines a communication system that allows a constant exchange of information.

Table 1 reports the quantity of emergency local plans adopted at urban level in Italy. Note that the national coverage rate (% coverage) is high. Nevertheless, a great part of these plans has to be updated according to recent guidelines [27]. However, adopting a plan alone does not represent a high level of preparedness. For instance, at EU level, the lack of standardization in current disaster training methods, represent a gap in the command, control, and communication levels [54].

Table 1: Coverage of emergency local plan at the urban level in Italy [54].

Regions/Autonomous provinces	Total towns (a)	Total town with a plan (b)	% coverage (a)/(b)
Abruzzo	305	301	99%
Basilicata	131	123	94%
Calabria	409	391	96%
Campania	550	486	88%
Emilia-Romagna	334	322	96%
Friuli Venezia Giulia	216	216	100%
Lazio	378	366	97%
Liguria	235	220	94%
Lombardia	1,544	1,209	78%
Marche	239	239	100%
Molise	136	136	100%
Piemonte	1,206	1,119	93%
Puglia	258	256	99%
Sardegna	377	297	79%
Sicilia	390	190	49%
Toscana	276	250	91%
Trentino AA – Bolzano (aut. prov.)	116	85	73%
Trentino AA – Provincia Trento (aut. prov)	210	210	100%
Valle d'Aosta	74	74	100%
Veneto	575	527	92%
Total	8,051	7,108	88%

5 CONCLUSIVE REMARKS

This paper has investigated the possibility to integrate planning for ordinary and extraordinary conditions. From the analysis of the literature review, it emerges a separation between planning approaches for ordinary and extraordinary conditions. The separation also concerns science that produces theories and laws for urban planning. This paper proposes a smart city planning approach to reach a convergence among theories, rules and implementation to integrate urban planning for ordinary and extraordinary conditions. In particular, the focus of this paper is on evacuation planning and a framework that integrates planning, training, experimentation and evaluation for improvement. The specific Italian case study shows a partial application of the proposed framework. The analysis of national and regional guidelines shows a set of limits, respect to the proposed framework. More integration between science and rules is needed. The exercises proposed by Italian guidelines could produce a great quantity of data and information useful to support the construction of transport system models available in the literature. This implies the necessity to introduce an integrated planning process for ordinary and extraordinary conditions. Further developments of this research concern the possibility to integrate TSM and ICT in an ITS for supporting integrated planning and decision-makers. These tools can be part of a control room in a smart city.

REFERENCES

[1] Bullock, J., Haddow, G. & Coppola, D.P., *Introduction to Emergency Management* 6th ed., Elsevier, 2017.

[2] UN, United Nations Sendai Framework for Disaster Risk Reduction 2015–2030. https://www.unisdr.org/files/43291_sendaiframeworkfordrren.pdf. Accessed on: 3 Oct. 2021.

[3] EU, European Union Civil Protection Mechanism (EUCPM). https://ec.europa.eu/echo/what/civil-protection/mechanism_en. Accessed on: 3 Oct. 2021.

[4] Joss, S., Sengers, F., Schraven, D., Caprotti, F. & Dayot, Y., The smart city as global discourse: Storylines and critical junctures across 27 cities. *J. Urban Technol.,* **26**, pp. 3–34, 2019.

[5] Fazzio, F., Olivieri, M., Parotto, R. & Pizzo, B., *Linee guida per la definizione della Struttura Urbana Minima nel PRG. Ricerca Por-Fesr 2007–2013*, Regione Umbria, Direzione Ambiente, Territorio e Infrastrutture. Dipartimento di Pianificazione Territoriale e Urbanistica, Università di Roma La Sapienza, 2010.

[6] French Government, Law on expropriation of 26 March, 1852. https://www.legifrance.gouv.fr. Accessed on: 3 Oct. 2021.

[7] UK Government, Law of 9 August 1844 for London and around, 1844. www.parliament.uk. Accessed on: 3 Oct. 2021.

[8] Italian Law n. 1150 of 17 August 1942, Legge Urbanistica.

[9] Taylor, N., *Urban Planning Theory Since 1945*. SAGE: London, 1998.

[10] Italian Law, n. 1444 of 1968, Standard Urbanistici.

[11] Musolino, G. & Panuccio, P., Planning instruments in Italy and the UK: Private and public spaces for emergency events in urban areas. *WIT Transactions on The Built Environment*, vol. 134, WIT Press: Southampton and Boston, 2013.

[12] Cirianni, F., Panuccio, P. & Rindone, C., A comparison of urban planning systems between the UK and Italy: Commercial development and city logistic plan. *WIT Transactions on The Built Environment*, vol. 130, WIT Press: Southampton and Boston, pp. 785–797, 2013.

[13] Di Salvo, G., Giuffré, M., Pellegrino, P. & Pizzo, B., Rischio sismico e pianificazione: dall'emergenza all'ordinario. La struttura urbana minima come pretesto per ripensare la città pubblica, INU – Biennale dello spazio pubblico, Dipartimento Design Tecnologia dell'Architettura Territorio e Ambiente, Università di Roma La Sapienza, 2011.
[14] Sarlo, A. (ed.), *Mitigazione del rischio sismico in aree urbane. Un'esperienza su Reggio Calabria*, Iiriti Editore: Reggio Calabria, Italy, 2004.
[15] Fera, G., *La città antisismica; storia, metodologie, prospettive*, Gangemi Editore: Reggio Calabria, Italy, 1990.
[16] Hall, P.G., *Urban and Regional Planning*, 4th ed., Routledge: London, 2002.
[17] Forester, J., *Critical Theory and Public Life*, MIT Press: Cambridge, 1985.
[18] Vitetta, A., Risk reduction in transport system in emergency conditions: A framework for network design problem. *WIT Transactions on the Built Environment*, vol. 206, WIT Press: Southampton and Boston, submitted for publication, 2021.
[19] Russo, F. & Chilà, G., Risk reduction in transport system in emergency conditions: A framework for demand analysis. *WIT Transactions on the Built Environment*, vol. 206, WIT Press: Southampton and Boston, submitted for publication, 2021.
[20] Musolino, G., Risk reduction in transport system in emergency conditions: A framework for supply analysis. *WIT Transactions on the Built Environment*, vol. 206, WIT Press: Southampton and Boston, submitted for publication, 2021.
[21] Di Gangi, M. & Belcore, O.M., Risk reduction in transport system in emergency conditions: a framework for decision support systems. *WIT Transactions on the Built Environment*, vol. 206, WIT Press: Southampton and Boston, submitted for publication, 2021.
[22] Friedmann, J., Planning theory revisited. *European Planning Studies*, **6**(3), pp. 245–253, 1998. DOI:10. 1080/09654319808720459.
[23] Faludi, A., *Planning Theory*, Pergamon Press: Oxford, 1973.
[24] Allmendinger, P., *Planning Theory*, 2nd ed., Palgrave Macmillan: Basingstoke, 2009.
[25] Italian Law n. 183 of 1989, Piano Assetto Idrogeologico.
[26] Italian Law n. 24 of 2020, Classificazione del rischio sismico delle costruzioni.
[27] Italian directive of President of the Council of Ministers 30 April 2021. Indirizzi per la predisposizione dei piani di protezione civile ai diversi livelli territoriali. Official Gazettes n. 160 of 6 July 2021.
[28] Italian Law n. 42 of 2004, Codice dei Beni Culturali e del Paesaggio.
[29] Russo, F., Panuccio, P. & Rindone, C., European plans for the smart city: from theories and rules to logistics test case. *European Planning Studies*, 2016. DOI: 10.1080/09654313.2016.1182120.
[30] Panuccio, P., Smart planning: From city to territorial system. *Sustainability*, **11**, p. 7184, 2019. DOI: 10.3390/su11247184.
[31] McEntire, D.A., The status of emergency management theory: Issues, barriers, and recommendations for improved scholarship. *Paper Presented at the FEMA Higher Education Conference,* June 8, Emmitsburg, MD, 2004.
[32] Russo, F. & Rindone, C., The planning process and logical framework approach in road evacuation: A coherent vision. *WIT Transactions on The Built Environment*, vol. 117, WIT Press: Southampton and Boston, pp. 415–425, 2011.
[33] Gkiotsalitis, K. & Cats, O., Public transport planning adaptation under the COVID-19 pandemic crisis: Literature review of research needs and directions. *Transp. Rev.*, **41**, pp. 374–392, 2021.

[34] Russo, F., Panuccio, P. & Rindone, C., The process of smart city definition at an EU level. *WIT Transactions on Ecology and the Environment*, vol. 191, WIT Press: Southampton and Boston, pp. 979–989, 2014.
[35] Hartama, D. et al., The planning of smart city to mitigate the impacts of natural disaster in North Sumatra. *1st International Conference on Information Technology in Disaster Risk Reduction (ITDRR)*, Sofia, Bulgaria, pp. 147–154, 2016. DOI: 10.1007/978-3-319-68486-4_12. hal-03213128.
[36] Yamashita, S., Watanabe, R. & Shimatani, Y., Smart adaptation activities and measures against urban flood disasters. *Sustainable Cities and Society*, **27**, pp. 175–184, 2016.
[37] Ullah, F., Qayyum, S., Thaheem, M.J., Al-Turjman, F. & Sepasgozar, S.M.E., Risk management in sustainable smart cities governance: A TOE framework. *Technological Forecasting and Social Change*, **167**, 2021.
[38] Wang, J., Lin, W. & Huang, Y.H., A performance-oriented risk management framework for innovative R&D projects. *Technovation*, **30**, pp. 601–611, 2010. DOI: 10.1016/j.technovation.2010.07.003.
[39] World Health Organization (WHO), A strategic framework for emergency preparedness. World Health Organization: Geneva, 2017. Licence: CC BY-NC-SA 3.0 IGO.
[40] Russo, F. & Chilà, G., Dynamic approaches to demand model in evacuation conditions. *WIT Transactions on The Built Environment*, vol. 111, WIT Press: Southampton and Boston, pp. 303–312, 2010.
[41] Russo, F. & Chilà, G., Safety of users in road evacuation: Modelling and DSS for demand. *WIT Transactions on Ecology and the Environment*, vol. 120, WIT Press: Southampton and Boston, pp. 465–474, 2009.
[42] Marcianò, F.A., Musolino, G. & Vitetta, A., A system of models for signal setting design of a signalized road network in evacuation conditions. *WIT Transactions on the Built Environment*, vol. 111, WIT Press: Southampton and Boston, pp. 313–323, 2010.
[43] Marcianò, F.A., Musolino, G. & Vitetta, A., Within-day traffic assignment and signal setting in road evacuation: A procedure with explicit path enumeration. *WIT Transactions on the Built Environment*, vol. 117, WIT Press: Southampton and Boston, pp. 403–414, 2011.
[44] Marcianò, F.A, Musolino, G. & Vitetta A., Signal setting design on a road network: Application of a system of models in evacuation conditions. *WIT Transactions on Information and Communication Technologies*, vol. 43, WIT Press: Southampton and Boston, pp. I443-I454, 2010.
[45] Marcianò F.A, Musolino G. & Vitetta A., Signal setting optimization on urban road transport networks: The case of emergency evacuation. *Safety Science*, **72**, pp. 209–220, 2015.
[46] Di Gangi, M., Watling, D. & Di Salvo, R., Modeling evacuation risk using a stochastic process formulation of mesoscopic dynamic network loading. *IEEE Transactions on Intelligent Transportation Systems*, 2020. DOI: 10.1109/TITS.2020.3038478.
[47] Di Gangi, M. & Polimeni, A., A mesoscopic approach to model route choice in emergency conditions. *Paper presented at the Springer Proceedings in Mathematics and Statistics*, **217**, pp. 547–555, 2017. DOI: 10.1007/978-3-319-67308-0_55.
[48] Russo, F. & Vitetta, A., Safety of users in road evacuation: General methodology and main results. *WIT Transactions on the Built Environment*, vol. 96, WIT Press: Southampton and Boston, pp. 763–772, 2007.

[49] Russo, F. & Rindone, C., Urban exposure: Training activities and risk reduction. *WIT Transactions on Ecology and the Environment*, vol. 191, WIT Press: Southampton and Boston, pp. 991–1001, 2014.
[50] Russo, F., Rindone, C. & Trecozzi, M.R., The role of training in evacuation. *WIT Transactions on Information and Communication Technologies*, vol. 44, WIT Press: Southampton and Boston, pp. 491–502, 2012.
[51] Italian Law n. 225 of 24 February 1992, Istituzione del Servizio nazionale della protezione civile.
[52] Italian Law 2012 n. 59, Disposizioni urgenti per il riordino della protezione civile.
[53] Italian Law 12 July 2012 n. 100, Conversione in legge, con modificazioni, del decreto-legge 15 maggio 2012, n. 59, recante disposizioni urgenti per il riordino della protezione civile.
[54] Khorram-Manesh, A. et al., Education in disaster management and emergencies: Defining a new European course. *Disaster Medicine and Public Health Preparedness*, **9**(3), pp. 245–255, 2015.
[55] Italian Department for Civil Protection Mappa dei piani di protezione civile comunali dati di dettaglio, 2020. https://servizio-nazionale.protezionecivile.gov.it/it/attivita/prevenzione/pianificazione-di-protezione-civile/mappa-piani-di-protezione-civile. Accessed on: 3 Oct. 2021.

RISK REDUCTION IN TRANSPORT SYSTEM IN EMERGENCY CONDITIONS: A FRAMEWORK FOR DECISION SUPPORT SYSTEMS

MASSIMO DI GANGI & ORLANDO MARCO BELCORE
Dipartimento di Ingegneria, Università degli Studi di Messina, Italy

ABSTRACT
Risk analysis and assessment allow analysts and decision-makers to measure whether the losses (economic and human lives) are tolerable and define appropriate actions. Moreover, planning tools and models applied to the emergency decision management (EDM) and spatial decision support system (SDSS) improve the collective interest in increasing the resilience for societal risks. Starting from recent improvements offered by the liberalisation for the GPS data and open data services, this paper aims to suggest an architecture for an open data Web SDSS framework. Furthermore, it integrates into its modular structure the possibility to analyse both networks and evacuees behaviours through the integration of the transportation system model (TSM) models with geographical information systems (GIS), information technology systems (ITS) and volunteered geographic information (VGI). The main advantage of such a system is coping with incoming and in-progress emergencies, utilising on-time data, connecting multiple databases and sources of information, and implementing a participatory process involving different stakeholders.
Keywords: decision support system, evacuation modelling, emergency management, GIS.

1 INTRODUCTION

The interest in Emergency Decision Managment (EDM) is a key issue in emergency management; it aims to increase the collective interest upon the need to increase the resilience for social risks.

Many researchers and stakeholders approached the problem by focusing on "risk analysis and assessment" as an indicator of managing emergency conditions [1]–[3]. Depending on the nature of the event, evacuation represents the instrument to limit the factor of risk exposure, as well as logistics operations (placing inventory management, rescue vehicles and teams) increase both response and recovery phases. The former approach will be of primary interest in this discussion, and therefore the following considerations will focus more on modelling and intervention strategies to limit exposure and reduce the risk factor. The present work is part of a collection of papers upon risk reduction in transport systems during emergency, the same are listed below:

- Risk reduction in transport system in emergency conditions: A framework for network design problem;
- Risk reduction in transport system in emergency conditions: A framework for demand analysis;
- Risk reduction in transport system in emergency conditions: A framework for supply analysis;
- Risk reduction in transport system in emergency conditions: A framework for evacuation planning.

The project represents a step forward respect to previous analyses conducted in Di Gangi [4], Russo and Vitetta [5] and Russo and Rindone [6], where authors focused on:

WIT Transactions on The Built Environment, Vol 206, © 2021 WIT Press
www.witpress.com, ISSN 1743-3509 (on-line)
doi:10.2495/SAFE210251

- calibrating, validating and test models and procedures to assess the effects of action to reduce risk in terms of exposure;
- constructing a prototype laboratory system of models;
- providing public administration with guidelines for planning and managing evacuation.

Main goal of the present consists in adapting planning emergency tools and models to the novelty given by communication, geographical information systems (GIS) based technology and information technology systems (ITS) and introduce the proposal for the creation of a new spatial decision support system (SDSS) framework. In Section 2, question marks at the basis of the research recovery increasing by social knowledge (RISK) are provided; in Section 3 it is described current knowledge on the theme; Section 4 is dedicated to some considerations and the proposal for a web decision support systems (DSS), focused on GIS; ITS technology, GIS and mobile tools, whose main feature will consist into the possibility to involve citizens as an active agent into evacuation process and obtain real-time information to help decision-makers.

2 RISK ASSESSMENT

Planning of a system for emergency conditions starts from a risk assessment (R) analysis in different disaster scenarios, from which it is possible to estimate the extent of damage and the number of victims. Risk assessment aims to identify and to analyse risks and understand the priorities for intervention in order to elaborate strategic actions to contain or mitigate them.

Since the last decade of the twentieth century, risk has been defined in international literature as the "undesirable consequence of a particular activity in relation to the probability of its occurrence" and conventionally expressed using a compact formulation.

$$R = P \cdot M, \tag{1}$$

where P is the probability of occurrence; M is the magnitude, which represents the extent of the consequences. The magnitude M can be further segmented into two elements:

$$M = V \cdot E, \tag{2}$$

where V is the vulnerability of the system, or the inability of the single element, or parts, of the system, to resist the event; E is the exposure of people and goods present in the system. The synthetic formulation of risk can be expressed as

$$R = P \cdot V \cdot E, \tag{3}$$

where each of its components may be characterised using both spatial and temporal variables. According to the nature of the event, in order to reduce the risk factor (see Fig. 1), it's possible ti implement different measures such as:

- prevention measures to reduce probability;
- protection measures to reduce magnitude, which can consist of a reduction of both vulnerability and exposure.

Nevertheless, the reduction of P is limited (in terms of probability) to some kind of manmade events (power failure, radiation leak, hazardous freight, etc.), whereas it results difficult on natural disasters. Besides, it is always possible to work on risk exposure and its factors; it represents the value of people, goods and land use at a specific point in time and space that might be affected during and after an event. According to the risk formulation, it is

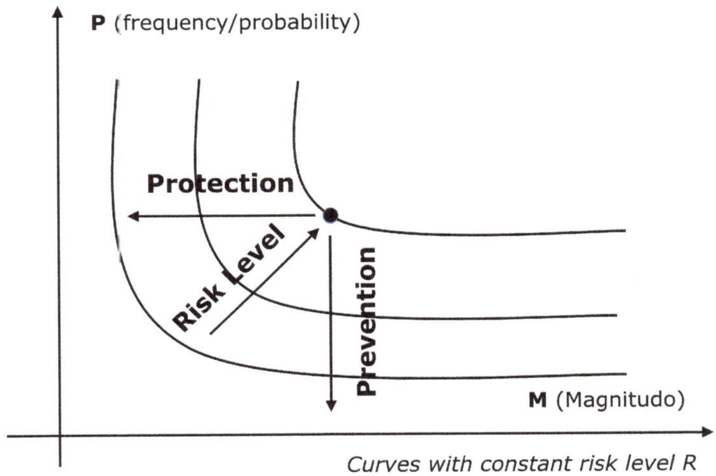

Figure 1: Possible measure for the social risk reduction [7].

important also to share between the individual level and the more interesting cardinal measure of social/societal risk whose formulation depends on $R : f(x, y, t)$.

Previous assumptions let the analyst focus also on important factors and resume a further classification of risk such [7]:

- the cause of the event, which can be natural (i.e. seismic, hydro-geological, volcanic, ...) or anthropic (i.e. accident with dangerous goods, nuclear event, ...);
- the spatial effect: punctual or diffuse;
- the temporal effect (impact), which can be immediate or delayed (each risk component may be associated to a reference period, which ranges from some years to few seconds);
- the effect on the transport system, which can be on the demand, network and demand – network interaction components.

For this reason, at the basis of our work we have to fix some clues and questions that will be part of our proposal:

- How can the risk be reduced by employing urban evacuation;
- How can transport system models may support operations;
- How transportation system model (TSM) interact with ITS during emergency;
- How can public administrations organise a decision support system to address an increasing recovery;
- How can citizens become a key part of the overall process;
- How to provide to evacuees the best social knowledge about safe destinations and routes;

to answer to the previous questions the proposal of a new framework will be introduced.

3 STATE OF THE ART

As stated before, the evacuation measures can be directly connected with recovery for a system, this implies how the risk reduction represents the basis for a resilient system, and the set of measures to implement is the result of a planning process.

Evacuation procedures and planning are defined according to the product-plans dimensions and interactions between the subjects involved. In Italy, the emergency planning process is governed by Law n. 225/92, which defines the Italian Civil Protection Service's competence in case of emergency conditions. At the different territorial scales (national, regional and local), the law identifies different public actors responsible for planning and program solutions to face emergencies. After introducing administrative decentralisation (Law n. 59/97, Legislative Decree n. 112/98), the Italian Regions and local authorities gained a central role in civil protection.

Nowadays, the legislation does not request the application of quantitative methods to verify evacuation procedures inside the local plans, and this lack in the law offers the researchers an opportunity. Indeed, the evaluation of risk index and safety conditions goes through a careful analysis of the performances that this can guarantee. From the mathematical point of view, the simulation of a system and the management of flows (goods, vehicles, and resources) correspond to the resolution of the canonical four-stage model (emission, destination, modal and route choice) and as highlighted in Fig. 2. The approach to be followed is strongly conditioned by the event, the effects that it can trigger on the transport systems, and the interactions among the different activities that persist in the area. Many authors dealt with emergency conditions both in the preventive and post-catastrophe phases [9] and following the principle of evacuation planning (EP) some guides have been drawn up such as the more generic cases developed by Russo and Rindone [6], [10] or more specific formulations as seen in Wei et al. [11] and Chen et al. [12], such as the tools for the mobilization of emergency teams [13] or even case studies for planning or verifying the evacuation of areas subject to disasters [14]–[16].

3.1 Planning an evacuation model

Thus, drawing up a study for an evacuation requires a complete survey on infrastructures, land use and the generation of scenarios for supply e demand and assignment interaction. As an output, the analyst obtains the exposure evaluation associated with flows and link performance. In Fig. 3 a generic evacuation plane framework is reported. The last step in the previous procedure involves risk acceptance and exposure as a direct consequence of the assignment analysis. The evaluated formulation can be used as an indicator whose threshold value may represent the success to reduce risk factors connected with emergency situations. As a matter of fact, Russo and Rindone [6] presented a planning tool for reducing the consequences of an emergency, the same was also calibrated and validated through on-field experimentation for a real case evacuation in a small town in Southern Italy. The output consisted in a multi-stage demand models calibration for the different user classes.

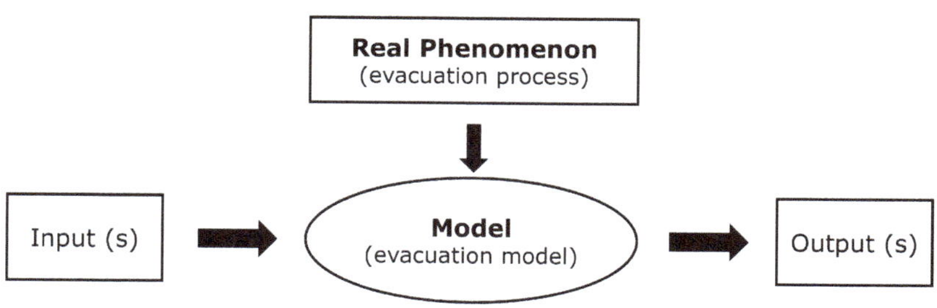

Figure 2: Generic evacuation modelling process [8].

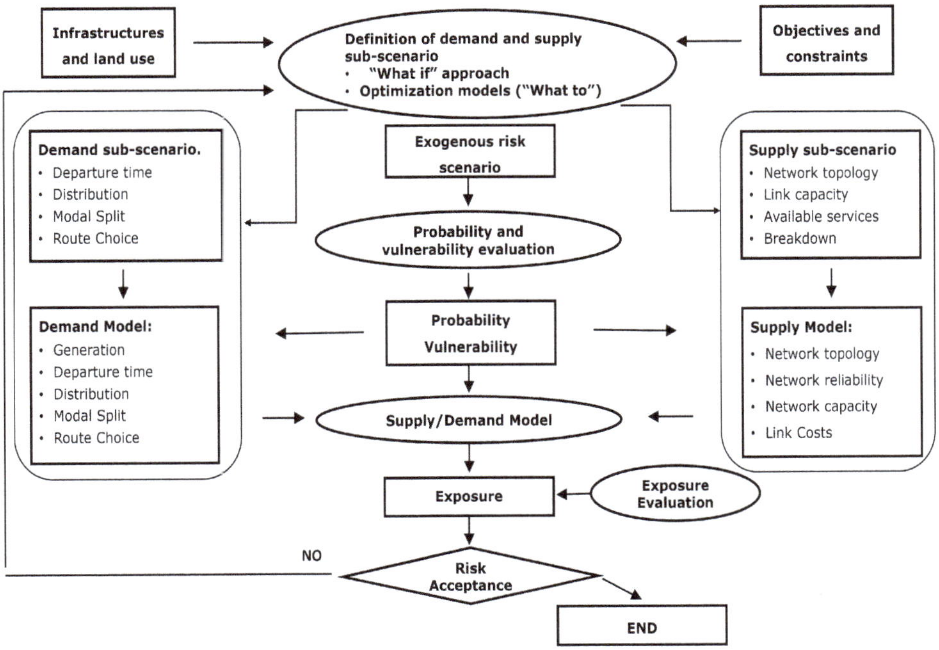

Figure 3: The procedure for a generic evacuation plan [5].

Urban evacuation studies have been carried out using methodologies based on TSM, and TSM simulate a transport system through a process based on the interaction between demand and supply system in terms of cost formulation (disutilities) and flows. Thus, the network model simulates performance and flows, the travel demand model simulates users' choices, the interaction between these to model is operated through assignment employing static (static traffic assignment (STA)) or dynamic interaction (dynamic traffic assignment (DTA)). The former simulates the transport system in stationary conditions, whereas DTA allows time evolution and progress. For a comprehensive review on existing DSS and their capability, the reader is suggested to read Vitetta et al. [17]. In general, exists various traffic assignment simulation tools on the market whose main limitation involves calibration and validation due to the complex behaviours to simulate [18]–[20].

Demand model for evacuation, when an exogenous event occurs to affect the system, have received little attention and classical algorithms do not cope with extraordinary conditions related to the event itself, one of the rare exceptions in this field is represented by Russo and Chilà [21]. In it authors proposed the resolution for the demand problem under emergency conditions, using real experimentation data from a small case study. As demonstrated by Russo and Chilà [22] demand models usually do not take into account information as an influencing factor on evacuees' behaviour, the authors reviewed some of the main important existing demand models and highlighted how only some could be capable of being adapted to emergency conditions and simulate evacuations, and as a consequence of this in the last three decades authors approached the problem mainly focusing on clear time evaluation, critical infrastructure and, the vulnerability of the network.

As a matter of fact when an exogenous event occurs users are not informed about congestion or network reliability, and since now TSM are based on two general conditions:(i) each person knows what to do; (ii) each person follows pre-order instructions; anyway in a real situation, furthermore during an emergency, this is not always true.

Usually, simulations are carried out employing different hypotheses on the characteristics of the flows and observed variables. Thus we share macroscopic with continuous flow representation and aggregate variables [23]; mesocopic, such as in Di Gangi and Velonà [24] where evacuees behaviours is represented by groups and their characteristics, as well as in micro-simulation that let the analyst to evaluate specifics characteristics about the evacuees intentions and reactions according to their profile [25], [26]. Concerning their nature, the macro model is adequate for application to extended areas where major needs consist of a relatively rapid execution and a scarce data granularity as input. Whereas mesoscopic and microscopic needs more data and their specification can benefit from the new opportunity offered by volunteered geographic information (VGI), satellite systems, long and short-range sensors, and the opportunity to implement variable messages and signal optimisation, as well as floating car data and in general all the information available from ITS function present along with the network. As demonstrated in Vitetta et al. [17] obtained data from direct observation are the basis of the calibration of a microscopic link model, as a response of the evacuees' choices, through the evaluation of performance indicators (total distance on the network, total travel time, average travel time and speed) and performance evolution indicators such as a number of vehicles on the network and number of the vehicles that reach the safe area. In further work, Vitetta et al. [27] underlined the importance of the use of TSM, to simulate classical one to one and many to one problems, and evacuation operation to organise evacuation for weak users and preserve their life. To achieve a better understanding a real-time monitoring is essential. The use of different data sources, real-time information and evacuees behaviour still today represents a novel proposal whose final objective will consist in reducing the existing gap between user optimum equilibrium and general optimum for the system. Furthermore, a real-time and mobile SDSS adoption will get the possibility to directly involve citizens as a source also. The continuous data stream in association with real-time simulation can provide enhanced situational awareness and dynamic feedback.

The opportunity to monitor evacuees using mobile and information and communication technology (ICT) infrastructure is an essential tool for the introduction of a new generation of collaborative SDSS. Gwynne et al. [28] the authors provided state of the art at the turn of the 1980s and 1990s, and it emerged how the models could not ignore a behavioural approach, and researchers should focus on the behaviour of displaced people who can deviate far from the mathematical approximations underlying the models as the same as in Vorst [29] where the author focused its work on user behaviour, highlighting how the psychological parameters, too often not evaluated, play a decisive role in understanding and therefore simulating evacuees' behaviour. Anyway many authors tried to give back a better understanding about people choices using different approaches to the problems and some of the most representative examples are offered by the use of rules and physical-mathematical approximations [30], the use of artificial intelligence models [31], probabilistic simulations [16] or agent models [8], [32], [33].

The importance of the commitment of ITS systems as tools for real-time monitoring of the effects produced, on networks, of evacuation and the impact of decisions taken by analysts and displaced persons is reiterated in Chilà et al. [34]. It represents an overview of a methodology that could be applied to emergency conditions integrated with ITS, and authors propose a general methodology to implement ITS into DSS. In it, thanks to the output from a small case study, models are calibrated and validated.

Finally, a frontier of great interest for research and public administrations lies in the possibility of integrating, in a single database, various sources of real-time data using sensors and short and long-range communications as well as alert and monitoring systems that are prepared by both central and international protocols as well as spontaneous user reports [32], [35]–[37].

3.2 Review on developed DSS framework approaches

We have further focused our review over the progress and intuitions from the academics ad practitioners during the last 40 years and emerged how the frameworks of the DSS moved from an analytical approach based on operative research (OR) dynamics and resolution based on scenario hypothesis and simulation to a more flexible and comprehensive use of GIS information and spatial analysis. Furthermore, in the last twenty years, the liberalisation of data and the progress in open data availability improved the possibility to develop SDSS and group decision support system (GDSS) through the employment of different decision-makers and a wider data source to elaborate solution [38].

Nowadays academics focuses on the development of SDSS whose essential elements are:

- a database capable of managing geo-referenced data;
- an interactive database capable to communicate with users, sensors and decision makers in real time;
- a mathematical model that simulates decision-making processes;
- a graphical interface that assists both the user and the decision maker.
- the opportunity to integrate multi criteria analyses and evaluate different layers of information;

This overview explicitly introduces the general objective at the basis of a SDSS for the simulation of a TSM as an instrument to support evacuation planning.

The literary review shed in light how only a few of the developed DSS nowadays integrate into their routine real evacuees' choices in real-time, follow the evolution of the event itself, and the collapse of arterial and bottlenecks, to do this the analyst has to estimate indicators, identify key parameters and cross-refer them with other databases like GIS or satellite, and implement ITS thanks to smart roads projects and VGI.

Most of them are evaluated for specific case study and emergencies, usually do not integrate dynamic assignment and micro simulation effects. The liberalisation of GPS data and geo-information, allowed faster progress on DSS workflow and the technologies supported by Web and Mobile, which now represent one of the frontiers to be explored [39] by integrating methodologies that support the use of Artificial Intelligence or interactive graphic interfaces, to evaluate different scenario effects employing geoprocessing tools and geocomputerized systems [38], [40]. The introduction of GIS information system let the analyst to evaluate qualitative and quantitative effects at the same time through the use of simulations. For a more in-depth discussion on GIS-based DSS, the reader is referred to Russo and Chilà [41], and as demonstrated by Zhou et al. [42], the interest of researchers focused on EDM that have at their basis make use of GIS to combine mathematical models and qualitative data offered by the geo-information and ICT, this parallel use of database and knowledge engine let the analysts to reduce the risks connected with the evolution of the event itself and some authors developed recently some example exploring the opportunities offered by the before mentioned coupling [43]–[45]. In Wei et al. [11], Zheng et al. [46], Zhu et al. [47], and Olyazadeh et al. [48] we found some proposal of generic architectures for the construction of a GIS-based DSS. The authors all agree to suggest modularity as a source of integration to enlarge the field of application and the possibility to combine different professions and the integration of open-data sources to enlarge the amount of information and to proceed to a universal standard to aggregate it in order to help the public administration to collect it with with a limited amount of money.

4 PROPOSAL FOR A MOBILE SDSS

So far, we focused our synopsis on the DSS focusing on the architecture to be associated with it and the simulation processes carried out by the analysis sub-models. It has been evident that many of the models developed so far have a more empirical or evaluation nature. Although included among the DSS they perform a more qualitative and informative task on disasters without providing an effective tool for the correct management of the emergency that makes them immediately usable as a support to the decision-maker in the phase of incipient danger or a few moments after the event. Thus, the broad review presented in the previous section has shown how the metadata, the management software (middleware), and sub-model analysis capable of interpreting data represent the cornerstones to work on in order to develop an innovative SDSS. It is, therefore, evident that institutions and agencies should be encouraged to use information technology, open-source software and open standards for EDM activities. It will result in a general system based on a geo-computerized environment that might be updated in real-time from different sources and which provides a central core of analysis that integrates the use of different solvers that collaborate – integrating, as foreseen in the objectives of the EU MUSTER Project [49].

4.1 Computational models

Play a crucial role in planning, and the development of simulation models is a significant challenge to grip with crisis management. Indeed those assist the decision-maker in drafting the appropriate plans; the use of DSS also allows you to plan not only the direct aspects but the evolution over time and the triggering of secondary phenomena not always directly related to the emergency. Upon it, a great gap is already represented by the limited possibility to directly evaluate the influence of heterogeneity into behaviours and the interference among different flows of people who operate different mode choices. It also emerged how dynamic assignment [50], and real-time applications [51], have been recently introduced and till now, the evolution over time has been extensively studied favouring a probabilistic approach. The introduction of the long and short-range sensors, floating car data, satellite and the VGI is giving impetus to a more straightforward definition of the problem. It follows that today a EDM must necessarily have a SDSS at its base, and the GIS component of the latter must be based on a rich and updated data package that is always kept active in the ordinary, and pre-disaster phases trough the creation of the spatial data infrastructures (SDI).

As a consequence of these statements, we can assume how the integration of ICT tools, and more generally ITS, allow public (e.g. municipalities, local administrations) and private (e.g. citizens) actors to increase their benefit while managing and using urban transport systems in emergency conditions. The information flow through mobile devices and web applications let the administrations increase the evacuees' consciousness and let them know what to do (e.g. which destination to reach, which path to undertake), modifying the way the evacuation will take place with respect to a scenario without ICT, and can be assumed that the introduction of such technologies will allow the decision-makers to shift the paradigm of the decisions undertaken by evacuees to a more sustainable system optimum (SO) model organized on cooperative behaviour, where evacuees interact with the system giving it back information upon behaviours and choices, letting the analyst monitor the demand and redefine information (i.e. suggested routes) according to the evolution of the emergency.

4.2 Collaborative system

The possibility to organise a user-friendly app, as evaluated in Astarita et al. [52] will involve the citizens to support the project and become a source of crowd-sourced information. A suitably developed App can allow the evacuees, familiar or not (i.e. tourists) with the urban area, to receive through their smartphone information about the nature of the alert, the safe destination and the route. The quality of the system and user interface will be assessed via experimentation of the user requirements for a successful disaster management system architecture. The DSS server must be easy to use during the management of incoming information about the event. Its main function consists of storing the evacuees' spatial-time data during their route; this detection could be operated using individual mobile devices and transmitted to the SDSS itself. On the other side, the system's repository will also collect and store data from official sources. Collected data from the App will be used as a source for the statistical analysis between the forecasted and the experienced behaviour. Thus the foretasted destinations and routes that the SDSS provides to the users, whose position is unknown for the analyst and detected by the SDSS at the "Alert" time instant, and the experienced destinations and routes selected and undertaken by the group of users along the (road) network, will be analysed.

The comparison of the forecasted vs experienced data will be carried out utilizing statistical analysis in order to verify the capacity of the TSM to simulate destination and route choice during a "real" evacuation, and this further validation will be part of the process to help municipalities to draw up efficient planning solutions during different kind of emergencies.

5 CONCLUSION

The introduction of this new proposal will positively impact different fields, both technological and social.

Technological, the use of the proposed methodologies by public administration and citizens could allow the integration between available traffic data. Many data can be shared between ordinary traffic conditions and emerging conditions. In the specific, both off-line and online data, obtained by operating sensors (e.g. IoT), can be used. The use on the city can make advancement on the installed sensor network that can be beneficial from the use of other social groups, particularly voluntary associations and no-profit organizations.

The results in terms of prototypal DSS can suggest to different start-up enterprises to work in this field to produces advancement of SDSS more linked to the specific territory. Furthermore, it will introduce the possibility to gives the possibility to customize the DSS for each city, developing the model used and upgrading the quality of the information provided to citizens, maintaining the same App for all the cities where applied and then easy to read. A consisting upgrade resides into the possibility to divide the DSS from the other parts gives the opportunity to link it with the main system of traffic control. These models allow the construction of procedures to be included in hardware and software tools for advancement user information other than the current ones that give the forecasted path conditioned to a choice destination and sometimes to the conditions of the network in terms of flow and time. The data obtained by testing experiences are highly useful for the calculation of statistics and forecasting by means of regressive statistical models to better organize the functioning of systems for next emergencies. This new class of models, together with the availability of data in real-time, could in the future lead to new smart objects to support the evacuation towards different places: from simple open square to the main hospital.

Social, from the citizens perspective, the ICT system gives positive economic impacts to the community reducing risk and increasing social knowledge, with no additional costs (for technicians, public administrators, citizens. Moreover, the system will provide to public administration the possibility to organize several trials with the social associations, using the ICT system as reference managing. Furthermore, the proposal for a new SDSS, whose main characteristics consist in its modularity, interoperability among different sources and actors, capable of involving citizens to reduce social risks, is in line with the agenda of many public administration who have to reach the goals subscribed into the UN Agenda 2015 (isnerire citazione) in terms of sustainability, for a safe and inclusive environment that has to reduce the number of deaths and the number of people affected by losses during an emergency thus, to centre specific goals into the area of "disaster-resilient societies" and "Smart mobility".

Finally, the potentiality for the proposed SDSS can be resumed as: (i) public administration: Major and Civil Protection during evacuations; (ii) citizens: residents (e.g. workers, students), and tourists; (iii) planners: the DSS allows testing, according to a "what-if" approach, the evacuation plan; (iv) design technicians: civil, industrial, and information technicians can use a developed version of App-Web server and DSS; (v) national and regional Civil Protection: the results can be diffused at a national or regional level to let citizens use the same App, but with specific DSS for each area.

REFERENCES

[1] Peckham, R., Haastrup, P. & Otway, H., A computer-based system for risk management support. *Decision Support Systems*, **4**(4), pp. 481–489, 1988.

[2] Van Raemdonck, K., Macharis, C. & Mairesse, O., Risk analysis system for the transport of hazardous materials. *Journal of Safety Research*, **45**, pp. 55–63, 2013.

[3] Russo, F. & Rindone, C., Urban exposure: Training activities and risk reduction. *WIT Transactions on Ecology and the Environment*, vol. 191, WIT Press: Southampton and Boston, pp. 991–1001, 2014.

[4] Di Gangi, M., Approaching the analysis of transport networks in emergency conditions for the design of evacuation plans. *WIT Transactions on Ecology and the Environment*, vol. 77, WIT Press: Southampton and Boston, 2004.

[5] Russo, F. & Vitetta, A., Risk evaluation in a transportation system. *International Journal of Sustainable Development and Planning*, **1**(2), pp. 170–191, 2006.

[6] Russo, F. & Rindone, C., Safety of users in road evacuation: Planning internal processes and guidelines. *WIT Transactions on the Built Environment*, vol. 96, WIT Press: Southampton and Boston, 2007.

[7] Russo, F. & Vitetta, A., Risk in anthropic environments: Methodologies for risk evaluation and exposition reduction. *Environmental Health Risk III*, **9**, p. 31, 2005.

[8] Tavares, R.M., Evacuation processes versus evacuation models: "quo vadimus"? *Fire Technology*, **45**(4), pp. 419–430, 2009.

[9] Russo, F. & Chilà, G., Safety of users in road evacuation: RP vs. SP surveys in demand analysis. *WIT Transactions on the Built Environment*, vol. 101, WIT Press: Southampton and Boston, pp. 703–713, 2008.

[10] Russo, F. & Rindone, C., Safety of users in road evacuation: The logical framework approach in evacuation planning. *WIT Transactions on the Built Environment*, vol. 101, WIT Press: Southampton and Boston, pp. 751–760, 2008.

[11] Wei, H., Zeng, Q.A., Hu, H., Wang, X. & Kukreti, A.R., Integrated urban evacuation planning framework for responding to human-caused disasters over a surface transportation network. *Transportation Research Record*, **2041**(1), pp. 29–37, 2008.

[12] Chen, X., Kwan, M.P., Li, Q. & Chen, J., A model for evacuation risk assessment with consideration of pre- and post-disaster factors. *Computers, Environment and Urban Systems*, **36**(3), pp. 207–217, 2012.

[13] Chang, M.S., Tseng, Y.L. & Chen, J.W., A scenario planning approach for the flood emergency logistics preparation problem under uncertainty. *Transportation Research Part E: Logistics and Transportation Review*, **43**(6), pp. 737–754, 2007.

[14] Chakraborty, J., Tobin, G.A. & Montz, B.E., Population evacuation: Assessing spatial variability in geophysical risk and social vulnerability to natural hazards. *Natural Hazards Review*, **6**(1), pp. 23–33, 2005.

[15] Di Gangi, M., Safety of users in road evacuation: Modelling and DSS for pedestrian outflow. *WIT Transactions on Ecology and the Environment*, vol. 120, WIT Press: Southampton and Boston, pp. 497–508, 2009.

[16] Di Gangi, M., Watling, D. & Di Salvo, R., Modeling evacuation risk using a stochastic process formulation of mesoscopic dynamic network loading. *IEEE Transactions on Intelligent Transportation Systems*, 2020.

[17] Vitetta, A., Musolino, G. & Marcianò, F., Safety of users in road evacuation: Supply and demand-supply interaction models for users. *WIT Transactions on the Built Environment*, vol. 96, WIT Press: Southampton and Boston, 2007.

[18] Stathopoulos, A. & Tsekeris, T., Economy of information in congested transport systems. *IET Intelligent Transport Systems*, **3**(3), pp. 336–344, 2009.

[19] Di Gangi, M., Modeling evacuation of a transport system: Application of a multimodal mesoscopic dynamic traffic assignment model. *IEEE Transactions on Intelligent Transportation Systems*, **12**(4), pp. 1157–1166, 2011.

[20] Marcianò, F.A., Musolino, G. & Vitetta, A., Signal setting optimization on urban road transport networks: The case of emergency evacuation. *Safety Science*, **72**, pp. 209–220, 2015.

[21] Russo, F. & Chilà, G., Safety of users in road evacuation: Demand models. *WIT Transactions on the Built Environment*, vol. 96, WIT Press: Southampton and Boston, 2007.

[22] Russo, F. & Chilà, G., Demand models in road evacuation: A synopsis of recent contributions. *WIT Transactions on the Built Environment*, vol. 116, WIT Press: Southampton and Boston, pp. 601–614, 2011.

[23] Mclean, M.A., Moeller, M. & Desrosiers, A., Clear: A model for the calculation of evacuation-time estimates in emergency planning zones. Technical report, Pacific Northwest Lab., 1983.

[24] Di Gangi, M. & Velonà, P., Safety of users in road evacuation: Pedestrian outflow models in a building. *WIT Transactions on The Built Environment*, vol. 96, WIT Press: Southampton and Boston, 2007.

[25] Sinuany-Stern, Z. & Stern, E., Simulating the evacuation of a small city: The effects of traffic factors. *Socio-Economic Planning Sciences*, **27**(2), pp. 97–108, 1993.

[26] Cova, T.J. & Johnson, J.P., Microsimulation of neighborhood evacuations in the urban–wildland interface. *Environment and Planning A*, **34**(12), pp. 2211–2229, 2002.

[27] Vitetta, A., Quattrone, A. & Polimeni, A., Safety of users in road evacuation: Algorithms for path design of emergency vehicles. *WIT Transactions on the Built Environment*, vol. 101, WIT Press: Southampton and Boston, pp. 727–737, 2008.

[28] Gwynne, S., Galea, E., Owen, M., Lawrence, P.J. & Filippidis, L., A review of the methodologies used in evacuation modelling. *Fire and Materials*, **23**(6), pp. 383–388, 1999.

[29] Vorst, H.C., Evacuation models and disaster psychology. *Procedia Engineering*, **3**, pp. 15–21, 2010.

[30] Santos, G. & Aguirre, B.E., A critical review of emergency evacuation simulation models, 2004.

[31] Pelechano, N. & Malkawi, A., Evacuation simulation models: Challenges in modeling high rise building evacuation with cellular automata approaches. *Automation in Construction*, **17**(4), pp. 377–385, 2008.

[32] Aguirre, B.E., El-Tawil, S., Best, E., Gill, K.B. & Fedorov, V., Contributions of social science to agent-based models of building evacuation. *Crowds in the 21st Century*, Routledge, pp. 163–180, 2015.

[33] Kasereka, S., Kasoro, N., Kyamakya, K., Goufo, E.F.D., Chokki, A.P. & Yengo, M.V., Agent-based modelling and simulation for evacuation of people from a building in case of fire. *Procedia Computer Science*, **130**, pp. 10–17, 2018.

[34] Chilà, G., Musolino, G., Polimeni, A., Rindone, C., Russo, F. & Vitetta, A., Transport models and intelligent transportation system to support urban evacuation planning process. *IET Intelligent Transport Systems*, **10**(4), pp. 279–286, 2016.

[35] Gunes, A.E. & Kovel, J.P., Using GIS in emergency management operations. *Journal of Urban Planning and Development*, **126**(3), pp. 136–149, 2000.

[36] Wilson, R.D. & Cales, B., Geographic information systems, evacuation planning and execution. *Communications of the IIMA*, **8**(4), p. 2, 2008.

[37] Abdalla, R. & Esmail, M., *WebGIS for Disaster Management and Emergency Response*, Springer, 2018.

[38] Sugumaran, V. & Sugumaran, R., Web-based spatial decision support systems (webSDSS): Evolution, architecture, examples and challenges. *Communications of the Association for Information Systems*, **19**(1), p. 40, 2007.

[39] Vizecky, K. & El-Gayar, O., Increasing research relevance in DSS: Looking forward by reflecting on 40 years of progress. *2011 44th Hawaii International Conference on System Sciences*, IEEE, pp. 1–9, 2011.

[40] Cova, T.J. & Church, R.L., Modelling community evacuation vulnerability using GIS. *International Journal of Geographical Information Science*, **11**(8), pp. 763–784, 1997.

[41] Russo, F. & Chilà, G., Safety of users in road evacuation: Modelling and DSS for demand. *WIT Transactions on Ecology and the Environment*, vol. 120, WIT Press: Southampton and Boston, pp. 465–474, 2009.

[42] Zhou, L., Wu, X., Xu, Z. & Fujita, H., Emergency decision making for natural disasters: An overview. *International Journal of Disaster Risk Reduction*, **27**, pp. 567–576, 2018.

[43] Mileu, N. & Queirós, M., Integrating risk assessment into spatial planning: Riskote decision support system. *ISPRS International Journal of Geo-Information*, **7**(5), p. 184, 2018.

[44] Zhang, Z., Hu, H., Yin, D., Kashem, S., Li, R., Cai, H., Perkins, D. & Wang, S., A cybergis-enabled multi-criteria spatial decision support system: A case study on flood emergency management. *International Journal of Digital Earth*, **12**(11), pp. 1364–1381, 2019.

[45] Heinzlef, C., Becue, V. & Serre, D., A spatial decision support system for enhancing resilience to floods: Bridging resilience modelling and geovisualization techniques. *Natural Hazards and Earth System Sciences*, **20**(4), pp. 1049–1068, 2020.

[46] Zheng, X., Zhong, T. & Liu, M., Modeling crowd evacuation of a building based on seven methodological approaches. *Building and Environment*, **44**(3), pp. 437–445, 2009.

[47] Zhu, L., Song, W. & Li, Q., Construction of emergency decision system based on GIS. *2009 Pacific-Asia Conference on Knowledge Engineering and Software Engineering*, IEEE, pp. 171–174, 2009.

[48] Olyazadeh, R., Aye, Z.C. & Jaboyedoff, M., Development of a prototype for spatial decision support system in risk reduction based on open-source web-based platform. *Conference Paper*, 2013.

[49] Gadomski, A.M., Bologna, S., Costanzo, G.D., Perini, A. & Schaerf, M., Towards intelligent decision support systems for emergency managers: The IDA approach. *International Journal of Risk Assessment and Management*, **2**(3–4), pp. 224–242, 2001.

[50] Ayfadopoulou, G., Stamos, I., Mitsakis, E. & Grau, J.M.S., Dynamic traffic assignment based evacuation planning for CBD areas. *Procedia-social and Behavioral Sciences*, **48**, pp. 1078–1087, 2012.

[51] Blaschke, T. & Schmidt, D., eMapBoard: From a collaborative situation mapping environment to a disaster management decision support system. *InterCarto–InterGIS 12 Berlin*, pp. 242–251, 2006.

[52] Astarita, V., Festa, D., Giofrè, G.P., Guido, G. & Stefano, G., Mobile for emergencies m4em: A cooperative software tool for emergency management operations. *Procedia Computer Science*, **134**, pp. 433–438, 2018.

Author index

www.ingramcontent.com/pod-product-compliance
Lightning Source LLC
LaVergne TN
LVHW062220200926
848484LV00006B/1442
* 9 7 8 1 7 8 4 6 6 4 4 3 5 *